ADVANCES IN BIOMEDICAL ENGINEERING

ADVANCES IN BIOMEDICAL ENGINEERING

Editor

PASCAL VERDONCK

ELSEVIER

Amsterdam • Boston • Heidelberg • London • New York • Oxford
Paris • San Diego • San Francisco • Singapore • Sydney • Tokyo

Elsevier
Radarweg 29, PO Box 211, 1000 AE Amsterdam, The Netherlands
Linacre House, Jordan Hill, Oxford OX2 8DP, UK

First edition 2009

British Library Cataloguing in Publication Data
A catalogue record for this book is available from the British Library

Library of Congress Cataloging-in-Publication Data
A catalog record for this book is available from the Library of Congress

ISBN: 978-0-444-53075-2

For information on all Elsevier publications
visit our web site at books.elsevier.com

Printed and bound in the United Kingdom

Transferred to Digital Print 2009

Contents

Preface *ix*
List of Contributors *xi*

1 Review of Research in Cardiovascular Devices **1**
Zbigniew Nawrat

1 Introduction 2
2 The Heart Diseases 8
3 The Cardiovascular Devices in Open-Heart Surgery 8
3.1 Blood Pumps 9
3.2 Valve Prostheses 23
3.3 Heart Pacemaker 34
4 The Minimally Invasive Cardiology Tools 34
5 The Technology for Atrial Fibrillation 38
6 Minimally Invasive Surgery 39
6.1 The Classical Thoracoscopic Tools 40
6.2 The Surgical Robots 43
6.3 Blood Pumps – MIS Application Study 49
7 The Minimally Invasive Valve Implantation 53
8 Support Technology for Surgery Planning 53
9 Conclusions 57

2 Biomechanical Modeling of Stents: Survey 1997–2007 **61**
Matthieu De Beule

1 Introduction 62
2 Finite Element Modeling of Stents 63
2.1 Finite element basics 63
2.2 Geometrical design and approximation 64
2.3 Material properties 65
2.4 Loading and boundary conditions 66
2.5 Finite element stent design 66
2.6 Effective use of FEA 68
3 Survey of the State of the Art in Stent Modeling: 1997–2007 68
3.1 Neglect of the balloon 69
3.2 Cylindrical balloon 74
3.3 Folded balloon 78
3.4 Summary 81
4 Alternative methods for biomechanical modeling of stents 84
4.1 FEM – Prolapse, flexibility and strut micromechanics 84
4.2 FEM – Self-expandable stents 85
4.3 CFD–drug elution and immersed FEM 87

5 Future Prospects 88
6 Conclusion 88

3 Signal Extraction in Multisensor Biomedical Recordings **95**
Vicente Zarzoso, Ronald Phlypo, Olivier Meste, and Pierre Comon

1 Introduction 96
1.1 Aim and scope of the chapter 96
1.2 Mathematical notations 97
2 Genesis of Biomedical Signals 98
2.1 A biomedical source model 98
2.2 Cardiac signals 101
2.3 Brain signals 105
3 Multi-Reference Optimal Wiener Filtering 109
3.1 Non-invasive fetal ECG extraction 109
3.2 Optimal Wiener filtering 110
3.3 Adaptive noise cancellation 112
3.4 Results 113
4 Spatio-Temporal Cancellation 115
4.1 Atrial activity extraction in atrial fibrillation 115
4.2 Spatio-temporal cancellation of the QRST complex in AF episodes 117
5 Blind Source Separation (BSS) 123
5.1 The isolation of interictal epileptic discharges in the EEG 123
5.2 Modeling and assumptions 125
5.3 Inherent indeterminacies 127
5.4 Statistical independence, higher-order statistics and non-Gaussianity 127
5.5 Independent component analysis 129
5.6 Algorithms 131
5.7 Results 133
5.8 Incorporating prior information into the separation model 136
5.9 Independent subspaces 138
5.10 Softening the stationarity constraint 138
5.11 Revealing more sources than sensor signals 138
6 Summary, Conclusions and Outlook 139

4 Fluorescence Lifetime Spectroscopy and Imaging of Visible Fluorescent Proteins **145**
Ankur Jain, Christian Blum, and Vinod Subramaniam

1 Introduction 146
2 Introduction to Fluorescence 146
2.1 Interaction of light with matter 146
2.2 The Jabłoński diagram 147
2.3 Fluorescence parameters 151

	2.4	Fluorescence lifetime	151
	2.5	Measurement of fluorescence lifetime	153
	2.6	Fluorescence anisotropy and polarization	155
	2.7	Factors affecting fluorescence	157
3	Fluorophores and Fluorescent Proteins		160
	3.1	Green fluorescent protein	161
	3.2	Red fluorescent protein	165
4	Applications of VFPs		166
	4.1	Lifetime spectroscopy and imaging of VFPs	167
5	Concluding Remarks		170

5 Monte Carlo Simulations in Nuclear Medicine Imaging **175**
Steven Staelens and Irène Buvat

1	Introduction		176
2	Nuclear Medicine Imaging		176
	2.1	Single photon imaging	177
	2.2	Positron emission tomography	178
	2.3	Emission tomography in small animal imaging	179
	2.4	Reconstruction	179
3	The MC Method		180
	3.1	Random numbers	180
	3.2	Sampling methods	181
	3.3	Photon transport modeling	182
	3.4	Scoring	183
4	Relevance of Accurate MC Simulations in Nuclear Medicine		184
	4.1	Studying detector design	184
	4.2	Analysing quantification issues	184
	4.3	Correction methods for image degradations	185
	4.4	Detection tasks using MC simulations	186
	4.5	Applications in other domains	186
5	Available MC Simulators		187
6	Gate		188
	6.1	Basic features	188
	6.2	GATE: Time management	192
	6.3	GATE: Digitization	193
7	Efficiency-Accuracy Trade-Off		194
	7.1	Accuracy and validation	194
	7.2	Calculation time	194
8	Case Studies		195
	8.1	Case study I: TOF-PET	195
	8.2	Case study II: Assessment of PVE correction	196
	8.3	Case study III: MC-based reconstruction	197
9	Future Prospects		200
10	Conclusion		200

6 Biomedical Visualization **209**
Chris R. Johnson and Xavier Tricoche

1 Introduction 210
2 Scalar Field Visualization 211
2.1 Direct volume rendering 211
2.2 Isosurface extraction 220
2.3 Time-dependent scalar field visualization 222
3 Vector Field Visualization 223
3.1 Vector field methods in scientific visualization 224
3.2 Streamline-based techniques 225
3.3 Stream surfaces 226
3.4 Texture representations 229
3.5 Topology 232
4 Tensor Field Visualization 234
4.1 Anisotropy and tensor invariants 235
4.2 Color coding of major eigenvector orientation 236
4.3 Tensor glyphs 236
4.4 Fiber tractography 239
4.5 Volume rendering 241
4.6 White matter segmentation using tensor invariants 244
5 Multi-field Visualization 245
6 Error and Uncertainty Visualization 250
7 Visualization Software 254
7.1 SCIRun/BioPSE visualization tools 255
7.2 map3d 258
8 Summary and Conclusion 263

Index 273

Preface

The aim of this book is to bridge the interdisciplinary gap in biomedical engineering education. No single medical device has ever been developed without an extensive interdisciplinary collaboration but, nevertheless, a wide gap still exists between different themes like biomaterials, biomechanics, extracorporeal circulation research, nanotechnology, and safety and regulations.

We have to evitate that, due to the growing complexity, we only gather with and communicate in our own expert groups with their specific research topics. Therefore it is, in my opinion, extremely important that a high quality book is available to cover these interdisciplinary islands by bringing representative research groups together. The aim of this book is a top review of current research advances in biomedical engineering but on an educational level, so it is also useful for teaching master's students, Ph.D. students, and professionals. Each chapter is complementary to reviews in specific scientific journals. The topics range from visualization technology and biosignal and imaging analysis over biomechanics and artificial organs to nanotechnology and biophotonics and cardiovascular medical devices. I thank all authors for their support and I sincerely hope this book will contribute to better engineering for health.

Prof. Pascal Verdonck
December 2007

List of Contributors

C. Blum
Biophysical Engineering Group, Institute for Biomedical Technology and MESA$^+$ Institute for Nanotechnology, University of Twente, PO Box 217, 7500AE Enschede, The Netherlands

I. Buvat
INSERM U678 UPMC, CHU Pitié-Salpêtrière, F-75634 Paris, France

P. Comon
Laboratoire I3S, UNSA-CNRS, Les Algorithmes, Euclide-B, BP 121, 2000 Route, des Lucioles, 06903 Sophia Antipolis Cedex, France

M. De Beule
Biophysical Engineering Group, Institute for Biomedical Technology and MESA$^+$ Institute for Nanotechnology, University of Twente, PO Box 217, 7500AE Enschede, The Netherlands

A. Jain
Biophysical Engineering Group, Institute for Biomedical Technology and MESA$^+$ Institute for Nanotechnology, University of Twente, PO Box 217, 7500AE Enschede, The Netherlands; Department of Biotechnology, Indian Institute of Technology Kharagpur, Kharagpur, India

C.R. Johnson
Scientific Computing and Imaging Institute, University of Utah, Salt Lake City, UT 84112, USA

O. Meste
Laboratoire I3S, UNSA-CNRS, Les Algorithmes, Euclide-B, BP 121, 2000 Route, des Lucioles, 06903 Sophia Antipolis Cedex, France

Z. Nawrat
Foundation for Cardiac Surgery Development & Medical University of Silesia, Poland

R. Phlypo
Biophysical Engineering Group, Institute for Biomedical Technology and MESA$^+$ Institute for Nanotechnology, University of Twente, PO Box 217, 7500AE Enschede, The Netherlands

S. Staelens
Biophysical Engineering Group, Institute for Biomedical Technology and MESA$^+$ Institute for Nanotechnology, University of Twente, PO Box 217, 7500AE Enschede, The Netherlands

V. Subramaniam
Biophysical Engineering Group, Institute for Biomedical Technology and MESA$^+$ Institute for Nanotechnology, University of Twente, PO Box 217, 7500AE Enschede, The Netherlands

X. Tricoche
Scientific Computing and Imaging Institute, University of Utah, Salt Lake City, UT 84112, USA

V. Zarzoso
Laboratoire I3S, UNSA-CNRS, Les Algorithmes, Euclide-B, BP 121, 2000 Route, des Lucioles, 06903 Sophia Antipolis Cedex, France

CHAPTER 1

Review of Research in Cardiovascular Devices

Zbigniew Nawrat

Contents

1. Introduction 2
2. The Heart Diseases 8
3. The Cardiovascular Devices in Open-Heart Surgery 8
   3.1. Blood Pumps 9
   3.2. Valve Prostheses 23
   3.3. Heart Pacemaker 34
4. The Minimally Invasive Cardiology Tools 34
5. The Technology for Atrial Fibrillation 38
6. Minimally Invasive Surgery 39
   6.1. The Classical Thoracoscopic Tools 40
   6.2. The Surgical Robots 43
   6.3. Blood Pumps – MIS Application Study 49
7. The Minimally Invasive Valve Implantation 53
8. Support Technology for Surgery Planning 53
9. Conclusions 57
Acknowledgments 58
References 58

Abstract

An explosion in multidisciplinary research, combining mechanical, chemical, and electrical engineering with physiology and medicine, during the 1960s created huge advances in modern health care. In cardiovascular therapy, lifesaving implantable defibrillators, ventricular assist devices, catheter-based ablation devices, vascular stent technology, and cell and tissue engineering technologies have been introduced. The latest and leading technology presents robots intended to keep the surgeon in the most comfortable, dexterous, and ergonomic position during the entire procedure. The branch of the medical and rehabilitation robotics includes the manipulators and robots providing surgery, therapy, prosthetics, and rehabilitation. This chapter provides an overview of research in cardiac surgery devices.

Keywords: Heart prostheses, valve prostheses, blood pumps, stents, training and expert systems, surgical tools, medical robots, biomaterials

Advances in Biomedical Engineering
ISBN 978-0-444-53075-2

1. INTRODUCTION

Remarkable advances in biomedical engineering create new possibilities of help for people with heart diseases. This chapter provides an overview of research in cardiac surgery devices. An explosion in multidisciplinary research, combining mechanical, chemical, and electrical engineering with physiology and medicine, during the 1960s created huge advances in modern health care. This decade opened new possibilities in aerospace traveling and in human body organ replacement. *Homo sapiens* after World War II trauma became not only the hero of mind and progress but also the creator of the culture of freedom. Computed tomographic (CT) scanning was developed at EMI Research Laboratories (Hayes, Middlesex, England) funded in part by the success of EMI's Beatles records. Modern medical imaging techniques such as CT, nuclear magnetic resonance (NMR), and ultrasonic imaging enable the surgeon to have a very precise representation of internal anatomy as preoperative scans. It creates possibilities of realizing new intervention methods, for instance, the very popular bypass surgery. It was a revolution in disease diagnosis and generally in medicine. In cardiovascular therapy, lifesaving implantable defibrillators, ventricular assist devices (VADs), catheter-based ablation devices, vascular stent technology, and cell and tissue engineering technologies have been introduced.

Currently, the number of people on Earth is more than 6 billion: increasingly lesser number of living organisms and about million increasingly more "intelligent" robots accompany them.

Robotics, a technical discipline, deals with the synthesis of certain functions of the human using some mechanisms, sensors, actuators, and computers. Among many types of robotics is the medical and rehabilitation robotics – the latest but rapidly developing branch at present, which includes the manipulators and robots providing surgery, therapy, prosthetics, and rehabilitation. They help fight pareses in humans and can also fulfill the role of a patient's assistant. Rehabilitation manipulators can be steered using ergonomic user interfaces – e.g., the head, the chin, and eye movements. The "nurse" robots for patients and physically challenged persons' service are being developed very quickly. Partially or fully robotic devices help in almost all life actions, such as person moving or consuming meals, simple mechanical devices, science education, and entertainment activities. HelpMate, an already existing robot-nurse, moving on the hospital corridors and rooms delivers meals, helps find the right way, etc.

On the one hand, robots are created that resemble the human body in appearance (humanoids), able to direct care; on the other hand, robotic devices are constructed – telemanipulators – controlled by the human tools allowing to improve the precision of human tasks. Robots such as ISAC (Highbrow Soft Arm Control) or HelpMate can replace several functions of the nurse, who will give information, help find the way, bring the medicines and the meal. In case of lack of qualified staff, to provide care for hospice patients at home, these robots will be of irreplaceable help.

Robotic surgery was born out of microsurgery and endoscopic experience. Minimally invasive interventions require a multitude of technical devices: cameras, light sources, special tools (offering the mechanical efficiency and tissue coagulation for preventing bleeding), and insufflations (thanks to advances in computer engineering, electronics, optics, materials, and miniaturization). The mobility of instruments is decreased [from seven, natural for human arm, to four degrees of freedom (DOFs)] due to the invariant point of insertion through the patient's body wall. Across the world, physicians and engineers are working together toward developing increasingly effective instruments to enable surgery using the latest technology. The leading technology presents robots intended to keep the surgeon in the most comfortable, dexterous, and ergonomic position during the entire procedure. The surgery is complex and requires precise control of position and force. The basic advantages of minimally invasive robot-aided surgery are safe, reliable, and repeatable operative results with less patient pain, trauma, and recovery time. Conventional open-heart surgery requires full median sternotomy, which means cracking of sternum, compromising pulmonary function, and considerable loss of blood.

Milestones in the Evolution of Cardiac Devices

1628	William Harvey, St Bartholomew's Hospital, London, presented his theory of the circulatory system. He described the function of the heart, arteries, and veins. It is considered to be one of the greatest advances in medicine.
1812	Julien-Jean Cesar LeGallois, a French physician, proposed the idea of artificial circulation.
1882	German von Schröder introduced the first bubble oxygenator.
1929	Werner Forssmann, a German surgeon, developed the technique of cardiac catheterization, the first to document right heart catheterization in humans using radiographic techniques (won the Nobel Prize in 1956).
1934	Dr Michael DeBakey invented the DeBakey pump (peristaltic).
1937	Artificial heart designed by the Soviet scientist W.P. Demichow was first successfully applied on the dog for 5.5 h.
1949	IBM developed the Gibbon Model I heart–lung machine, delivered to Jefferson Medical College, Philadelphia, PA, USA. It consisted of DeBakey pumps and film oxygenator.
1952	Paul Zoll developed the first cardiac pacemaker.
1952	Charles Hufnagel sewed an artificial valve into a patient's aorta.
1953	Dr John H. Gibbon, Jr, Jefferson Medical College Hospital, Philadelphia, PA, USA, first successfully applied extracorporeal circulation in an 18-year-old female with an atrial septal defect.
1953	Dr Michael DeBakey, Baylor University, Houston, TX, USA, implanted a seamless, knit Dacron tube for surgical repairs and/or replacement of occluded vessels or vascular aneurysms.
1957	Wild et al. reported the use of ultrasound to visualize the heart noninvasively.
1957	Dr C. Walton Lillehei and Earl Bakken, an electronic engineer, developed the first portable pacemaker. Bakken later formed the Medtronics Corporation.
1957	Drs William Kolff and Tetsuzo Akutsu at the Cleveland Clinic implanted the first artificial heart in a dog. The animal survived for 90 min.

1958	Dr Mason Sones, a cardiologist at the Cleveland Clinic Foundation, developed coronary angiography.
1960s	Semm et al. developed laparoscopic instrumentation.
1960	Dr Albert Starr, an Oregon surgeon, developed the Starr–Edwards heart valve. One of the most successful heart valves produced until the late 1970s.
1967	René Favaloro, an Argentine surgeon in the United States, performed the first coronary bypass operation using the patient's native saphenous vein as an autograft.
1967	Christiaan Barnard performed the first heart transplantation.
1968	A. Kantrowitz et al. performed the first clinical trial in a man with intra-aortic balloon pumping.
1969	Dr Denton Cooley, Texas Heart Institute, Houston, TX, USA, implanted a total artificial heart (TAH) designed by Domingo Liotta. The device served as a "bridge" to heart transplantation until a donor heart was found, for 64 h. The heart transplant functioned for an additional 32 h until the patient died of pneumonia.
1971	White – ECMO on newborn babies using veno-venous bypass for up to 9 days.
1975	A. Gruentzing developed the first balloon catheter.
1975	Dr Willem Kolff, University of Utah, designed a nuclear-powered artificial heart (Westinghouse Corporation).
1975	BioMedicus BioPump (Centrifugal) introduced for clinical applications.
1975	Computerized axial tomography, the "CAT scanner", was introduced.
1977	Newer generations of mechanical prostheses included the monoleaflet (Medtronic-Hall) and the bileaflet (St Jude Medical).
1979	The Jarvik TAH was designed using a flexible four-layer diaphragm and a structural design that fits in the human chest. This design was a larger 100 cc version of today's CardioWest TAH–t, which is 70 cc.
1981	Dr Denton Cooley implanted another pneumatically driven artificial heart designed by Dr Akutsu. This artificial heart was used for 27 h as a "bridge" to cardiac transplantation.
1982	Dr William DeVries implanted the Jarvik 7 into Barney Clark, DDS. Dr Clark lived for 112 days.
1984	Baby girl Faye's native heart, Loma Linda Medical Center, was explanted and replaced with a baboon heart. She survived for 3 weeks.
1984	First human implant and successful bridge-to-transplant – a Novacor® LVAS.
1985	The FDA gave approval for Hershey Medical Center to perform six PennState artificial heart implants as bridges to human heart transplantations. This heart is no longer used with human subjects.
1985	At the University of Arizona, Dr Jack Copeland implanted a prototype TAH in a patient who had rejected a recently transplanted heart.
1986	The first atherectomy devices that remove material from the vessel wall were introduced.
1987	Introduction of the first use of coronary stent (by 1997, more than 1 million angioplasties had been performed worldwide).
1990	First LVAS patient discharged home with a Novacor LVAS.
1990–1992	The FDA had withdrawn the Investigational Device Exemption (IDE) from Symbion for the clinical study of the Jarvik TAH. Symbion

	subsequently donated the TAH technology to University Medical Center (UMC), Tucson, AZ, USA, which reincorporated the company and renamed it CardioWest.
1994	First FDA-approved robot for assisting surgery [automated endoscopic system for optimal positioning (AESOP) produced by Computer Motion (CM; Goleta, CA, USA)].
1994	FDA approved the pneumatically driven HeartMate® LVAD (Thoratec Corporation, Burlington, MA, USA) for bridge to transplantation (the first pump with textured blood-contacting surfaces).
1994	HeartMate LVAS has been approved as a bridge to cardiac transplantation.
1996	REMATCH Trial (Randomized Evaluation of Mechanical Assistance for the Treatment of Congestive Heart failure, E. Rose principal investigator) initiated with HeartMate® VE (Thoratec Corp.). Results published in 2002 showed mortality reduction of 50% at 1 year as compared to patients receiving optimal medical therapy.
1998	Simultaneous FDA approval of HeartMate VE (Thoratec Corporation, Burlington, MA, USA) and Novacor LVAS (World Heart Corporation, Ontario, Canada), electrically powered, wearable assist systems for bridge to transplantation, utilized in more than 4000 procedures to 2002. Till now, we can estimate 4500 HeartMate XVE, more than 440 IVAD (Implantable Ventricular Assist Device) and more than 1700 Novacor in this kind of blood pump.
1998	First clinical application of next-generation continuous-flow assist devices. DeBakey (MicroMed Inc.) axial-flow pump implanted by R. Hetzer, G. Noon, and M. DeBakey.
1998	Carpentier and Loulmet performed first in the world endoscopic operation of single bypass graft between left internal thoracic artery and left anterior descending (LITA–LAD) and first operation inside the heart – mitral valve plastic and atrial septal defect closure was performed using surgical robot da Vinci (Intuitive Surgical, Sunnyvale, CA, USA).
1998	Mohr and Falk bypass surgery and mitral valve repairs in near endoscopic technique (da Vinci).
1999	First clinical application of a totally implantable circulatory support system. LionHeart LVAS implanted in a 67 year-old male recipient by R. Koerfer and W. Pae.
1999	D. Boyd first totally endoscopic coronary artery bypass graft (E-CABG) using Zeus robot (Computer Motion, CA, USA, currently intuitive surgical, not in the market).
2000	Physicians in Houston US have realised the implantation and have got the first patient in the clinical investigation of the Jarvik 2000 Heart. Jarvik Heart, Inc. and the Texas Heart Institute began developing the Jarvik 2000 Heart in 1988.
2000	THI was granted permission by the Food and Drug Administration to evaluate the Jarvik 2000 Heart as a bridge to transplantation in five patients.
2000	The FDA gave permission to extend the study. Patients have been sustained for more than 400 days with this device. (http://www.texasheartinstitute.org/Research/Devices/j2000.cfm).
2001	The AbioCor was implanted in Robert Tools by cardiac surgeons Laman Gray and Robert Dowling on July 2, 2001, at Jewish Hospital in Louisville, Kentucky. L. Robert Tools aged 59 years with the artificial heart survived 5 months. He died because of the blood clot.

2001	Doctors Laman Gray and Robert Dowling in Louisville (KY, USA) implanted the first autonomic artificial heart – AbioCor (Abiomed, Inc., Danvers, MA, USA). The FDA approved the Abiocor for commercial approval under a Humanitarian Device Exemption in September, 2006. ABIOMED is also working on the next generation implantable replacement heart, the AbioCor II. Incorporating technology both from ABIOMED and Penn State, the AbioCor II is smaller and is being designed with a goal of 5-year reliability. The BVS 5000 was the first extracorporeal, or outside the body, ventricular assist device on the market and is still the most widely used bridge to recovery device with systems located in more than 700 institutions throughout the world. Abiomed also offers the Impella 2.5 – a minimally invasive, percutaneous ventricular assist device that allows the heart to rest and recover (http://www.abiomed.com/products/faqs.cfm).
2001	Drs Laman Gray and Robert Dowling in Louisville (KY, USA) implanted the first autonomic artificial heart – AbioCor (Abiomed Inc., Danvers, MA, USA).
2001	The first transatlantic telesurgery – Lindbergh operation – surgeon from New York operated patients in Strasburg using Zeus® system.
2001	The first full implantable TAH Lion Heart (the Texas Heart Institute in Houston, Abiomed in Danvers, US) was used. The first-ever human implant of the LionHeart™ left ventricular assist system took place October 26, 1999 at the Hearzzentrum NRW in Bad Oeynhausen, Germany. Eight patients have lived with the device for more than 1 year, and four patients have lived with the device for more than 2 years. The Food and Drug Administration approved the first series of US clinical trials for the Arrow LionHeart™ heart assist device in February 2001. Penn State Milton S. Hershey Medical Center implanted the device in the first US patient later that month (http://www.hmc.psu.edu/lionheart/clinical/index.htm).
2001	The first totally implantable TAH LionHeart (the Texas Heart Institute in Houston, TX, USA; Abiomed in Danvers, MA, USA) was used.
2002	FDA approved the HeartMate VE LVAD for permanent use (Thoratec Corp.).
2002	Novacor LVAS became the first implanted heart assist device to support a single patient for longer than 5 years.
2002	The first percutaneous aortic valve replacement was performed by Alain Cribier in April of 2002 [xxxx]. To
2004	The CardioWest TAH-t becomes the world's first and only FDA-approved temporary total artificial heart (TAH-t). The indication for use is as a bridge to transplant in cardiac transplant patients at risk of imminent death from nonreversible biventricular failure. SynCardia Systems, Inc. (Tucson, AZ, USA) is the manufacturer of the CardioWest™ TAH-t. It is the only FDA- and CE-approved TAH-t in the world. It is designed for severely ill patients with end-stage biventricular failure. The TAH-t serves as a bridge to human heart transplant for transplant eligible patients who are within days or even hours of death. A New England Journal of Medicine paper published on August 26, 2004 (NEJM 2004; 351: 859–867), states that in the pivotal clinical study of the TAH-t, the 1-year survival rate for patients receiving the CardioWest TAH-t was 70% versus 31% for control patients who did not receive the device. One-year and 5-year survival rates after transplantation among patients who had received a TAH-t as a bridge to human heart transplant were 86% and 64%. The highest bridge to human heart transplant rate of any heart device is 79%

	(New England Journal of Medicine 2004; 351: 859–867). Over 715 implants account for more than 125 patient years on the TAH-t (http://www.syncardia.com).
2005	A total of about 3000 cardiac procedures were performed worldwide using the da Vinci system. This includes totally endoscopic coronary artery bypass grafting (TECAB), mitral valve repair (MVR) procedures, ASD closure, and cardiac tissue ablation for atrial fibrillation.
2006	The next-generation pulsatile VAD, the Novacor II, entered a key phase of animal testing with the first chronic animal implant.
2006	The FDA approved the first totally implantable TAH – developed by AbioCor – for people who are not eligible for heart transplantation and who are unlikely to live more than a month without intervention.
2006	Have experience with human implantation: Edwards Lifesciences (>100 patients) and CoreValve (>70 patients). (www.touchbriefings.com/pdf/2046/Babaliaros.pdf)
2006	US FDA approval to start pilot clinical trial for Impella 2.5 (Abiomed, Inc., Danvers, MA, USA). To date the Impella 2.5 has been used to support over 500 patients during high-risk PCI, post PCI, and with AMI with low cardiac output.
2007	More than 4,500 in 186 centers advanced heart failure patients worldwide have been implanted with the HeartMate XVE. Longest duration of support (ongoing patient on one device): 1,854 days, age range: 8–74 (average 51). The REMATCH (Randomized Evaluation of Mechanical Assistance for the Treatment of Congestive Heart Failure) clinical trial demonstrated an 81% improvement in two-year survival among patients receiving HeartMate XVE versus optimal medical management. A Destination Therapy study following the REMATCH trial demonstrated an additional 17% improvement (61% vs. 52%) in one-year survival of patients receiving the HeartMate XVE, with an implication for the appropriate selection of candidates and timing of VAD implantation.
2008	State of art in 2008, based on the experience almost 20–30 years shows among others: – future of long-term heart support methods will be connected with rotary pumps, the stop of activity in pulsating pump technology is observed (rotary pumps are smaller and comfortable in the implantation especially it is possible introduce mini invasive surgical methods) – the mono-leaflet disk valve are applied more and more seldom (they require during the implantation the definite orientation in the heart) in the comparison with bi-leaflet – the new kind of biological valve prostheses based on tissue engineering method get into the clinical experiment – the multiple companies have engineered percutaneous heart valves (PHV), primarily for aortic stenosis
2008	There have been 867 unit shipments worldwide – 647 in the United States, 148 in Europe and 72 in the rest of the world (http://www.intuitivesurgical.com). Devices for "robotic surgery" are designed to perform entirely autonomous movements after being programmed by a surgeon. The *da Vinci* Surgical System is a computer-enhanced system that interposes a computer between the surgeon's hands and the tips of microinstruments. The system replicates the surgeon's movements in real time.

2. THE HEART DISEASES

The human biological heart has two sets of pumping chambers. The right atrium receives oxygen-depleted blood from the body, which is pumped into the lungs through the right ventricle. The left atrium receives aerated blood from the lungs, which is pumped out to the body through the left ventricle. With each heart beat, the ventricles contract together. The valves control the direction of blood flow through the heart.

Congestive heart failure, a condition in which the heart is unable to pump the blood effectively throughout the body, is one of the leading causes of death. This disease is caused by sudden damage from heart attacks, deterioration from viral infections, valve malfunctions, high blood pressure, and other problems. Although medication and surgical techniques can help control symptoms, the only cure for heart failure is heart transplantation. Artificial hearts and pump assist devices have thus been developed as potential alternatives.

Ischemic heart disease is caused by progressive atherosclerosis with increasing occlusion of coronary arteries resulting in a reduction in coronary blood flow. Blood flow can be further decreased by superimposed events such as vasospasm, thrombosis, or circulatory changes leading to hypoperfusion. Myocardial infarction (MI) is the rapid development of myocardial necrosis caused by a critical imbalance between the oxygen supply and the oxygen demand of the myocardium. Coronary artery bypass grafts (CABGs) are implanted in patients with stenosed coronary arteries to support myocardial blood flow.

Valvular heart disease is a life-threatening disease that affects millions of people worldwide and leads to valve repairs and/or replacements.

In the USA and Europe alone, with more than 600 million inhabitants and more than 6 million patients with heart failure, the prevalence of advanced heart failure, constituting 1–10% of the heart failure population, is estimated to total between 60,000 and 600,000 patients. More than 700,000 Americans die each year from heart failure, making it the number one cause of death in the U.S., as well as worldwide. About half of these are sudden cardiac deaths, which occur so quickly that there is not enough time for intervention with a cardiac assist or replacement device. For the remaining half, heart transplantation is one of the few options available today. Though hundreds of thousands are in need, only about 2,000 people in the U.S. will be able to receive donor hearts every year. This consistent shortage in the supply of donor hearts in the U.S. demonstrates the need for an alternative to heart transplantation. The total potential market for the artificial heart is more than 100,000 people in the U.S. each year. (http://www.abiomed.com).

3. THE CARDIOVASCULAR DEVICES IN OPEN-HEART SURGERY

Cardiac surgery is surgery on the heart and/or great vessels. This surgery is a complex procedure requiring precise control of position and force. Conventional open-heart surgery requires full median sternotomy, which means cracking of sternum, compromising pulmonary function, and considerable loss of blood.

The repair of intracardiac defects requires a bloodless and motionless environment, which means that the heart should be stopped and drained of blood. Hence, the patient requires the function of the heart and lungs provided by an artificial method.

Modern heart–lung machines can perform a number of other tasks required for a safe open-heart operation. This system preserves the patient's own blood throughout the operation and the patient's body temperature can be controlled by selectively cooling or heating the blood as it flows through the heart–lung machine. Medications and anesthetic drugs can be administered via separate connections. The disadvantages include the formation of small blood clots, which increase the risk of stroke, pulmonary complications, and renal complications. The machine can also trigger an inflammatory process that can damage many of the body's systems and organs. Those risks push today's biomedical engineers to improve the heart–lung machine and oxygenator, while surgeons are developing advances that would eliminate the need for the machine altogether. One such advance is minimally invasive surgery (MIS).

The surgeons have begun to perform "off-pump bypass surgery" – coronary artery bypass surgery without the aforementioned cardiopulmonary bypass. The surgeons operate on the beating heart stabilized to provide an (almost) still work area. One of the greatest challenges in beating-heart surgery is the difficulty of suturing or sewing on a beating heart. A stabilization system makes it possible for the surgeon to work on the patient's beating heart carefully and, in the vast majority of cases, eliminates the need for the heart–lung machine.

3.1. Blood Pumps

The human heart is a pump that is made of muscle tissue. A special group of cells called the sinus node is located in the right atrium. The sinus node generates electrical stimuli that make the heart contract and pump out blood. The normal human heart beats about 75 times per minute (i.e., about 40 million times a year) – i.e., the heart pumps 5 l of blood per minute. The normal systemic blood pressure is 120/80 mmHg. The mechanical power (calculated by multiplying the pressure by the flow rate) of the human heart is about 1.3 W. However, to provide this mechanical power, the heart requires 10 times much higher rate of energy turnover, owing to its low mechanical efficiency (less than 10%).

However, the development in biotechnology can open the opportunity for tissue engineering (a branch of biotechnology) – a prospect of saving people with extremely complex or irreversible failure heart will still be realized using mechanical heart support devices.

During the last half-century, various blood pumps were introduced into clinical practice, which can partially support or replace the heart during open-heart surgery or considerably for a longer time period until heart recovers or until transplantation is performed. Several millions of people owe their health and lives to these devices. According to the American Heart Association, an estimated 5 million Americans are living with heart failure and more than

400,000 new cases are diagnosed every year. About 50% of all patients die within 5 years.

The first operation on a beating heart was performed by Gibbon in 1953 using a peristaltic pump. Since then, there has been rapid development in mechanical assist devices, leading to the realization of today's mechanical aid system for blood circulation, based on the requirements using the following:

- intra-aortic balloon pump (IABP)
- continuous blood flow devices – roll and centrifugal rotary pumps
- pulsating flow blood pumps – pneumatic or electro-control membrane pumps

The assumed time of blood pump using in the organism influences the construction and material used in its design. According to this criterion, the pumps can be divided into following categories:

- short-term (during the operation or in sudden rescue operations)
- medium-term – over several weeks or months (as a bridge to heart transplantation or treatment)
- long-term – several years (already at present) and permanently (in the intention) (as the target therapy)

The extracorporeal circulation (*perfusion*) and the controlled stopping of heart action make it possible to perform the open-heart operation. The peristaltic pumps are in common use, where the turning roll locally tightens the silicon drain to move a suitable volume of the blood. The disadvantages of this procedure, such as the damage to blood components, are eliminated or reduced by pump construction and material improvement. The patients with both heart and lung failure are assisted by the system consisting of pump and oxygenator. The method of time oxygenation (supply oxygen to the blood) for extracorporeal blood – extracorporeal membrane oxygenation (ECMO) – was applied the first time in 1972 by Hill. The blood pump, membrane oxygenator, heat exchanger, and a system of cannulas (tubes) for patient's vascular system connection are the elements of the system. Inner surface of all these parts are covered with heparin. The Extracorporeal Life Support Organization Registry has collected (since 1989) data on more than 30,000 patients, most of whom have been neonates with respiratory failure [7]. The currently available centrifugal pump VADs are BioMedicus BioPump (Medtronic, Minneapolis, MN, USA), CentriMag® (Levitronix, Zürich, Switzerland), RotaFlow® (Jostra, Hirrlingen, Germany), and Capiox® (Terumo, Ann Arbor, MI, USA). These pumps have been available since 1989 to support neonates and older children with postoperative cardiac failure but competent lung function.

Centrifugal pumps benefit from the physical phenomena (centrifugal, inertia force) of blood acceleration during temporal rotational movement. These pumps consist of a driving unit and an acrylic head driven by a magnetic couple. Input and output blood flow channels are perpendicular to each other. The output velocity of the blood on the conical rotor depends on the input rotary speed, preload, and afterload. Based on vortex technology, these pumps use turbine spins of 10,000–20,000 rpm to create a flow of 5–6 l/min and have generally been applied for temporary assistance of stunned myocardium of the left ventricle. The construction of working pump and its environment conditions influence the blood hemolysis during its use. According to the results of in vitro experiment, with the pump working as VAD and extracorporeal circulation (CPB and ECMO), the use of the pump as VAD caused the least degree of hemolysis, and the hemolysis of pumps in the time to ECMO strongly depends on the kind of oxygenator used. Pumps with the conical rotor caused a greater degree of hemolysis working with small flows and large pressures (ECMO). On the contrary, a lesser degree of hemolysis was observed for pumps with the flat rotor, regardless of its purpose.

The *intra-aortic balloon* was introduced as a tool for coronary circulation assistance in works of Maulopulos, Topaz, and Kolff in 1960. In 1968, Kantrowitz was the first to prove the clinical effectiveness of intra-aortic counterpulsation. It became the classic method in the 1970s, thanks to Datascope Company for its first device (System 80) for their clinical applications. The balloon (30–40 ml) placed in the aorta, synchronically with the heart, is filled up and emptied by raising the diastolic pressure and the perfusion of coronary arteries. As a result, the myocardial contractibility is improved, and hence, the cardiac output is bigger. This is one of the basic devices in cardiosurgery units. The balloon, rolled up on a catheter, is introduced through the artery. Kantrowitz introduced to cardiosurgery also another type of pump called the *aortic patch*, which has been in clinical use for many years, and uses the natural localization of heart valve. The balloon, which is sewn to cut the descending aorta, is connected through the skin with the pneumatic driver unit. The aortic blood flow is improved by the pulsate balloon filling and the aortic valve blocked the backflow to the ventricle.

New constructions of TAHs and VADs offer new hope for millions of heart patients whose life expectancy is greatly reduced because the number of patients waiting for a transplant far exceeds the donated hearts available. An artificial heart or VAD is made of metal (typically titanium–aluminum–vanadium alloy), plastic, ceramic, and animal parts (valve bioprostheses). A blood-contacting diaphragm within the pump is made from a special type of polyurethane that may also be textured to provide blood cell adherence.

The VAD (Figure 1) is an extracorporeal pneumatic blood pump, invented to assist a failure left or right ventricle, or the whole heart until it recovers or until a replacement by transplantation is performed. A mechanical heart assist is used to support the heart during failure caused by the ischemia of the heart muscle, the small cardiac output, cardiomyopathy, or heart valve diseases. The role of these aid pumps is to support the life function of the brain and other organs and heart treatment (thanks to the devices that partly take over its role) by improving hemodynamic conditions. It is fitted with inflow and outflow cannulas for connecting the VAD with heart and vascular system in parallel with the biological heart ventricle being assisted. The VAD consist one or two ventricles placed outside the body (failing heart is left in place) connected to the control unit.

The TAH (Figure 2) is designed for emergency replacement of irreversible damaged natural heart. Because the natural heart is completely removed, the TAH must ensure the hemodynamic, regulatory, and control functions of the circulatory system.

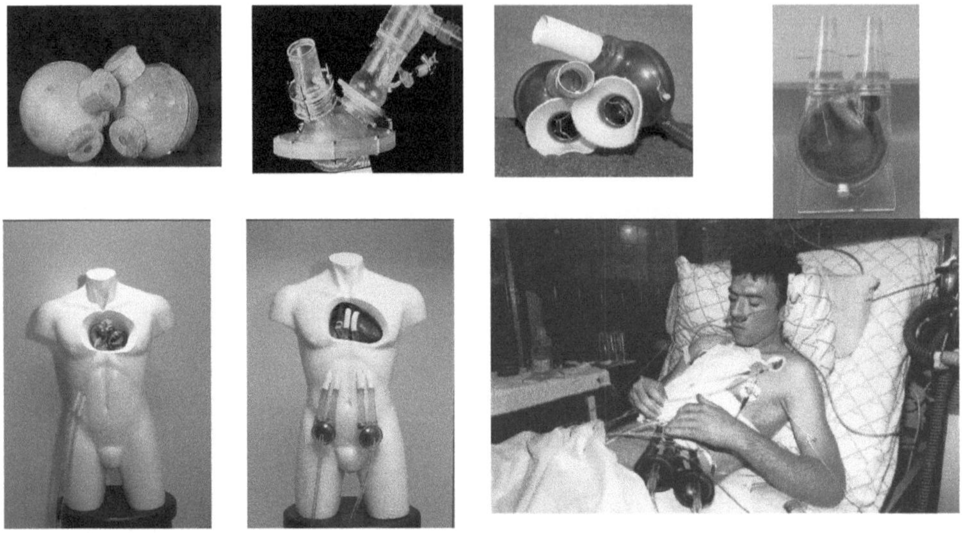

Figure 1 The steps in POLTAH design process (model after cadaver study, transparent model for laser visualization investigation, first polyurethane model for starting animal experiments). Nonsymmetrical shape of POLVAD proposed and designed creates a very good internal flow condition. The first successful POLVAD implantation was performed in 1993 in T. Gruszczyński. Until now, more than 200 POLVAD clinical implantations have been performed, the longest time successful application being over 200 days. Pneumatically driven artificial heart (POLTAH) and ventricular assist device (POLVAD). (See color plate 1).

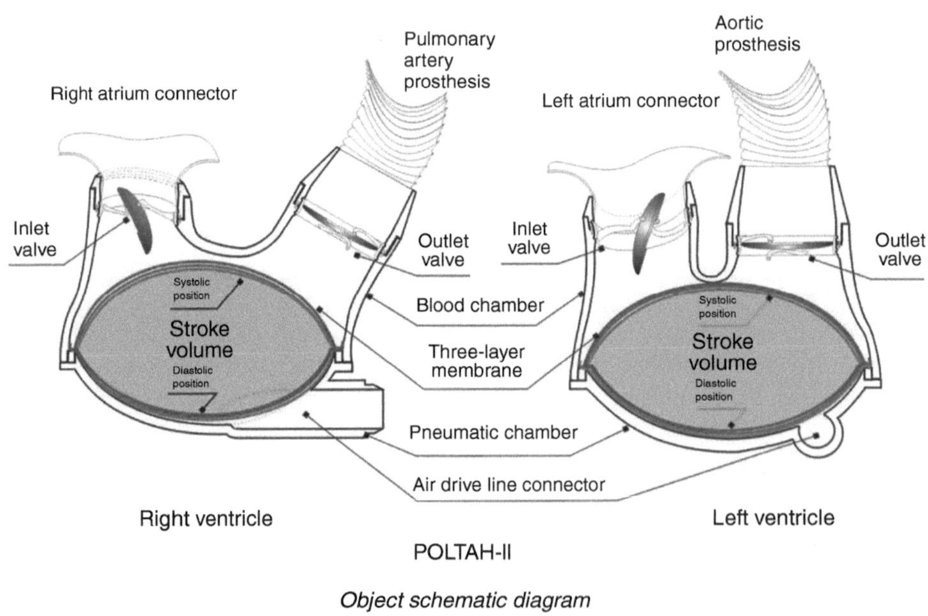

Figure 2 Schematic diagram of an artificial heart.

If the total replacement artificial heart is to mimic the function of the natural human heart, it meet the following requirements [1]:

1. Two separate pumps duplicating the left and the right heart
2. Output range of 5–10 l/min from each side
3. Aortic arterial pressure range of 120–180 mmHg
4. Pulmonary arterial pressure range of 20–80 mmHg
5. Physical dimensions permitting easy surgical insertion within the pericardium
6. Low weight to minimize restraints
7. A minimum of noise and vibration
8. A heat output of less than 25 W
9. A low degree of hemolysis
10. Inert to body chemicals

The artificial heart, designed by the Soviet scientist W.P. Demichow, was first successfully applied on a dog for 5.5 h in 1937. Several study groups in the United States in the 1950s performed the first clinical experiments. In 1957 T. Akutsu and W. Kolff (Figure 3) made the pneumatically driven artificial heart. The artificial heart constructed by D. Liotta with the same drive type successfully replaced the patient heart for 64 h, before heart transplantation in 1969. On 4 April 1969, Dr Denton A. Cooley performed the first human implantation of a TAH when he used a device developed by Dr Domingo Liotta to sustain the life of Mr Haskell Karp. After having undergone resection of a left ventricular aneurysm, the 47-year-old patient developed cardiac failure. The Liotta heart supported him for nearly 3 days until a suitable donor heart was available and successfully transplanted. This case demonstrated the feasibility of bridging to transplantation with a mechanical circulatory support system.

In 1982 William DeVries (UT, USA) implanted the artificial heart designed by R. Jarvik, cooperating in that time with W. Kolff, called *Jarvik* 7 in a patient for 112 days. The Jarvik heart was pneumatically driven (driver unit called COMDU) through a drain carried out by the skin. Four disk valves from Medtronic-Hall were fixed in the polyurethane body. The inner surface of ventricle was covered with a smooth layer of polyurethane (Biomer, Ethicon Inc.) and a four-layer membrane made from Biomer that separated the pneumatically driven part from the part

Figure 3 William Kolff in Kampen and in Gent (photo taken by the author).

having contact with the blood. There were two types of these VADs: Jarvik 7-70 and Jarvik 7-100 with 70 and 100 ml of stroke volume, respectively. Since 1993, these types of the artificial heart after a little modification have been clinically used [now under the name CardioWest (the technology transferred from the firm Symbion to CardioWest Technologies Inc., Tuscon, AZ, USA currently **Syncardia Systems**, Inc. Tucson, AZ, USA) and is marked C-70 and C-100, respectively]. The pump can operate at 30–199 bpm and also has variable systolic duration, delay, and stroke volume. By 1989, such devices had largely become a bridge to human heart transplantations. In a pivotal clinical study [2], these patients were successfully transplanted 79% of the time. The 1-year and 5-year survival rates after heart transplantation among these patients were 86 and 64%, respectively.

In Europe, the hospital La Pitié in Paris has the biggest experiences in implanting the artificial heart in patients. From April 1986 to June 2000, there had been 120 implantations of pneumatically driven Jarvik VADs (with the trade name TAH *Symbion* and at present TAH CardioWest). The total time of the implantation was more 3000 days (with the longest being 607 days implantation). Another type of pneumatically driven TAH named *PennState* was made at the Pennsylvania State University, where ventricles were also made from Biomer, but sacks were used instead of membrane-type construction. PennState VAD with the Björk–Shiley (B–S) disk valves was considerably not as often applied as the Jarvik heart.

As a next stage, the electrically powered PennState artificial heart was constructed. In this device, a DC brushless motor turns over the helical (screw) drive, which causes compression and expansion of segment polyurethane sack resulting in an 8 l/min volume flow with 16–22 W of consumed power. The total animal experimentation time is about 5 years at present, the longest trial being more than 5 months. Searching for the biocompatible and durable biomaterials in 1969, Y. Nose made the steel heart covered with Teflon (so-called *Green Heart*), which he implanted in the ram for 50 h. There were also trials to use properly prepared heart pericardium for the internal structure of the ventricle. Electrohydraulic artificial heart (Japan) *EHTAH* is equipped with ellipsoidal ventricle membranes made from nonthrombogenic, segmented polyurethane, 21 and 23 mm B–S valves and vascular Dacron grafts (Medox Medical, USA). *Baylor* TAH (Baylor College of Medicine, USA) is the electromechanical pump (the rotation of the screw causes the movement of membrane). The membrane is made from polyolefin rubber, the ventricle body from epoxy Hysol (Dexter Electronic Materials, USA) or at present from coal fibers, which is lighter and stronger. In the laboratory of Kolff, small ventricles with semielastic shell were worked out. They were made from Pellethane 80AE/2363 (Dow Chemical, Japan) by the vacuum forming and glued by dielectric welding. Valves are made from Isoplast, covered with Pellethane, and glued. W. Kolff sent the sets of ventricles to many centers in the world, for realization experiments, to Poland as well. In Zabrze comparative in vitro investigations were carried out with these ventricles to detect the extent of artificially induced thrombogenesis.

AbioCor implantable replacement heart (Abiomed, Inc., Danvers, MA, USA) is a fully implantable, electrically driven artificial heart. The rotary pump forces the movement of liquid, which by means of a second motor is directed to the left or right ventricle causing cramp of polyurethane sacks with two polyurethane valves. It does not require a compliance chamber (the internal hydraulic platform was rather applied). The normal flow is 4–8 l/min up to a maximum of 10 l/min. Internal implanted batteries enable 30–60 min of work. The pump regulates automatically its work frequency in the range 60–140 bpm. As a pump biomaterial tested for the last 20 years, the elastomer Angioflex was applied. After successful laboratory investigations on implantation in animals, the clinical experiment begun. On 3 July 2001, Drs Laman Gray and Robert Dowling in Louisville (KY, USA) implanted the first autonomic artificial heart (AbioCor). It consists of two ventricles separated by the pumping mechanism. Four original polyurethane valves are used to direct the flow (identical valves like these had been applied for many years in the Abiomed BVS pump). Energy from the external battery is wirelessly transferred through the skin port. The final goal of AbioCor applications is to reach the 5-year, statistical survival of patients with the replaced natural heart. Unlike the Jarvik 7, the AbioCor is powered by electrical energy that is transmitted from a battery across the skin to an internal coil and a backup battery. Because an opening in the skin is not needed to allow passage for tubes or wires, the risk of infection is greatly reduced. In addition, the external battery pack is designed to be worn on a belt or suspenders, enabling the patient to be mobile. From 2001 to 2004, the patients who had received the heart and survived the operation lived for 5 months, on average, the longest being not more than 17 months. In 2006, the AbioCor was approved for use in patients who do not qualify for a heart transplantation if their life expectancy as a result of heart failure is less than a month; the device is also approved as a temporary measure for patients awaiting a transplant.

The Abiomed Company designed AbioCor based on experience with application blood assist pumps. *Abiomed BVS 500* (Abiomed, Inc., Danvers, MA, USA) is pneumatically driven, two-chamber pump one atrium without any power support and the other pneumatically driven ventricle for pumping freely (means not pneumatically driven, under the natural blood pressure) filled atrium with the freely filled atrium, vertically fixed to the patient's bed. Relatively long cannulas are used for establishing the patient's circulatory system connection. It is relatively easy to adjust, with the full ejection mode (80 ml). As a first, the company introduced the polyurethane valves to the pump (had been in clinical use for more than 10 years). Cannulas are made from Angioflex, which is a nontoxic and durable material (can withstand 100,000 beats/day). Endings of the cannulas sewn to the atrium have a special armature and the shape preserving collapsing the atrium walls or the blocking of the cannula inlet. From the output side, the arterial cannulas have special Dacron collars facilitating their sewing.

In Europe, the most developed construction of implantable artificial heart is located in Aachen and in Berlin. In Poland, apart from pneumatically driven artificial heart POLTAH, in the Artificial Heart Laboratory of Foundation for Cardiac Surgery Development in Zabrze, promising construction of implantable electrohydraulic pump project is being led. According to plan, this project will give as a result a totally implantable VAD with all driven and control units inside the patient's body.

In 1966, the pump designed by M. DeBakey, who was also the constructor of the first roll pump applied to the perfusion of the heart during surgery, was a first successfully implanted membrane VAD for left heart ventricle support. The future of implantable pumps is based on the electric drive. The first VAD of this type was

Novacor (Novacor Division, World Heart Corporation, Ontario, Canada) with electromagnetic driving system. The blood flows in the rhythmically compressed sack, made from Biomer (polyurethane) having two biological valves (made from pericardium 25 mm or porcine 21 mm from the Edwards Laboratories, Baxter Inc.) working as the one-way flow valves. The controller provides electrical energy, via the percutaneous lead, to actuate the driver during the ejection cycle. The driver's balanced, symmetrical solenoid (electromagnet) converts this electrical energy into mechanical energy, stored in a pair of springs. The spring forces act directly on the blood pump's dual symmetrically opposed pusher plates, pressurizing the blood within the pump, and propelling it into the aorta to support systemic circulation. The ventricle body was made from polyester resin, strengthened by glass fibers. It is located in the left upper square of the stomach. The blood flows into the ventricle from Dacron cannula placed in the apex of the left ventricle and is ejected to the system through the Dacron cannula as well to the side wall of the ascending aorta. It works synchronically with the heart. The pump/drive unit can eject from all-fill volumes (from 25 ml to its maximum capacity of 70 ml), to match the stroke volume received from the left ventricle. A small residual blood volume, proportional to afterload (systemic pressure), remains in the blood pump at the end of ejection. The system can thereby adapt to the natural heart's function without depending on an ECG signal or other physiological measurement. Currently, WorldHeart's Novacor LVAS has been implanted in nearly 100 medical centers worldwide in more than 1700 patients, and is the first mechanical circulatory support device to support a single patient for more than 6 years. Also, 172 primarily bridge-to-transplant patients have been supported by Novacor LVAS for more than 1 year, 45 have been supported for more than 2 years, 24 for more than 3 years, 11 for more than 4 years, and 1 for more than 6 years. Only 1.4% of the pumps have needed replacement. No patient deaths have been attributed to Novacor LVAS failure.

Another type of the implantable VAD used for several years as a pneumatically and electrically driven pump is *HeartMate VAD* (HE HAS; Thermo Corporation, Burlington, MA, USA (http://www.thoratec.com/vad-trials-outcomes)). *Heart-Mate* is the family of ventricles. First FDA on the mechanical VAD (Pneumatic HeartMate® I) in the United States was designed as the platform to the transplantation. Electrically driven HeartMate received FDA in 1998. All ventricles were made from the especially rough surface facilitating the formation of natural endothelium. This VAD is implanted in abdominal space between the heart apex and the ascending aorta and the power supply is transferred by the wire through the skin. The second pipe (through the same skin port) is used to deaerate the chamber or in emergency states allows an external person to manually drive the pump. The volume flow is up to 11 l/min. Closed in titan body motor through the cam system causes the pulsator to move up and down the synthetic sack filled with the blood, closed at two sides by biological pig valves (25 mm, Medtronic). This company also introduced different solution, assuring biocompatibility of membrane. Instead of focusing on achieving the maximal smoothest of polyurethane bag, they worked out material with built-in network of the fibers that facilitates the creation of natural endothelium. Currently, similarly to R. Jarvik, this company works out the artificial heart with turbine axial pump.

PennState LionHeart, *LionHeart LVD LVAS* (Arrow International, Inc., Reading, PA, USA), is an implantable pulsating pump that has developed since 1994. The stroke volume is automatically controlled, resulting in the 3–7 l/min of flow. A set of external batteries (3.5 kg weight) is adequate for 6 h of work. It was first clinically used in Germany in October 1999.

Thoratec® VAD (Thoratec Laboratories, Inc., Pleasanton, CA, USA) is an external pump, applied to the short and long support comparatively often. The new implantable version has streamlined shapes, body made of polished titanium, and rest made from polyurethane and the sack through which the blood will flow from Tharalon. It is supplied pneumatically from the mobile, handy driver, connected through the skin by 9 mm drain. This relatively small VAD is also suitable for the biventricular assistance, weighs 339 g, and has an external volume of 252 ml.

The Thoratec PVAD (Paracorporeal Ventricular Assist Device) provides short-to-intermediate, uni- or biventricular support with the option and benefit of home discharge. With more than 20 years of clinical use, the PVAD has been well established as a Bridge-to-Transplantation or Post-Cardiotomy Recovery option for advanced heart failure patients of any age or size. Around 69% of PVAD patients were successfully supported to cardiac transplantation or device removal for myocardial recovery. More than 4,000 patients implanted at over 240 medical centers in 26 countries, the longest duration of support: 1,204 days (858 days discharged to home), smallest patient: 17 kg (BSA 0.73 m^2).

The IVAD (Implantable Ventricular Assist Device) provides intermediate-to-chronic support and is the only biventricular, implantable, and home-dischargeable VAD. It is indicated for Post-Cardiotomy Recovery and Bridge-to-Transplantation. Patients implanted: More than 440 patients at 95 medical centers in 9 countries, longest duration of support: 735 days, smallest patient: BSA 1.31 m^2.

As Thoratec's first-line intermediate-to-chronic left ventricular assist device, the HeartMate II has been extensively studied as Bridge-to-Transplantation for advanced heart failure. Over 1,200 worldwide patients implanted, longest duration of support (ongoing patient on one device): 3.6 years, transplanted, recovered, or supported to 180 days: 80%. World Heart registry (http://www.worldheart.com).

In the nearly 100 medical centers worldwide that use Novacor® LVAS, it is renowned for its reliability, durability and predictability of wear. Of the more than 1,700 Novacor® LVAS recipients to date, 172 primarily Bridge-to-Transplant patients have been supported by Novacor® LVAS for more than one year. Among these recipients, 45 have been supported for more than two years, 24 for more than three years, 11 for more than four years and 1 for more than six years. Only 1.4% of the pumps have needed replacement. No patient deaths have been attributed to Novacor® LVAS failure. In Germany, the 1,500th patient was implanted with Novacor® LVAS in 2004.

Medos-HIA VAD from Aachen is made from biocompatible polyurethane (project led by Helmut Reul). It is similar to natural trileaflet valves also made from polyurethane, but with the valve cusp optimization resulting in a conical shape (natural are spherical). Ventricles are made in the full row of sizes, from 60 to 9 ml (pediatric ventricle). Medos was clinically used in more than 100 cases. The investigations carried out in 1994 allowed to guarantee the minimum time of valve functioning of half-year (18 million of cycles). Since 1990, in Aachen, the HIA TAH has been developed, which is based on electromechanical driving unit. The inclination of the plate on which the membrane is fixed about 2 cm allows to reach the 65 ml stroke volume. Brushless motor rotates in only one direction (similar to the case of HeartMate, but the construction of the cam unit is different). The pump allows to reach above 10 l/min (with the frequency of 140 bpm under physiological conditions).

HeartSaver VAD (WorldHeart Corporation, Ottawa, ON, Canada) is the pulsate pump with hydraulic drive (project led by Tofy Mussivand). In 2000, it was almost ready for the starting of clinical experiments. Different types were prepared: the axial pumps for continuous flow and VADs and TAHs for pulsating pumps. All models are driven electrohydraulically (rotor). The external volume of the VAD is approximately 530 ml, the membrane is made from smooth polyurethane, and Medtronic-Hall mechanical valves are used, which according to plan will be replaced with the less thrombogenic biological Carpentier–Edwards valves. The working liquid in this pump is the silicon oil, which is currently not in use.

Berlin Heart is a series of external, pneumatically driven, ventricle assist devices, with the 50, 60, or 80 ml ejection, equipped with B–S mechanical valves or the polyurethane valves. Berlin Heart VADs have comfortable silicone cannulas with the wire armature, which renders cannulas a suitable shape. This type of VADs had been successfully in vitro tested for 5 years (more than 210 million cycles), mainly in Berlin. Berlin Heart VAD was used in Poland by Z. Religa's team in first successful mechanical support of heart as a bridge to transplantation.

The pneumatically driven membrane-type Polish ventricle assist devices *POLVAD* (U-shaped) and artificial heart *POLTAH* (spherical) were developed in Zabrze, Poland. Until now, more than 200 VAD implantations have been performed. First successful bridge to transplantation was constructed by Z. Religa's team. The implantation of BVAD produced by Berlin Heart LVAD was performed in May 1991. The patient T. Gruszczyński was awaiting heart transplantation. After 11 days of support, the heart transplantation was performed and patient returned home. Unfortunately, rejection occurred and the patient needed a second, new heart. In the meantime, author (physicist) designed and B. Stolarzewicz (chemist) performed a new Polish ventricular assist pump named POLVAD. After tests, POLVAD was ready for clinical applications. It happened that the first patient to use POLVAD was the same patient T. Gruszczyński (the heart was rejected after 2 years). Fortunately, the POLVAD worked very well, and after 21 days, the heart transplantation was performed. Unfortunately, the patient died during surgery (second heart transplantation).

In pulsate blood pumps, the part producing mechanical energy is separated from the blood by a biocompatible and durable membrane. Membrane-type VADs have it fixed to the hard or half-elastic body. They divide the artificial chamber in two almost even parts. However, in the ventricle with sack, the membrane is shaped into this bag hung in stiff chamber. The artificial ventricles can be driven pneumatically, hydraulically, or electromechanically. The trials of the nuclear power source introduction failed. The VADs are used for supporting patient's life as a bridge to transplantation or to natural heart recovery. At the end of this procedure, the cannulas connecting VAD to the patient's circulatory system are operationally removed. The TAHs are used in these cases, when heart transplantation is the only rescue for the patient or as a target therapy.

Distinct from abovedescribed pulsate pumps, the Jarvik 2000 and MicroMed–DeBakey are the currently used implantable axial pumps with continuous flow. The first axial flow pump to be introduced into clinical practice for intermediate to long-term treatment of end-stage heart failure in adults was the DeBakey VAD®. *MicroMed–DeBakey VAD* is the rotor, axial pump (flow up to 5 l/min), connected by the cable through the skin to the driver unit held on the patient's belt. It is implanted between the heart apex and aorta (ascending or descending). The DeBakey VAD is 30 mm × 76 mm, weighs 93 g, and is approximately 1/10th the size of pulsatile products on the market. In 2000, the first successful usage was performed in Vienna.

Jarvik 2000 is a small (2.5/5.5 cm) rotor pump (the electromagnetic rotor covered by titanium layer). It fits directly into the left ventricle, which may eliminate problems with clotting. The outflow graft connects to the descending aorta. The device itself is nonpulsatile, but the natural heart continues to beat and provides a pulse. The rotor has ceramic bearings, which are washed by the blood (smearing and the receipt of warmth). The angular velocity of 9000–16,000 rpm ensures 3–6 l/min of output flow with the aortic pressure of 80 mmHg and power consumption of 4–10 W. The system is powered by external batteries remotely through skin using electromagnetic field created by a set of coupled coins or using the wire carried out through the "port" in the skull safety plugged with the pyrolytic carbon. On 25 April 2000, physicians in Houston, TX, USA, have realized the implantation and introduced the first patient in the clinical investigation of the Jarvik 2000. For lifetime use, the Jarvik 2000 has also had proven successful in treating a target population of patients suffering from chronic heart failure due to a prior heart attack or cardiomyopathy. Many have been rehabilitated to a dramatically improved life at home, and in some cases patients have even returned to work. So far, the Jarvik 2000 FlowMaker® has been used to treat more than 200 patients in the United States, Europe, and Asia. Of those, roughly 79% received the Jarvik 2000 as a bridge to transplantation and 21% as a permanent implant, with a number of patients in each group being terminally ill, near-death cases. Nearly 70% of those patients were supported successfully. Several surgeons reports have described placement of continuous flow devices without cardiopulmonary bypass. Frazier described a patient in whom he placed this pump while briefly fibrillating the heart and placement of the Jarvik 2000 with an anterior, intraperitoneal approach without bypass. This technique is attractive in reoperative situations.

The University of Pittsburgh–Thermo Cardiosystems *HeartMate® II* is an axial rotary pump with the stone bearing with a volume of 89 ml, weighs 350 g, and the estimated time of use is 5–7 years. As intermediate-to-chronic left ventricular assist device, the HeartMate II (now Thoratec's) has been extensively studied as Bridge-to-Transplantation for advanced heart failure. Over 1,200 worldwide patients implanted, longest duration of support (ongoing patient on one device): 3.6 years, transplanted, recovered, or supported to 180 days: 80%. The third-generation,

HeartMate III, will be the centrifugal pump with the rotor levitating in the magnetic field (without mechanical bearings).

There have been several scientific groups working in the field of artificial heart in Japan, Australia, Austria, Argentina, France, Germany, Poland, Czech Republic, Russia, and so on. But from the market point of view, there is not easy business. Currently, the strongest company offering the wide range of products in the area of blood pumps is Thoratec. Thoratec Corporation is engaged in the research, development, manufacturing, and marketing of medical devices for circulatory support, vascular graft, blood coagulation, and skin incision applications. The Thoratec VAD system is the device that is approved for left, right, or total heart support and that can be used both as a bridge to transplantation and for recovery from open-heart surgery. More than 4300 of these devices have been used in the treatment of over 2800 patients worldwide. With the introduction of the Implantable Ventricular Assist Device (IVAD™), Thoratec delivers the first and only implantable VADs for left, right, and biventricular support for bridge to transplantation and for postcardiotomy recovery. The HeartMate® XVE Left Ventricular Assist System (LVAS) is now FDA approved as a long-term permanent implant, called destination therapy. In addition, the accompanying Thoratec TLC-II® Portable VAD Driver provides hospitals with the first mobile system that allows these univentricular or biventricular VAD patients to be discharged home to await cardiac transplantation or myocardial recovery. The company is also a leader in implantable LVADs. Its air-driven and electric HeartMate LVAD, which has been implanted in more than 4100 patients worldwide, are implanted alongside the natural heart and take over the pumping function of the left ventricle for patients whose hearts are too damaged or diseased to produce adequate blood flow.

WorldHeart is a developer of mechanical circulatory support systems (e.g., Novacor) with leading next-generation technologies. The Levacor is a next-generation rotary VAD. It is the only bearingless, fully magnetically levitated implantable centrifugal rotary pump with clinical experience. An advanced, continuous-flow pump, the Levacor uses magnetic levitation to fully suspend the spinning rotor, its only moving part, inside a compact housing.

WorldHeart's Novacor II LVAS is a next-generation, pulsatile VAD. It can be fully implanted without a volume displacement chamber, thereby reducing the risk of complications by eliminating the need to perforate the skin. The operation of the pump drive unit is very interesting. When the pusher plate is driven to the right (pumping stroke), the prechamber expands, filling from the left ventricle. Simultaneously, the pumping chamber is compressed, ejecting blood into the body. When the pusher plate returns to the left (transfer stroke), the prechamber is compressed while the pumping chamber expands; blood transfers from the prechamber to the pumping chamber, with no inflow or outflow. Because the total volume of the two chambers remains constant as one fills and the other empties, the system can operate without a volume compensator or venting through the skin. WorldHeart's Novacor II LVAS is not currently available. In 2005, WorldHeart conducted the first animal implant of the Novacor II LVAS – ahead of schedule. WorldHeart Company currently focuses on new products – especially for pediatric patients. The PediaFlow VAD is an implantable, magnetically levitated blood pump based on

WorldHeart's proprietary rotary VAD MagLev™ technology. In its pediatric configuration, the device is designed to provide a flow rate from 0.3 to 1.5 l/min. The PediaFlow VAD is being developed to provide medium-term (less than 1 year) implantable circulatory support to patients from birth to 2 years of age with congenital or acquired heart disease.

Pneumatic pulsatile VADs have been available in pediatric sizes since 1992. At Herzzentrum Berlin, VADs are used lasting from several days to 14 months in 70 infants and children with myocarditis and cardiomyopathy, leading to a notable rise in survival in the past 5 years. It is possible to discharge 78% of the infants under 1 year old [3].

Several types of VADs have been used in children and adolescents whose body surface area is greater than 1.2 m^2 – that is, generally in children older than 5 years. The Thoratec VADs (Thoratec Laboratories, Inc.) have been available since the early 1980s for adult use, but they can also be implanted in older children and adolescents. Several other adult-size VADs, such as the Novacor (Baxter Healthcare Corporation, Irvine, CA, USA), have been applied in adolescents, and a version of the axial flow DeBakey VAD (MicroMed Technology Inc., The Woodlands, TX, USA) has been developed that is suitable for use in children and has been implanted in several patients. Two miniaturized extracorporeal, pneumatically driven VADs designed specifically for smaller children and infants have been introduced in Europe so far: the Berlin Heart Excor® (Berlin Heart AG, Berlin, Germany) in 1992 and the Medos HIA device (Medos Medizintechnik AG, Stolberg, Germany) in 1994. The first reported implantation of a Medos VAD as a bridge to transplantation in a child took place in 1994. Only the extracorporeal, pneumatically driven Berlin Heart Excor and the Medos HIA pulsatile systems have so far proven successful in children of all ages. The pediatric version of the Berlin Heart Excor VAD is mounted with trileaflet polyurethane valves and is available with pump sizes of 10, 25, 30, 50, 60, and 80 ml. The 10-ml pumps are suitable for neonates and infants with body weight of up to 9 kg (body surface area 0.43 m^2), and the 25 and 30 ml pumps can be used in children up to the age of 7 years (weight 30 kg and body surface area of about 0.95 m^2); adult-sized pumps can be implanted in older children. The adult pump has a stroke volume of 80 ml and tilting disk valves. Pediatric-sized pumps are suitable for children with a body weight of 3–9 kg. This pump has a stroke volume of 10 ml and polyurethane trileaflet valves [3].

Several other pulsatile devices developed for the adult population are used in school-aged children: the HeartMate I (Thoratec Laboratories, Inc.), Toyobo (National Cardiovascular Center Tokyo, Japan), Abiomed® BVS 5000 (Abiomed Inc., Delaware, MA, USA), and Novacor (World Heart Corporation, Ontario, Canada). Two continuous flow rotary VADs that use axial flow or centrifugal flow – the Incor® VAD (Berlin Heart) and the DeBakey VAD – have been introduced into routine clinical care. Some further devices are still being subjected to clinical trials: Jarvik 2000 (Jarvik, New York, NY, USA), HeartMate II, Duraheart® (Terumo Kabushiki Kaisha Corporation, Shibuya-ku, Japan), VentrAssist (Ventracor, Chatswood, NSW, Australia), and CorAid® (Cleveland Clinic, Cleveland, OH, USA) [3].

The Incor device (Berlin Heart AG) is 146 mm long and 30 mm wide and weighs 200 g. MicroMed modified the adult pump to fit children and in 2004 received FDA humanitarian device exemption status, enabling implantation of the DeBakey VAD Child pump for persons, aged 5–16, awaiting heart transplantation.

Berlin Heart AG products are Incor, Excor cannulas, and driving units. In June 2002, the worldwide first implantation of the Incor device took place in the German Heart Center DHZB. By July 2005, no less than 300 Incor devices had been implanted (two patients have been living with the device for 3 years). Excor is an extracorporeal, pulsatile VAD. Various types and sizes of blood pumps and a wide range of cannulas allow us to meet all clinical needs and treat all patients, regardless of their age.

The VentrAssist, which is made by the Australian company Ventracor, has a moving part – a hydrodynamically suspended impeller. It has been designed to have no wearing parts or cause blood damage. It weighs just 298 g and measures 60 mm in diameter, making it suitable for both children and adults. VentrAssist also has an advantage over its one competitor, Incor made by the German company Berlin Heart. The VentrAssist is less likely to damage red blood cells because it moves the blood more slowly with a bigger impeller.

The term VAD has been applied to a wide variety of mechanical circulatory support systems designed to unload the heart and provide adequate perfusion of the organs.

Short-term circulatory support with an LVAD may be indicated for patients with end-stage heart failure (of any etiology) who are awaiting a donor heart for transplantation, and for patients with a severe acute heart failure syndrome from which myocardial recovery is anticipated (such as acute myocarditis). An LVAD is sometimes used if weaning from cardiopulmonary bypass after cardiac surgery fails. In the active arm of a nonrandomized controlled study, 78% (32/41) of patients survived for a mean of 215 days with LVAD support. In another comparative study, 81% (13/16) of patients survived to transplantation (duration of support not stated). One case series showed that at 30 days of bridging to transplantation with an LVAD, survival was 83%, falling to 19% after 24 months' support.

In a nonrandomized controlled trial, posttransplant survival of patients bridged on LVAD support was 66% (21/32) at 41 months, compared with 67% (98/146) of patients at 36 months who had a transplant without circulatory support, although patients in the latter group were significantly older. One case series of 243 patients in whom LVADs were used to bridge to transplantation reported actuarial post-transplant survival of 91% at 1 year, 70% at 5 years, and 40% at 10 years. Results from case series included in a systematic review showed that between 60% (12/20) and 83% (5/6) of patients survived to transplantation or were still alive awaiting transplantation on LVAD support. Of the total cases of bridge to recovery reported, 58% (7/12) of patients survived to final follow-up; successful explantation of the device or weaning from support was achieved in all these patients [4].

The randomized controlled study showed a reduction of 48% in the risk of death from any cause in the group that received left VADs as compared with the medical therapy group [relative risk 0.52; 95% confidence interval (CI) 0.34–0.78; $P = 0.001$]. Randomly assigned 129 patients with end-stage heart failure were ineligible for cardiac transplantation to receive a left VAD (68 patients) or optimal medical management (class IV heart failure). The rates of survival at 1 year were

52% in the device group and 25% in the medical therapy group ($P = 0.002$), and the rates at 2 years were 23% and 8% ($P = 0.09$), respectively. The frequency of serious adverse events in the device group was 2.35 (95% CI 1.86–2.95) times that in the medical therapy group, with a predominance of infection, bleeding, and malfunctioning of the device. The quality of life was significantly improved at 1 year in the device group. The use of a left VAD in patients with advanced heart failure resulted in a clinically meaningful survival benefit and an improved quality of life. A left VAD is an acceptable alternative therapy in selected patients who are not candidates for cardiac transplantation [5].

All mechanical circulatory support systems are associated with a wide range of possible complications: bleeding, infection, device malfunction, hemolysis, peripheral ischemia, and perforation of a ventricle or the aorta, of which bleeding and thromboembolic complications are the most frequent and most serious. Infections, hemolysis, pulmonary edema, and multiorgan failure have also been reported. It was also noted that implantation of an LVAD could unmask previously subclinical right ventricular dysfunction. Damage to the blood pump equipment rarely causes harm to the patient. As examples, one Jarvik 2000 patient broke a cable connector when he slammed it in a car door; another accidentally cut his cable with scissors while changing his bandage; yet another patient lost connection to his battery and controller when a purse snatcher grabbed his shoulder bag and ran off with it. Fortunately, in all these cases, the patient suffered no harm. Their own natural hearts were able to sustain them until they could connect to their backup equipment. But it can be noticed that the longest-running Jarvik 2000 FlowMaker patient has been supported by the device longer than any patient in the world with any other type of mechanical heart, either TAH or VAD – i.e., more than 4 years.

In September 2006, the FDA approved the first totally implantable TAH (AbioCor) for people who are not eligible for a heart transplant and who are unlikely to live more than a month without intervention. It is a big success, but it also means that many problems are awaiting to be solved in laboratories (Figure 4) in the near future.

3.2. Valve Prostheses

Valvular heart disease is a life-threatening disease that affects millions of people worldwide and leads to approximately 250,000 valve repairs and/or replacements every year. Almost 90,000 Americans a year need surgery for valve disease. Malfunctioning of a native valve impairs its efficient fluid mechanical/hemodynamic performance. Human heart valves act as check valves, controlling the direction of blood flow through the heart. The aortic valve is between the left ventricle and the aorta and the mitral valve is between the left atrium and the left ventricle. It opens and closes to control the blood flow into the left side of the heart. The normal aortic valve area is 3.0–4.0 cm^2. In general, severe aortic stenosis has been defined as a valve whose area was reduced to 0.75–1.0 cm^2. In general, mean transvalvular pressure gradients greater than 50 mmHg represent severe aortic stenosis, while mean gradients less than 25 mmHg suggest mild aortic stenosis. Mitral stenosis

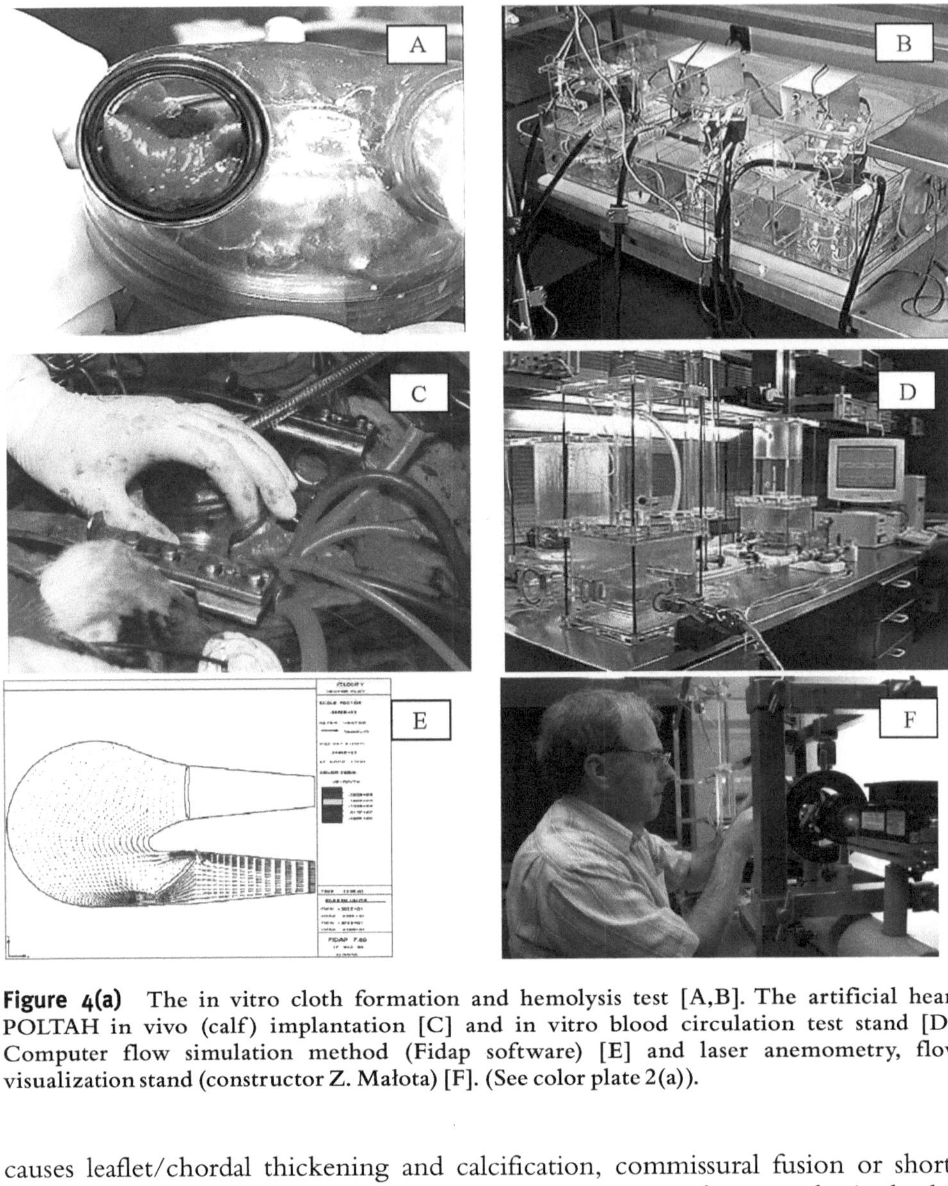

Figure 4(a) The in vitro cloth formation and hemolysis test [A,B]. The artificial heart POLTAH in vivo (calf) implantation [C] and in vitro blood circulation test stand [D]. Computer flow simulation method (Fidap software) [E] and laser anemometry, flow visualization stand (constructor Z. Małota) [F]. (See color plate 2(a)).

causes leaflet/chordal thickening and calcification, commissural fusion or shortening, chordal fusion, or a combination of these processes. The normal mitral valve area (MVA) is 4.0–5.0 cm^2. Accordingly, mild mitral stenosis is defined as a condition with an MVA of 1.5–2.5 cm^2 and a mean gradient at rest less than 5 mmHg. Moderate and severe mitral stenoses are defined as conditions with an MVA 1.0–1.5 and less than 1.0 cm^2, respectively, with mean gradients greater than 5 mmHg [6].

Valve repair is now the method of choice for surgical treatment of valve disease, thanks to improvements in techniques over the past 40 years. Starr and associates first reported a technique for aortic repair in 1960. In the early 1980s, surgeons who

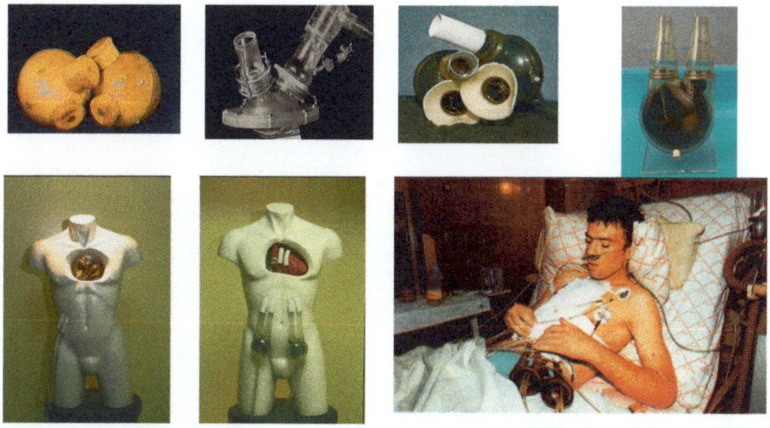

Plate 1 The steps in POLTAH design process (model after cadaver study, transparent model for laser visualization investigation, first polyurethane model for starting animal experiments). Nonsymmetrical shape of POLVAD proposed and designed creates a very good internal flow condition. The first successful POLVAD implantation was performed in 1993 in T. Gruszczyński. Until now, more than 200 POLVAD clinical implantations have been performed, the longest time successful application being over 200 days. Pneumatically driven artificial heart (POLTAH) and ventricular assist device (POLVAD). (See Figure 1 of chapter 1, p. 12).

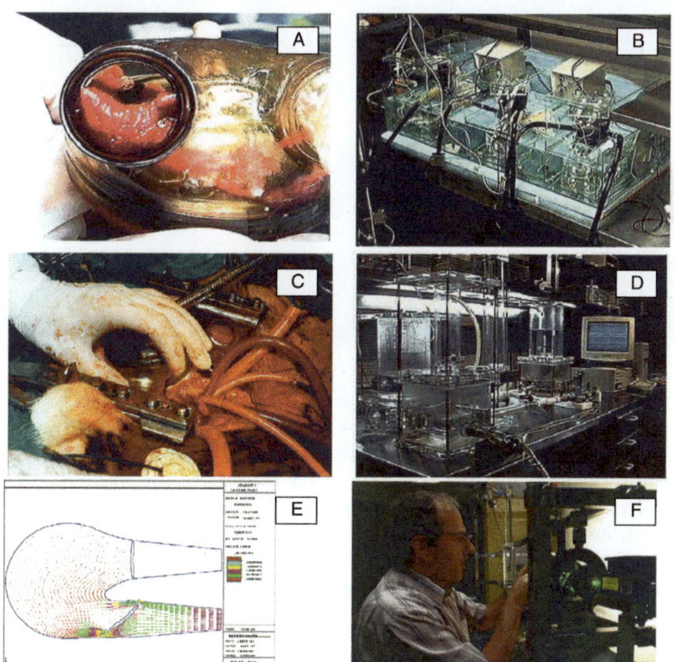

Plate 2(a) The in vitro cloth formation and hemolysis test [A,B]. The artificial heart POLTAH in vivo (calf) implantation [C] and in vitro blood circulation test stand [D]. Computer flow simulation method (Fidap software) [E] and laser anemometry, flow visualization stand (constructor Z. Małota) [F]. (See Figure 4a of chapter 1, p. 24).

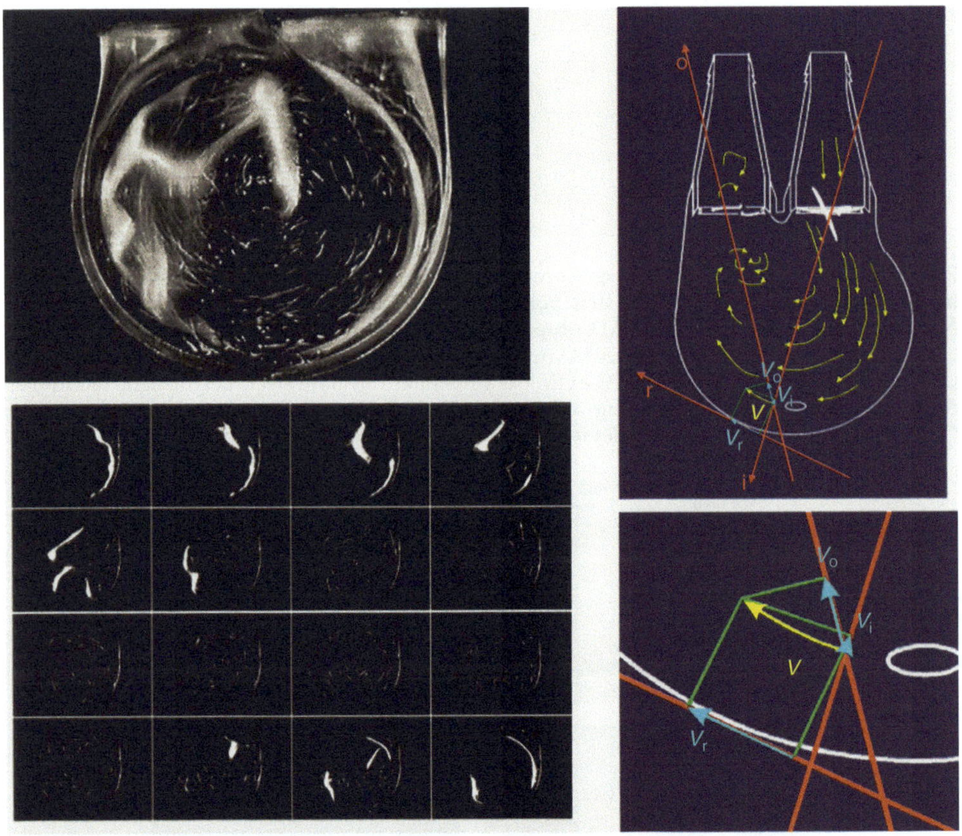

Plate 2(b) One of the problems that are not satisfactorily solved was the interpretation of laser flow visualization pictures. Proposed by the author in the 1990s, functional analyzing method (FAM) method involves monitoring the quality of a flow system. The artificial heart is a pump that should cause fluid flow from the inlet (atrial) to the outlet channel and provide blood with a determined amount of energy to overcome the load, in order to create clinically desired pressure-flow state in the circulatory system. Both goals must be achieved, while conserving the safe transportation of morphotic blood elements, which is essential for long-term blood pumps. Therefore, it must minimize the undesired phenomena occurring at the blood–artificial surface interface. (See Figure 4b of chapter 1, p. 25).

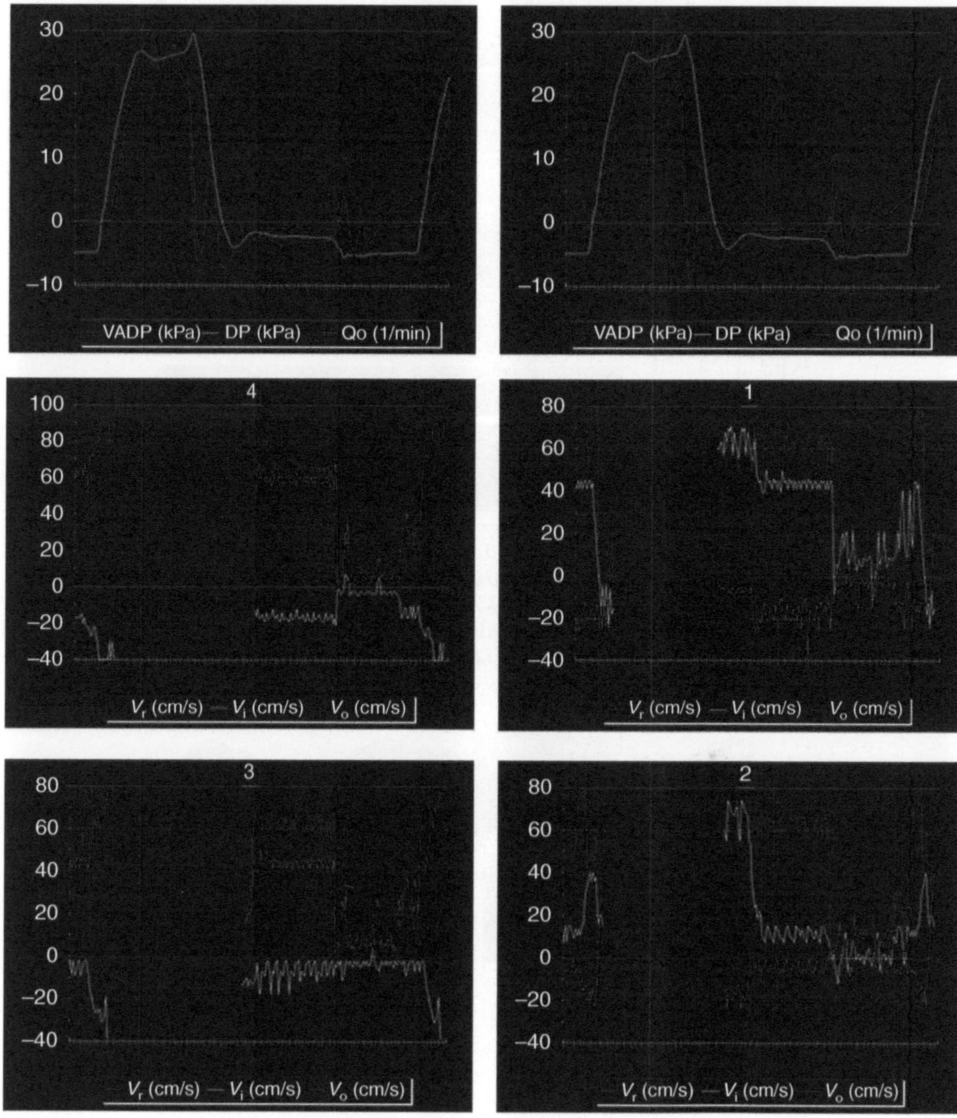

Plate 2(c) The chamber was divided into areas where, for a given section, and for every local flow velocity, vectors are assigned with its components in the direction of an outlet V_o and inlet V_i channel as well as the component V_r tangent to the chamber's wall. The results depend on the phase of work cycle. The analysis of time changes of the first two components allows to track down the dynamics of flow structure organization during an entire cycle, to detect inertial effects and passage time, while the tangent component is responsible for a good washing of wall vicinity areas. The full analysis makes possible the classification of obtained pictures (since digital picture recording is suitable for computer analysis and gives results that are easy to interpret), so that the conclusions regarding the causes of phenomena and their regulation may be drawn easily. (See Figure 4c of chapter 1, p. 26).

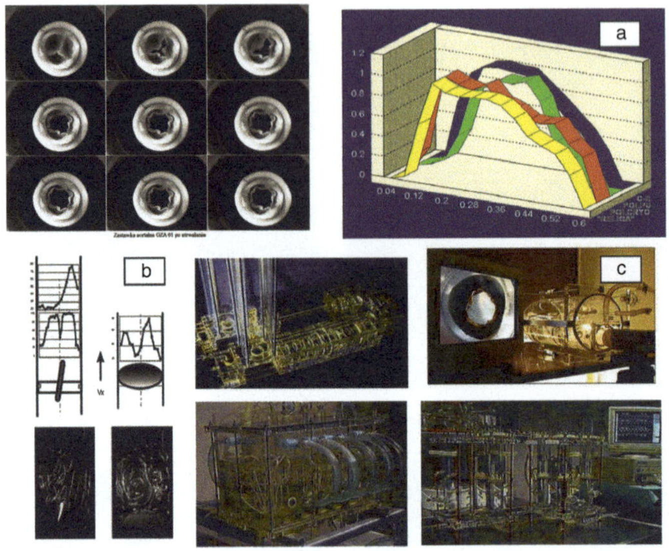

Plate 3 Results of valve test tester designed and introduced in Institute Heart Prostheses, FCSD. The opening area valve test and comparison in one cycle of working biological Religa, Polcryo, and synthetic Polpu prototypes with Carpentier–Edwards bioprotheses (a). The laser visualization and laser anemometer test results of Polish disk valve prototype (b). Several testing apparatus used in our laboratories (c). (See Figure 5 of chapter 1, p. 27).

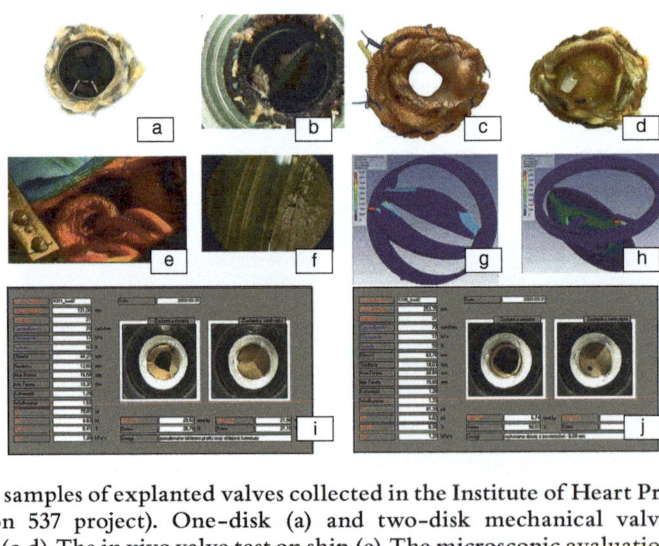

Plate 4 The samples of explanted valves collected in the Institute of Heart Prostheses, FCSD (COST Action 537 project). One-disk (a) and two-disk mechanical valve (b) and two bioprostheses (c,d). The in vivo valve test on ship (e). The microscopic evaluation of valve's disk (f) and modeling of mechanical valves – equivalent elastic strain values (g,h). The effect of biological heart valve prosthesis damage on hemodynamic efficiency has been investigated and analyzed: improper functioning due to leaflet calcification – the opening area of "calcified" (leaflets partially glued (i)) valve decreased by 44%, while the pressure gradient grew up to 22 mmHg. Perforation of a leaflet (j) – significant perforation (hole area 6.68 mm^2) of leaflet caused increase in back-leakage of 147% to 8.38 (6.8 ml). (See Figure 7 of chapter 1, p. 31).

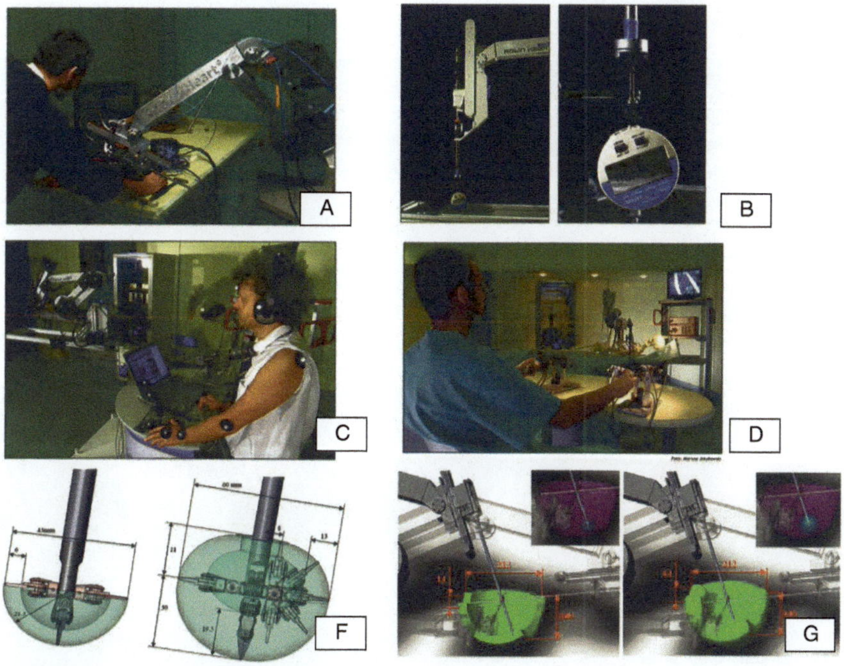

Plate 5c Evaluation of Robin Heart robot is carried out using the following measurements: vibration (new accelerometer sensors) [A]; linear movement (using digital micrometer) [B]; system of external trajectory recording (several digital cameras, markers, and image analysis methods) [C]; and physical [D] and computer surgery simulation (*bottom*: general range of robot mobility and workspace equipped with [F] standard laparoscopic tool and [G] Robin Heart I instrument). (See Figure 10c of chapter 1, p. 45).

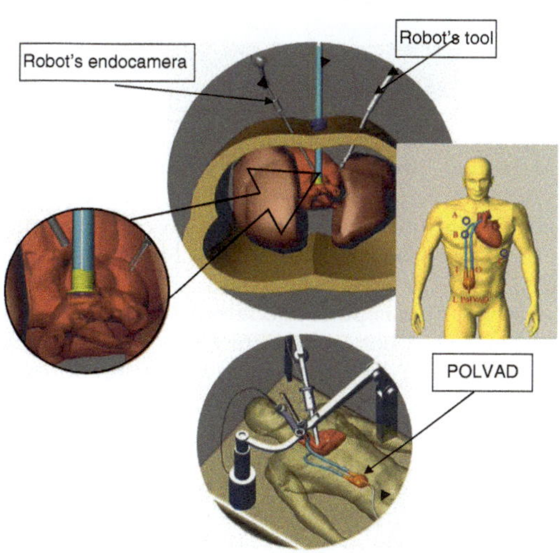

Plate 6 Semiautomatic tools for robotically assisted POLVAD implantation. (See Figure 11 of chapter 1, p. 50).

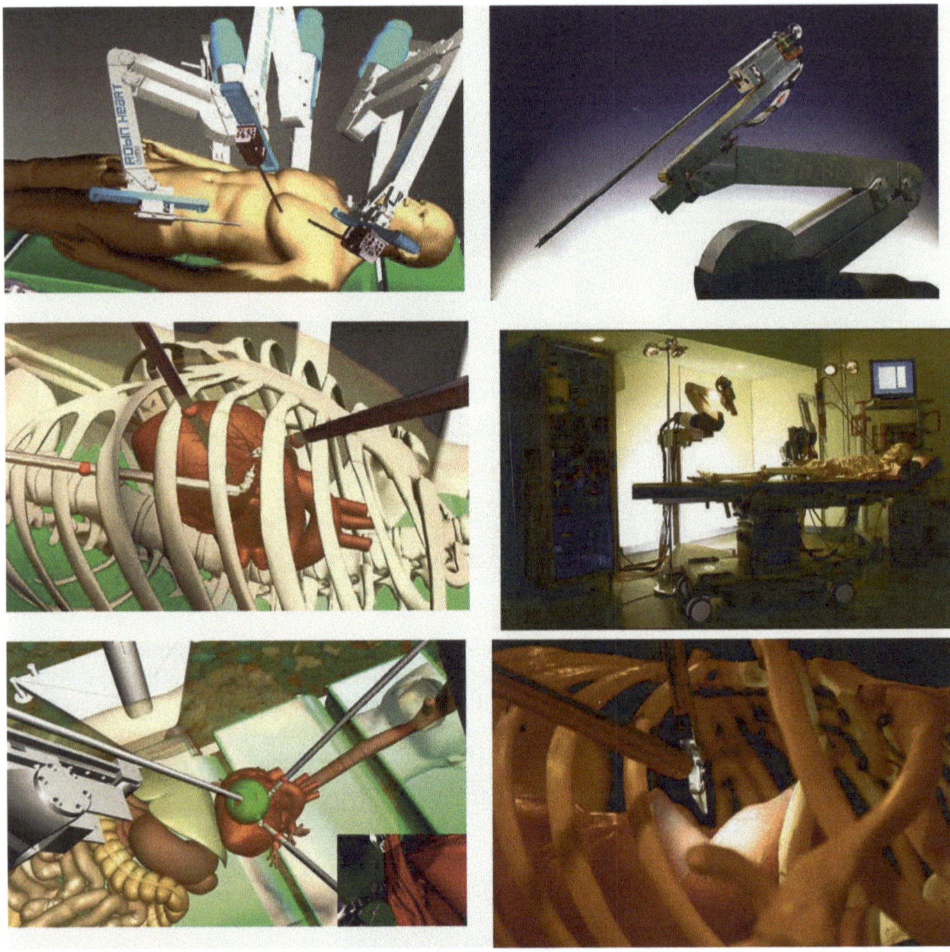

Plate 7 The virtual and the real condition for testing the Robin Heart robot. Using a virtual reality (VR) technology, an interactive model of surgery room equipped with a Robin Heart system was created using EON Professional software. This computer modeling method allows for an advanced procedure training and will be used as a low-cost training station for surgeons in the future. The model allows for a better understanding process of less invasive surgery treatment and a robot behavior. This type of modeling and a computer-aided design (CAD) technique use an accurate CAD robot model in a VR software together with a precise reflection of workspace geometry. This approach gives a surgeon easy and intuitive way to understand the technical details and use it to optimize and plan medical process. The next step in FCSD research work will be establishing the surgery workshops based on the newest technology, and some new projects using affordable semiautomatic robotic tools. Presented model of operating room in VR environment has been successfully used during Surgery Workshop in FCSD (May 2006). This system is intuitive for a user and gives them a very realistic 3D visualization. (See Figure 14 of chapter 1, p. 56).

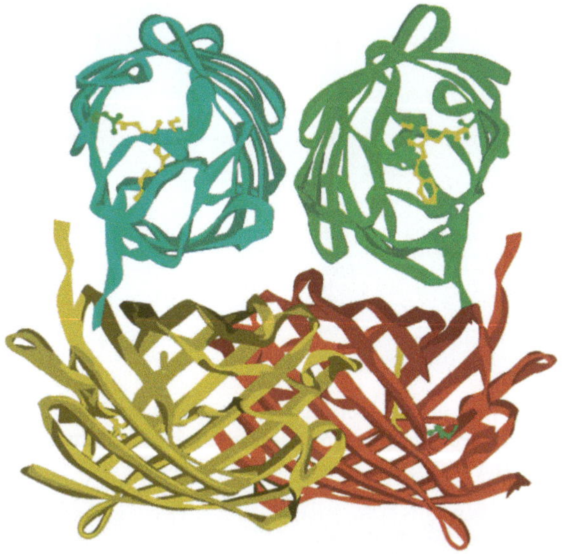

Plate 8 Schematic of X-ray structure of the DsRed protein. Structure generated with Accelrys DS ViewerPro Suite v5.1 and X-ray data from the Protein Data Bank [65]. (See Figure 13 of chapter 4, p. 165).

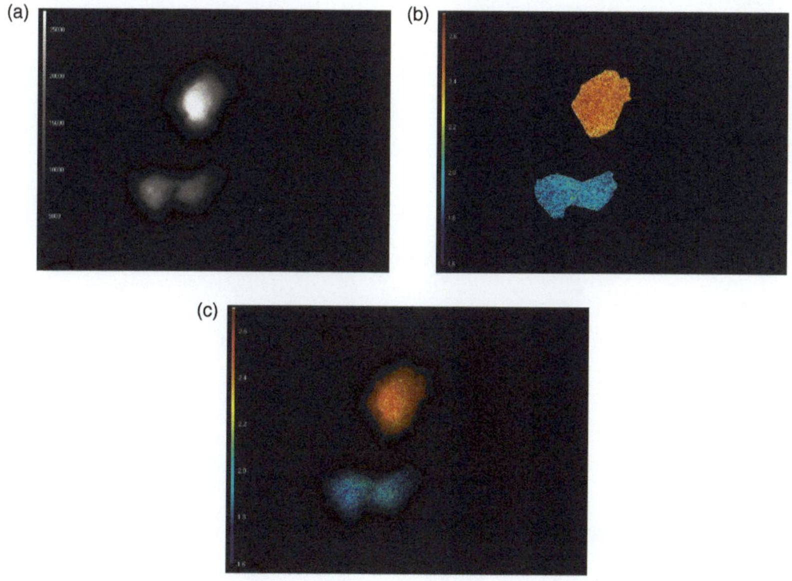

Plate 9 Fluorescence lifetime images of mammalian cells transfected with CFP alone and a CFP–YFP fusion construct. Panel (a) fluorescence intensity image, Panel (b) lifetime image, and Panel (c) overlay of fluorescence intensity and lifetime images. Change in CFP lifetime was selectively monitored by using a CFP band pass filter. The upper cell is transfected with CFP alone and yields a lifetime of 2.5 ns. For the cells expressing the fusion construct, the lifetime is substantially lowered (1.8 ns) due to FRET from CFP (donor) to the YFP (acceptor) molecule. Images courtesy of Lambert Instruments. (See Figure 16 of chapter 4, p. 169).

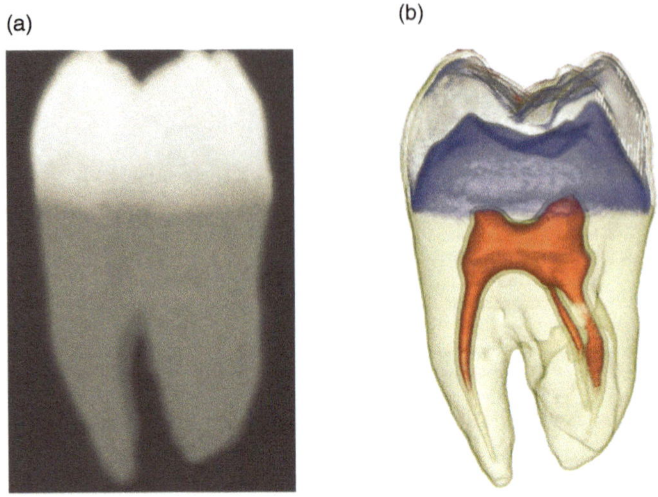

Plate 10 (a) Maximum intensity projection (MIP) volume rendering of a tooth from CTdata, and (b) A full volume rendering of the same data using multi-dimensional transfer functions with SCIRun (Section 7.1). (See Figure 3b of chapter 6, p. 213).

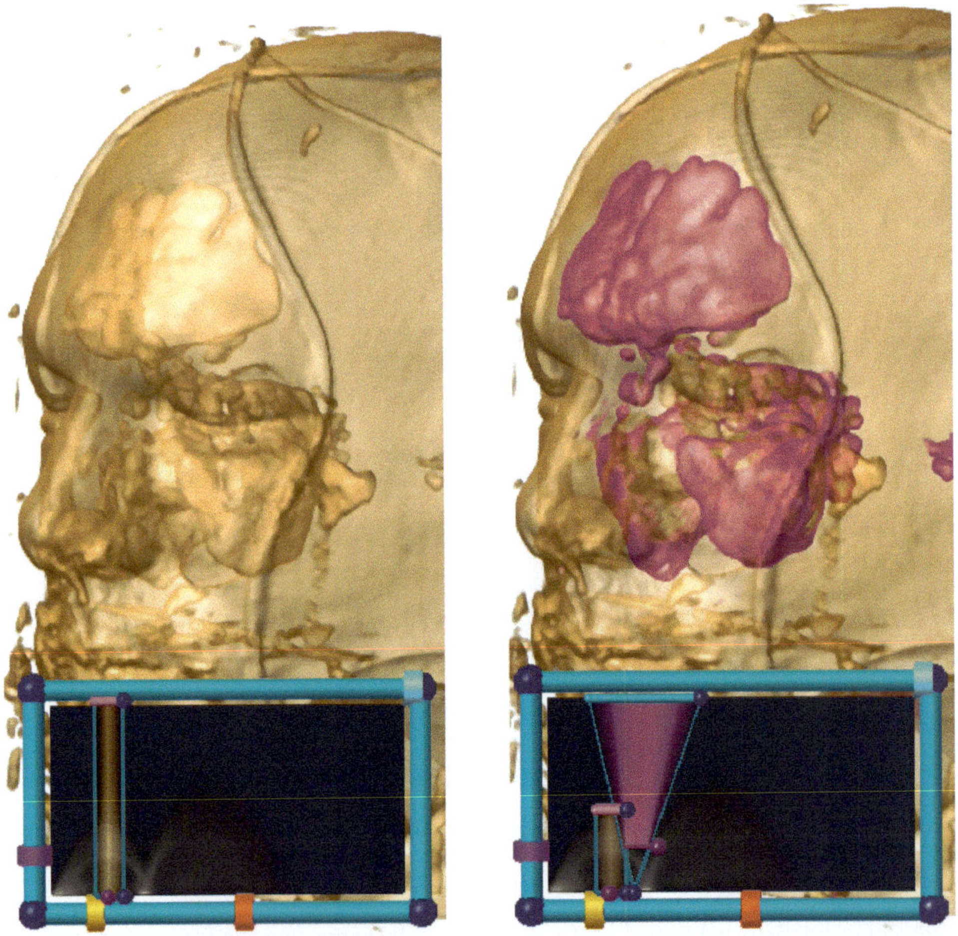

Plate 11 A comparison of scalar (left) and multi-field (right) volume renderings of the Visible Male CTdata. (See Figure 7 of chapter 6, p. 216).

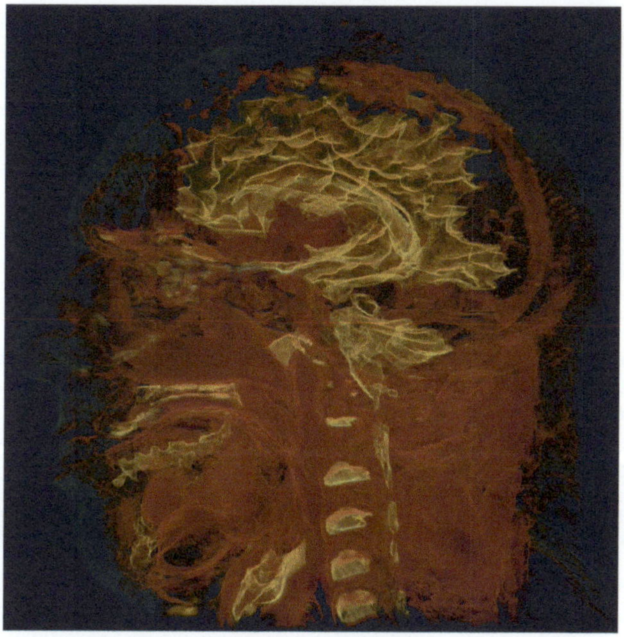

Plate 12 Example of 4D multi-field volume classified using a Gaussian transfer function. The dataset is the Visible Male Color Cryosection, courtesy of the NLM NIH. The analytical integration of the transfer function along rays creates a high-quality representation. In particular, muscles and boundaries between white and gray matter can be precisely isolated. Image courtesy of Joe Kniss [7]. (See Figure 8 of chapter 6, p. 218).

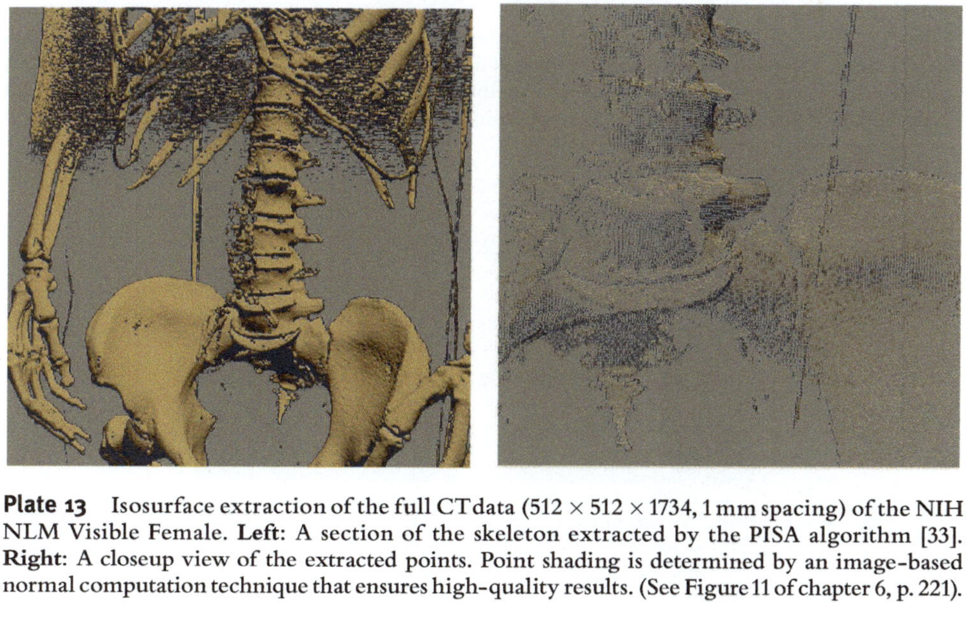

Plate 13 Isosurface extraction of the full CT data ($512 \times 512 \times 1734$, 1 mm spacing) of the NIH NLM Visible Female. **Left**: A section of the skeleton extracted by the PISA algorithm [33]. **Right**: A closeup view of the extracted points. Point shading is determined by an image-based normal computation technique that ensures high-quality results. (See Figure 11 of chapter 6, p. 221).

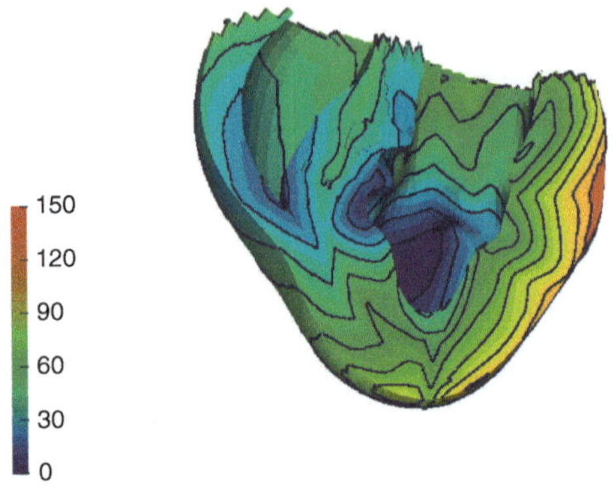

Plate 14 BioPSE visualization of the isochrones for a cardiac activation simulation. The front of the heart has been clipped away to reveal the transmural activation times. (See Figure 12 of chapter 6, p. 222).

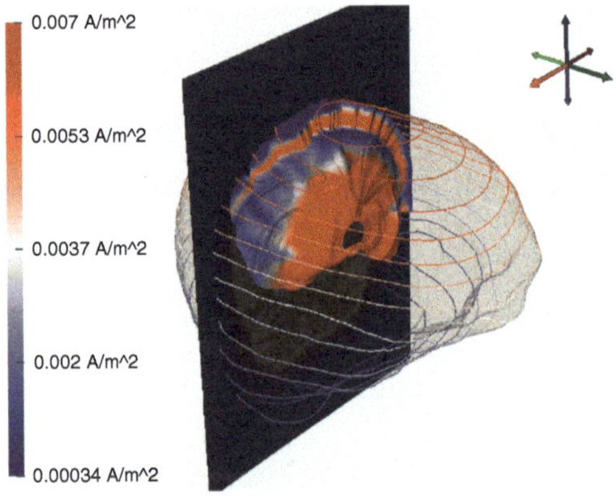

Plate 15 Visualization of the potential distribution through the cranium due to a dipolar source simulation. Potentials are color-mapped onto a stream-surface within the head volume, and evenly spaced isopotential contours are plotted on the scalp surface. Data courtesy of Scott Makeig, UCSD and Greg Worrell, Mayo Clinic. (See Figure 13 of chapter 6, p. 223).

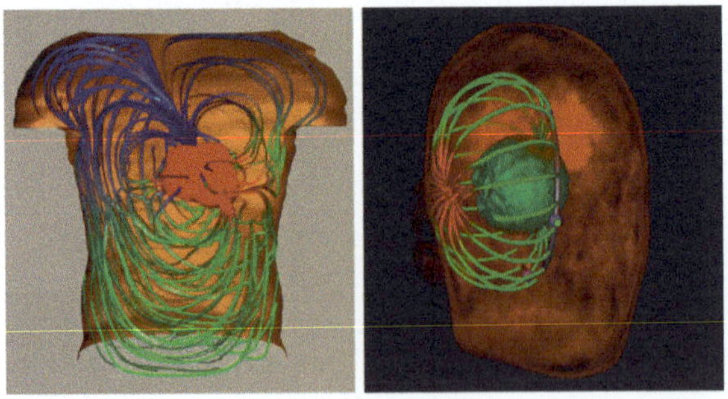

Plate 16 Applications of streamlines to Finite Element biomedical datasets. **Left**: The bioelectric field in the torso visualized through streamlines seeded randomly around the epicardium. **Right**: Use of a circular rack to seed streamlines regularly around a dipolar source in the brain. (See Figure 14 of chapter 6, p. 225).

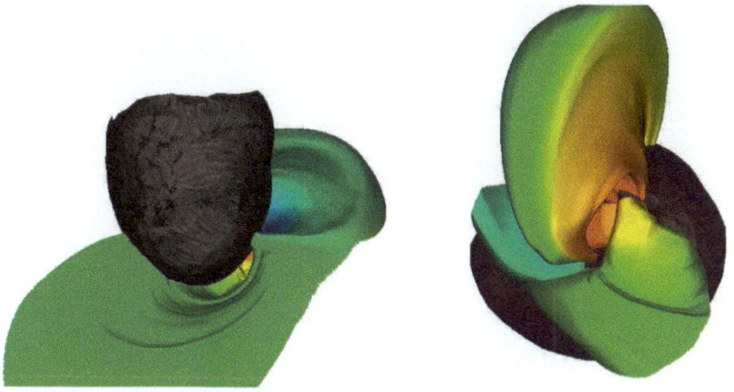

Plate 17 Visualization of the bioelectric field in the direct vicinity of epicardium with high-quality stream surfaces. The image shows the technique applied to the cardiothoracic current in a Finite Element forward computation. The surfaces capture the geometry of the current induced by the dipole equivalent cardiac source. They also provide an effective representation of the interconnections that exist between different regions on the epicardium. The seeding curves correspond to isocontours of the electric potential selected close to local extrema. A rainbow color map is used along each seeding curve to visualize flow stretch. (See Figure 17 of chapter 6, p. 228).

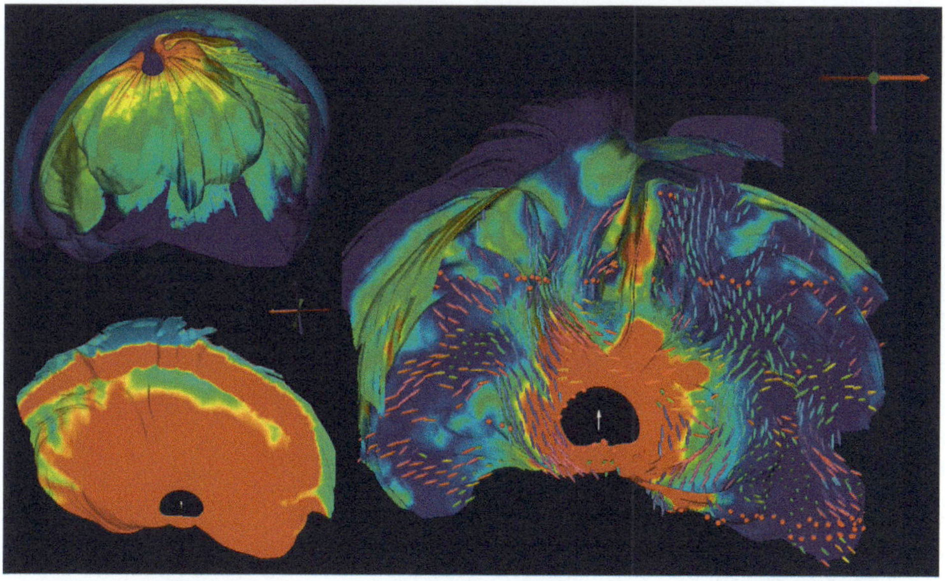

Plate 18 Stream surface visualization of bioelectric field induced by a dipolar source in left thalamus. **Left top**. Stream surfaces seeded along isocontour of electric flux on sphere enclosing the source. Culling is used to address occlusion. White matter has anisotropic conductivity. **Left bottom**. Stream surface started along circle contained in coronal slice and centered around source location. White matter is assumed isotropic. Color coding corresponds to magnitude of electric field. **Right**. Similar image obtained for anisotropic white matter. Glyphs visualize major eigenvector of conductivity tensor. Color coding shows magnitude of return current. (See Figure 18 of chapter 6, p. 228).

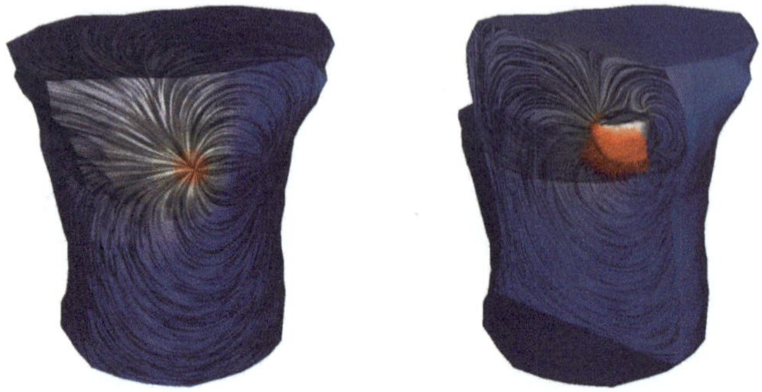

Plate 19 **Left**. LIC representation of the current on the boundary of a Finite Element grid used in forward computation. The color-coded electric potential (negative values in blue, positive values in red) are superimposed on the texture. **Right**. Two LIC textures computed over cutting planes combined by transparency. The geometry of the heart remains opaque for context. (See Figure 22 of chapter 6, p. 232).

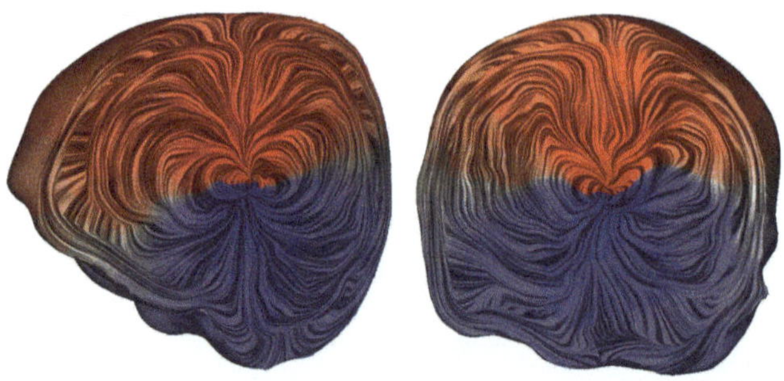

Plate 20 LIC applied to coronal and sagittal clipping planes reveals details of the dipolar source and its interaction with the surrounding anisotropic tissue. Rather than a typical smooth, symmetric dipolar pattern, the electric current is clearly diverted by the presence of white matter tracts that lie close to the source. The field also changes direction very rapidly as it approaches the skull just beneath the surface of the head. (See Figure 23 of chapter 6, p. 232).

Plate 21 Topology of bioelectric field on epicardium. The images show an LIC representation of the potential gradient on the surface enhanced by the depiction of the associated topological graph. Green points correspond to potential minima, blue points mark potentials maxima. Left and middle images show an anterior view of the epicardium. Right image shows a view of the torso from the left. (See Figure 25 of chapter 6, p. 233).

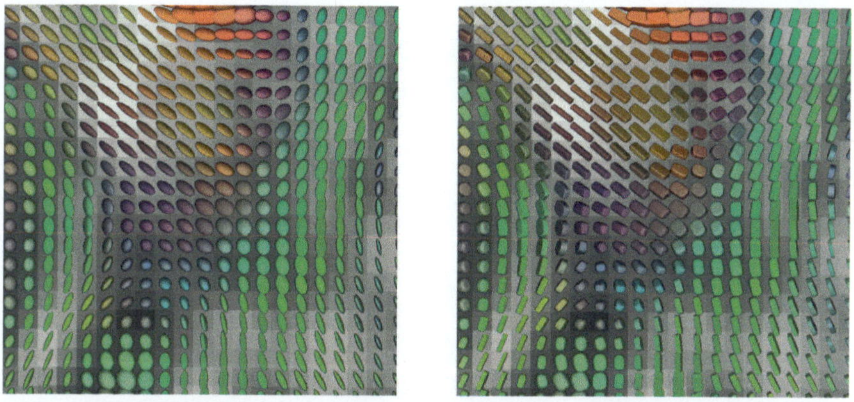

Plate 22 Ellipsoidal (left) and superquadric (right) glyph visualizations of an axial slice of a DT-MRI scan. (See Figure 27 of chapter 6, p. 237).

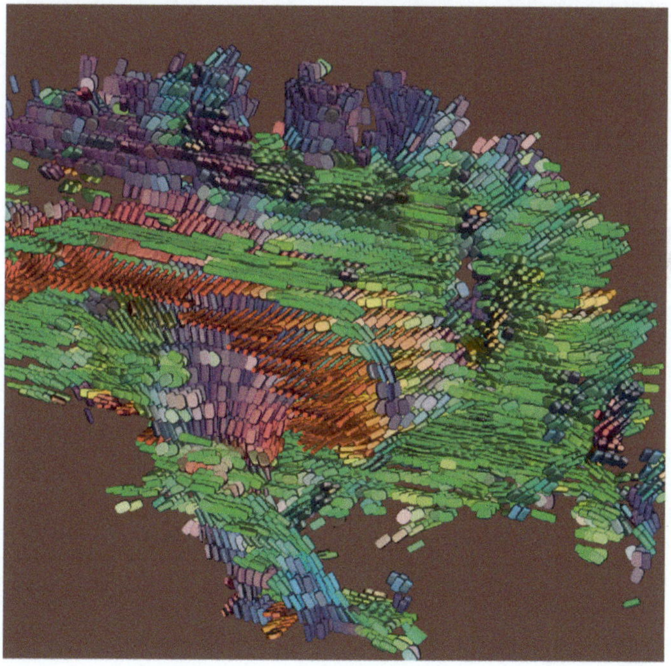

Plate 23 Visualization of half a brain DT-MRI volume using superquadrics glyphs. Glyphs are depicted only in anisotropic voxels. Red indicates left/right fiber orientation, green indicates anterior/posterior, and blue indicates superior/inferior. Image courtesy of Gordon Kindlmann. (See Figure 28 of chapter 6, p. 238).

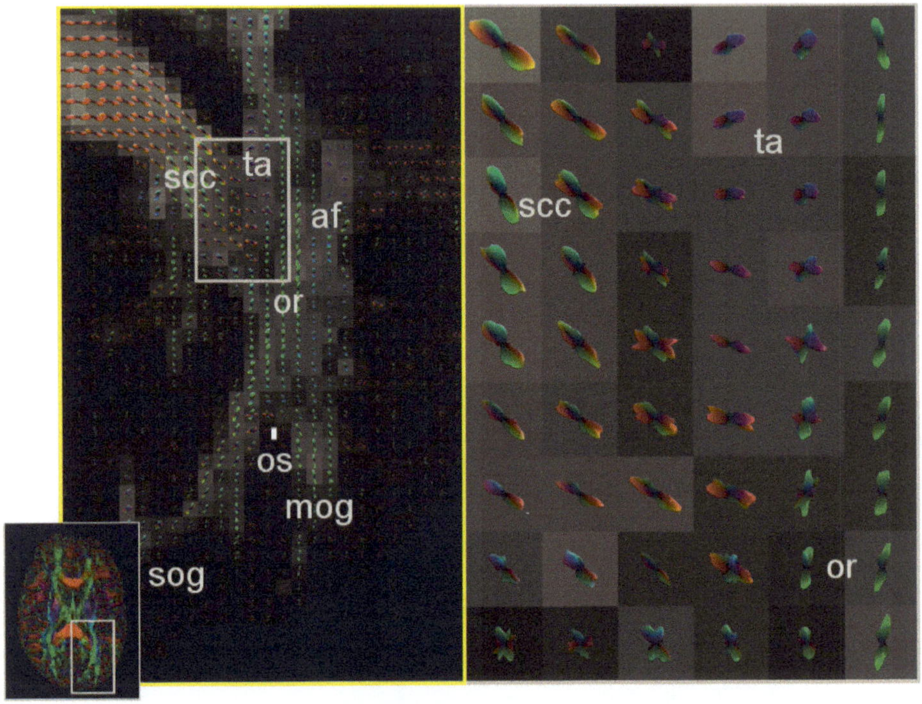

Plate 24 Q-ball image showing intravoxel fiber crossing in the major forceps. Figure courtesy of D.S. Tuch. (See Figure 29 of chapter 6, p. 239).

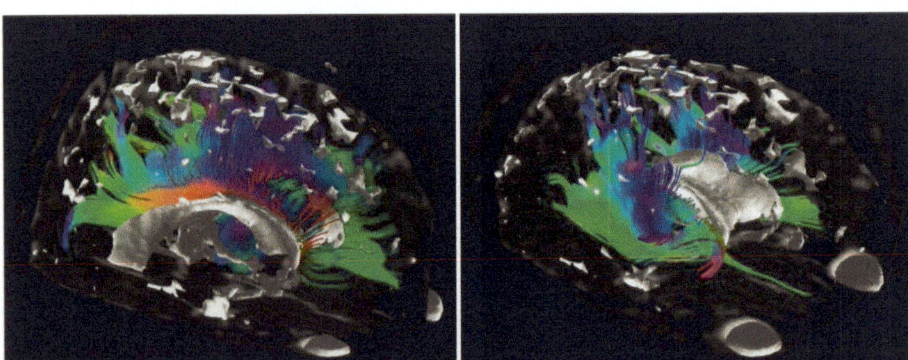

Plate 25 Brain structures: neural pathways computed using the MLS algorithm in the corpus callosum (left) and corona radiata (right) shown together with isotropic brain structures – ventricle, eye sockets, and pockets of CSF on the top of the brain. Cutting planes show isotropic c_s values. Images courtesy of Leonid Zhukov and Barr [99]. (See Figure 30 of chapter 6, p. 240).

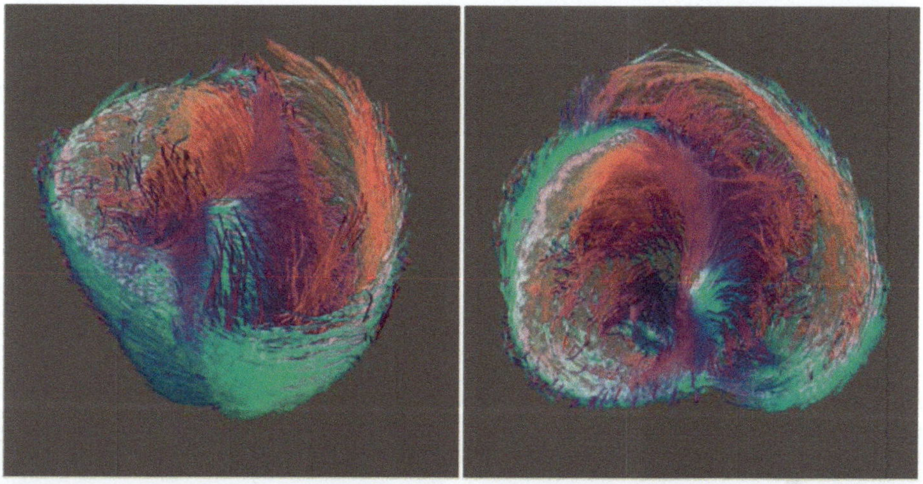

Plate 26 Reconstruction of heart muscle fibers using the MLS algorithm. The color coding changes smoothly from clockwise to counterclockwise spiral oriented fibers. Horizontal parts (very small pitch angle) of the fibers are shown in white. This coloration is consistent with observations of some heart researchers, who have described a systematic smooth variation in pitch and direction of heart muscle fibers from endocardium to epicardium. Images courtesy of Leonid Zhukov and Barr [100]. (See Figure 31 of chapter 6, p. 241).

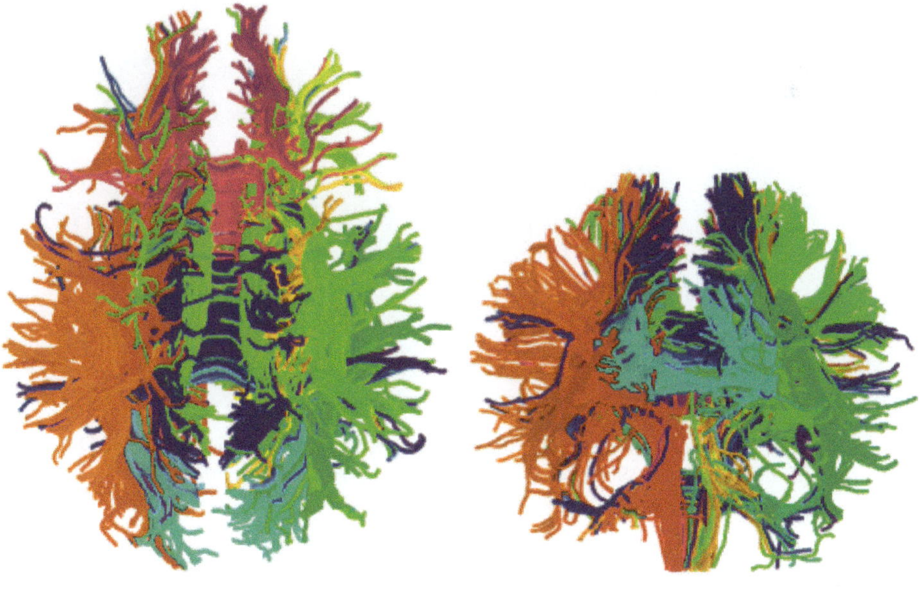

Plate 27 **Left**: Axial view of a segmentation obtained from recursive bipartitioning of the white matter fiber traces. The color of fiber traces indicate cluster membership. **Middle**: coronal view. Images courtesy of Carl-Fredrik Westin [107]. (See Figure 32 of chapter 6, p. 242).

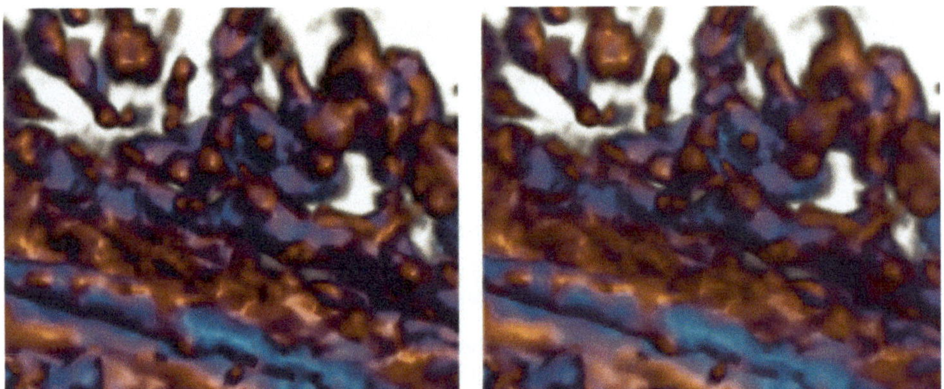

Plate 28 Mixing between lit-tensor and opacity gradient shading for a portion of brain data. Images from Kindlmann and Weinstein [111]. (See Figure 34 of chapter 6, p. 243).

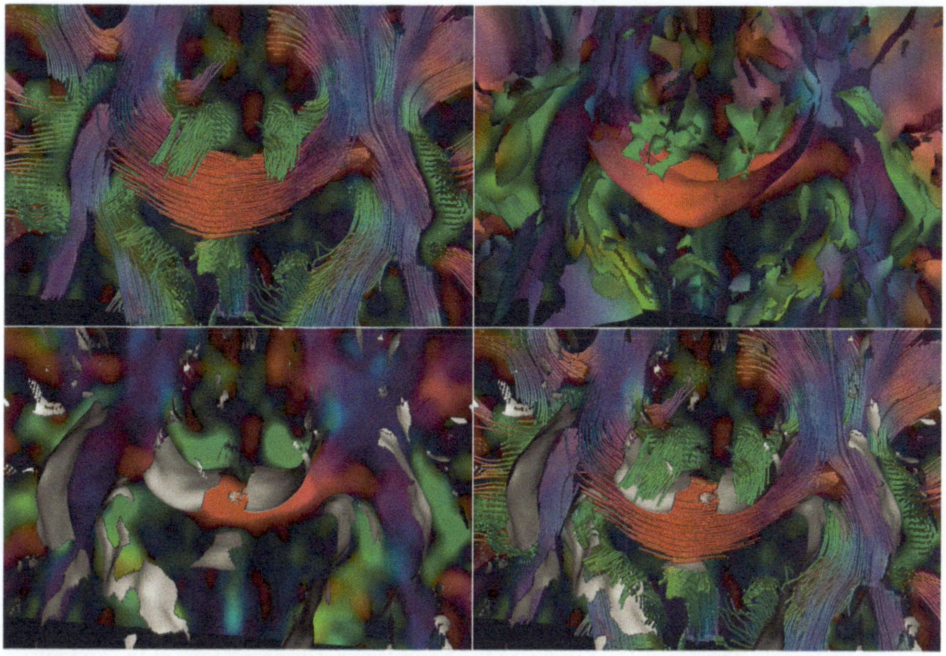

Plate 29 Anisotropy creases near the corpus callosum. **Top left**: RGB map with fibers. **Top right**: Ridge surfaces. **Bottom left**: Valley surfaces. **Bottom right**: Valleys with fibers. From Kindlmann [114]. (See Figure 35 of chapter 6, p. 244).

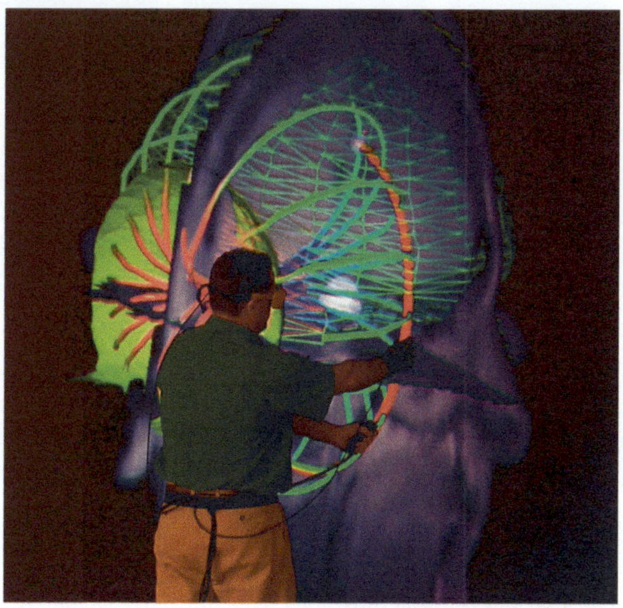

Plate 30 Electric current flow within the brain arising from a localized source, visualized in an immersive environment. (See Figure 37 of chapter 6, p. 246).

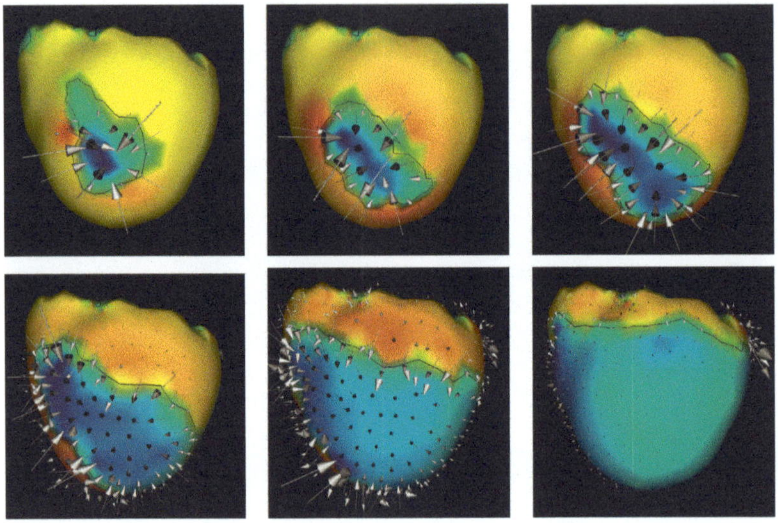

Plate 31 Combined scalar and vector visualization I. The image shows the electrical potential and the electrical current over the cardiac surface during excitation, at 10, 20, 30, 40, 50, and 60 ms, respectively. The black lines indicate the boundary between regions of in- and outflow of the electrical current. Data and image courtesy of Bruno Taccardi and Frank Sachse of the Cardiovascular Research and Training Institute. (See Figure 39 of chapter 6, p. 248).

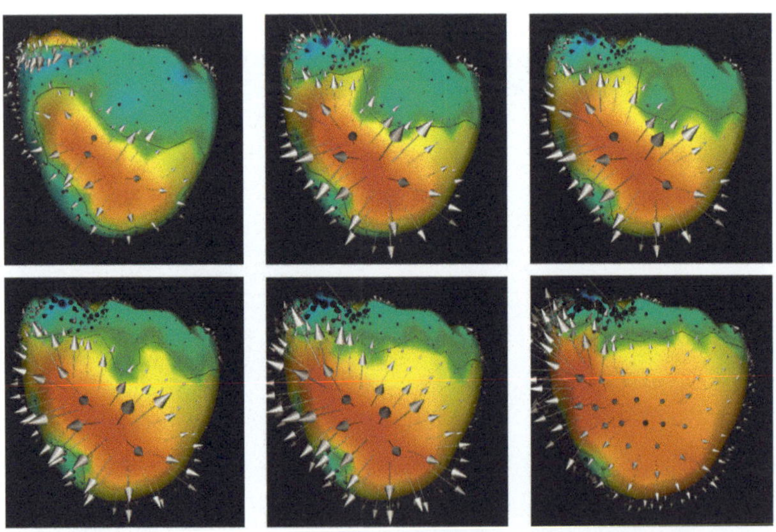

Plate 32 Combined scalar and vector visualization II. Electrical potential and electrical current over the cardiac surface are shown during repolarization, at 110, 130, 150, 170, 190, and 210 ms, respectively. Data and image courtesy of Bruno Taccardi and Frank Sachse of the Cardiovascular Research and Training Institute. (See Figure 40 of chapter 6, p. 249).

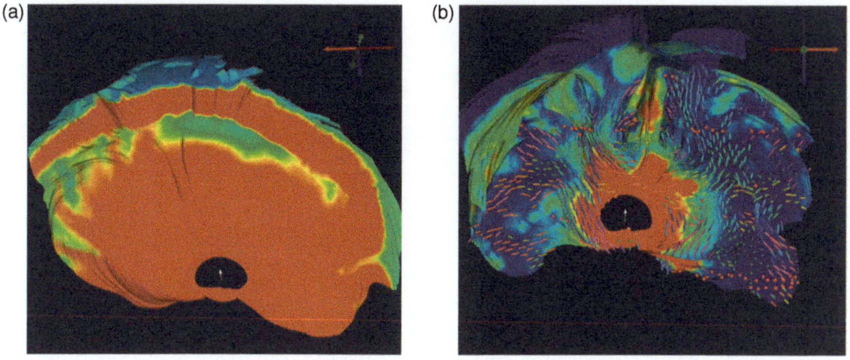

Plate 33 Visualization of return current surfaces from an EEG simulation using a deep thalamic source. The two visualizations show the results of the simulation using an isotropic model (a) and a model including anisotropic white matter (b). (See Figure 41 of chapter 6, p. 249).

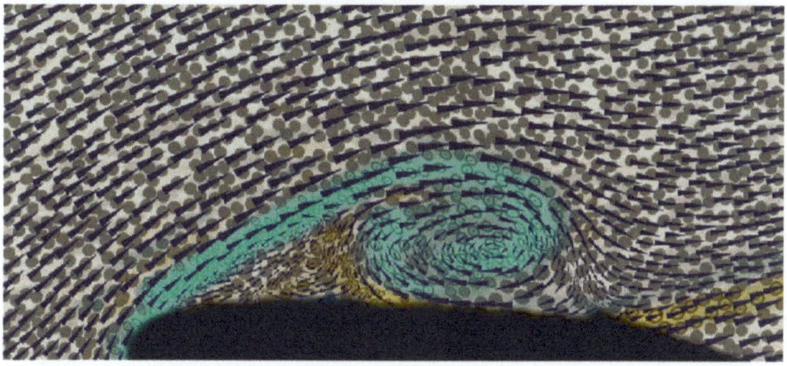

Plate 34 Visualization of experimental 2D flow past an airfoil, indicating velocity, speed, vorticity, rate of strain, divergence, and shear. Image courtesy R.M. Kirby and H. Marmanis and D.H. Laidlaw. (See Figure 42 of chapter 6, p. 250).

Plate 35 Particle tracing using blurring to show the uncertainty in the path. Used with permission of the author. (See Figure 43 of chapter 6, p. 251).

Plate 36 An isosurface of a synthetic dataset (in gray) is bracketed with volume-rendered regions (red and green) indicating uncertainty around the isovalue. (See Figure 46 of chapter 6, p. 253).

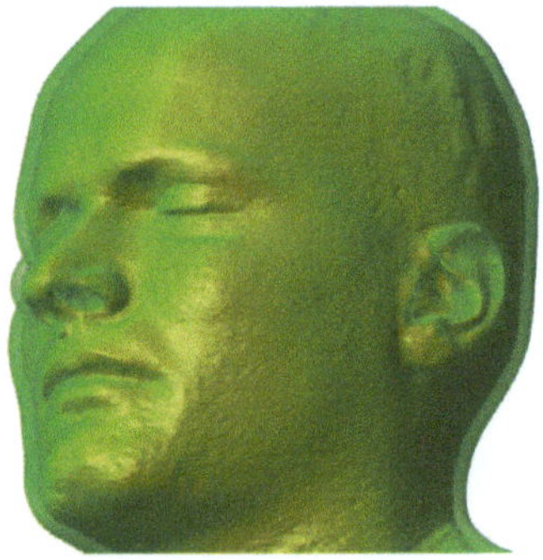

Plate 37 An isosurface of an MRI dataset (in orange) is surrounded by a volume-rendered region of low opacity (in green) to indicate uncertainty in surface position. (See Figure 47 of chapter 6, p. 253).

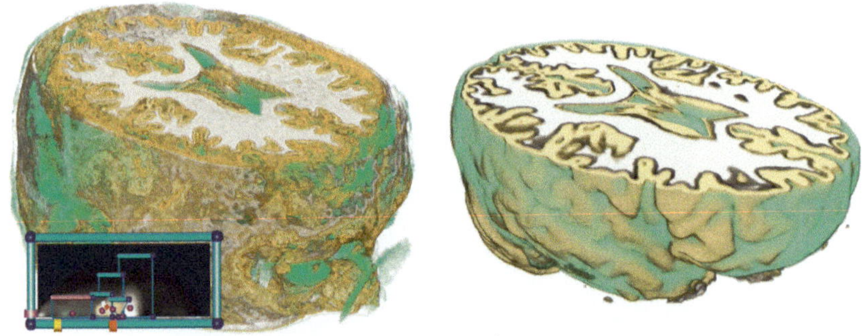

Plate 38 A Comparison of transfer function-based classification versus data-specific probabilistic classification. Both images are based on T1 MRI scans of a human head and show fuzzy classified white matter, gray matter, and cerebro-spinal fluid from Kniss et al. [124]. **Left**: Results of classification using a carefully designed 2D transfer function based on data value and gradient magnitude. **Right**: Visualization of the data classified using a fully automatic, atlas-based method that infers class statistics using minimum entropy, non-parametric density estimation [125]. (See Figure 48 of chapter 6, p. 254).

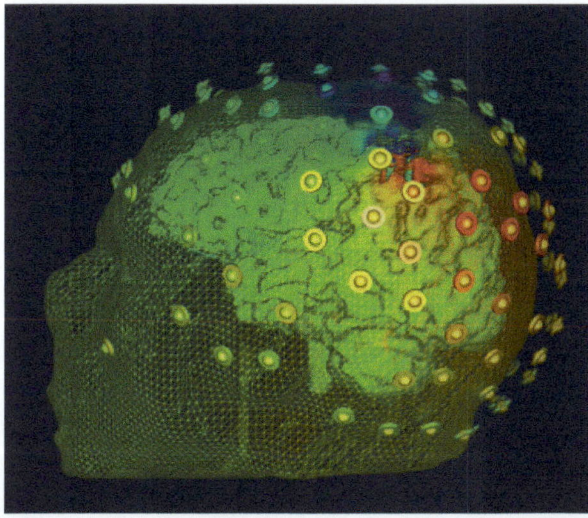

Plate 39 Visualization of simulation results of an EEG simulation localizing a neural source. (See Figure 50 of chapter 6, p. 257).

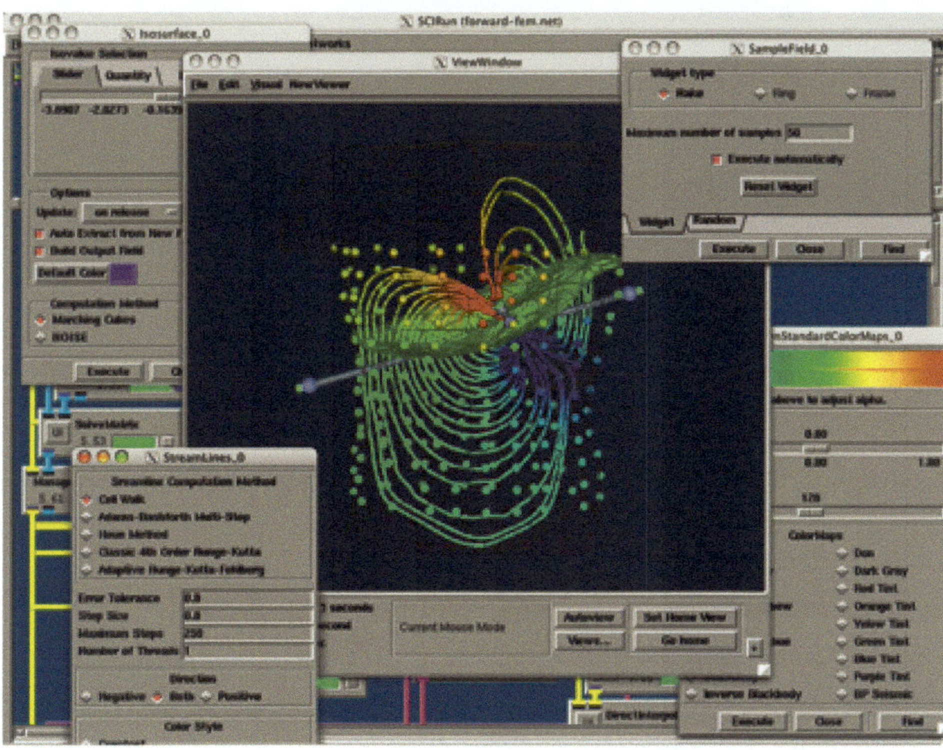

Plate 40 BioPSE dataflow interface for a bioelectric field application. (See Figure 51 of chapter 6, p. 258).

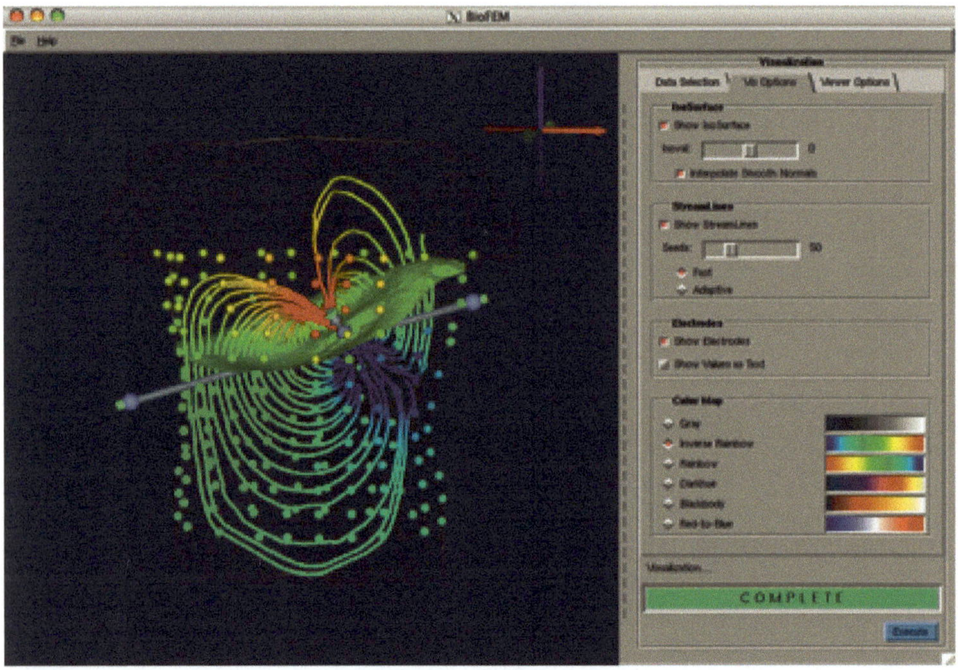

Plate 41 The BioFEM PowerApp custom interface. Though the application is functionality equivalent to the dataflow version shown in Figure 51, this PowerApp version provides an easier-to-use custom interface. Everything is contained within a single window; the user is lead through the steps of loading and visualizing the data with the tabs on the right; and generic control settings have been replaced with contextually appropriate labels; and application-specific tooltips (not shown) appear when the user places the cursor over any user interface element. (See Figure 52 of chapter 6, p. 259).

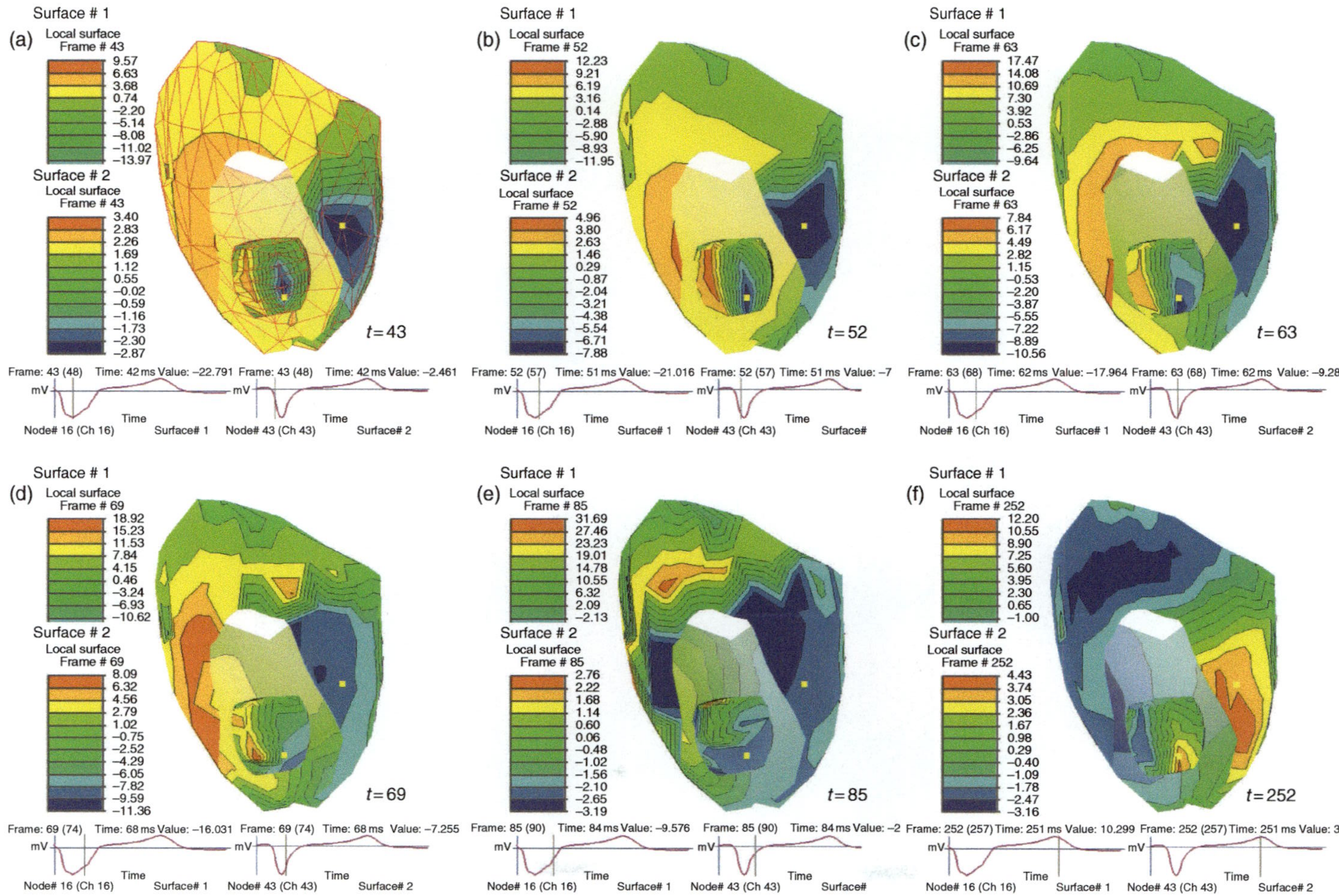

Plate 42 Sequence of potentials from the outer and inner surfaces of the heart. Images show the use of clipping planes to reveal data from two concentric surfaces simultaneously. The first image in the sequence shows the underlying measurement meshes for both surfaces (in red) and the time values marked in each panel indicate the time after stimulation of the heart from the epicardium. (See Figure 54 of chapter 6, p. 262).

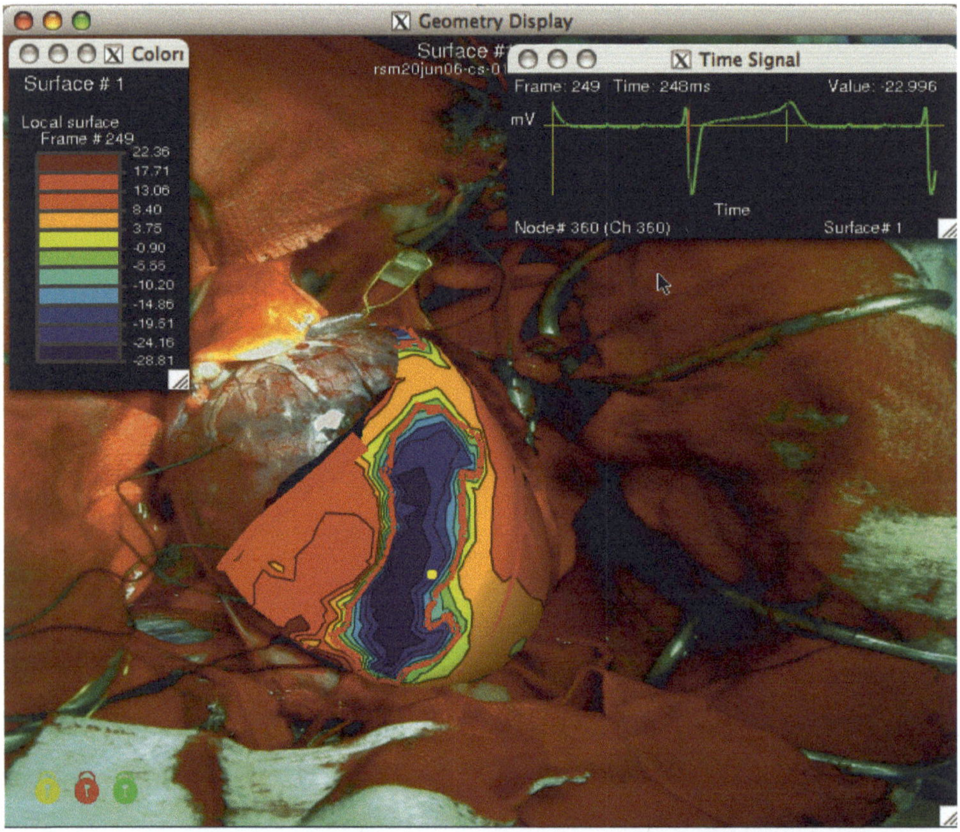

Plate 43 Visualization of data from the surface of the heart superimposed on an image of the same heart during the experiment. Color coding is for surface potentials with the think red contour indicating the location of activation. The time signal in the upper right corner of the figure shows a vertical red mark indicating the time instant captured in the surface map. (See Figure 55 of chapter 6, p. 263).

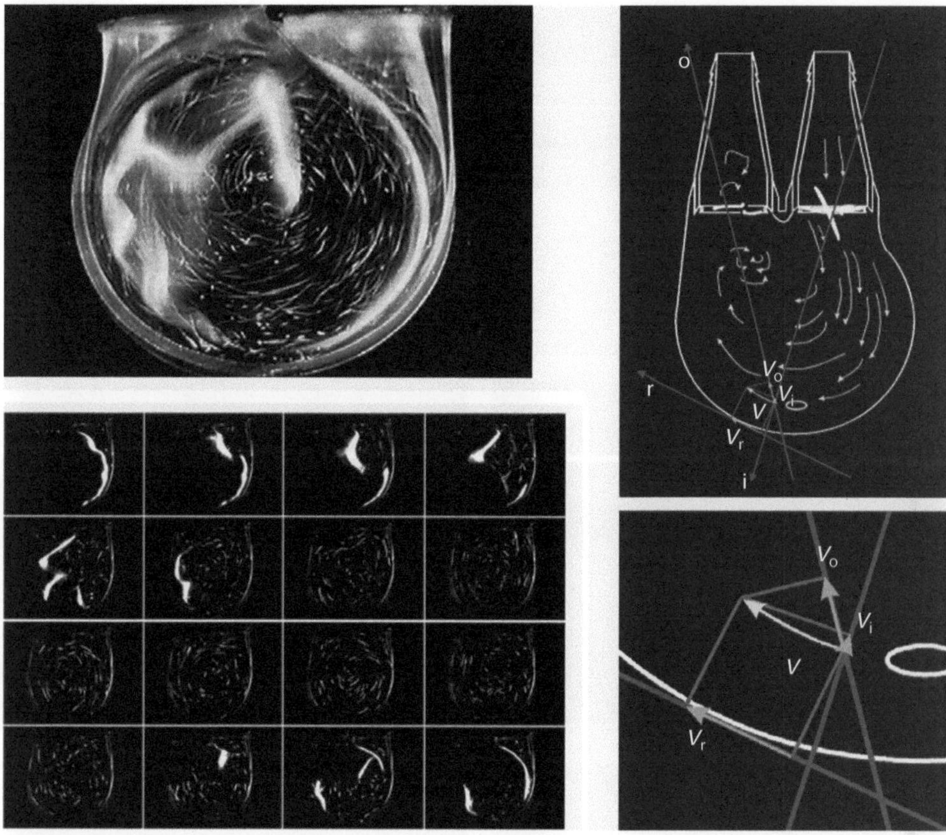

Figure 4(b) One of the problems that are not satisfactorily solved was the interpretation of laser flow visualization pictures. Proposed by the author in the 1990s, functional analyzing method (FAM) method involves monitoring the quality of a flow system. The artificial heart is a pump that should cause fluid flow from the inlet (atrial) to the outlet channel and provide blood with a determined amount of energy to overcome the load, in order to create clinically desired pressure-flow state in the circulatory system. Both goals must be achieved, while conserving the safe transportation of morphotic blood elements, which is essential for long-term blood pumps. Therefore, it must minimize the undesired phenomena occurring at the blood–artificial surface interface. (See color plate 2(b)).

performed percutaneous balloon valvotomy became more involved in aortic valve repair after annular disruptions, and other balloon-induced injuries caused acute insufficiency in young patients requiring immediate repair. Balloon valvotomy may be appropriate for children and adolescents with congenital aortic stenosis, but not for adults with calcific aortic stenosis.

The application of mitral and tricuspid valve repair method allows to maintain the natural anatomy of the heart valve. The surgeon repairs the tissue of the damaged valve and usually implants an *annuloplasty ring* to provide extra support to the valve.

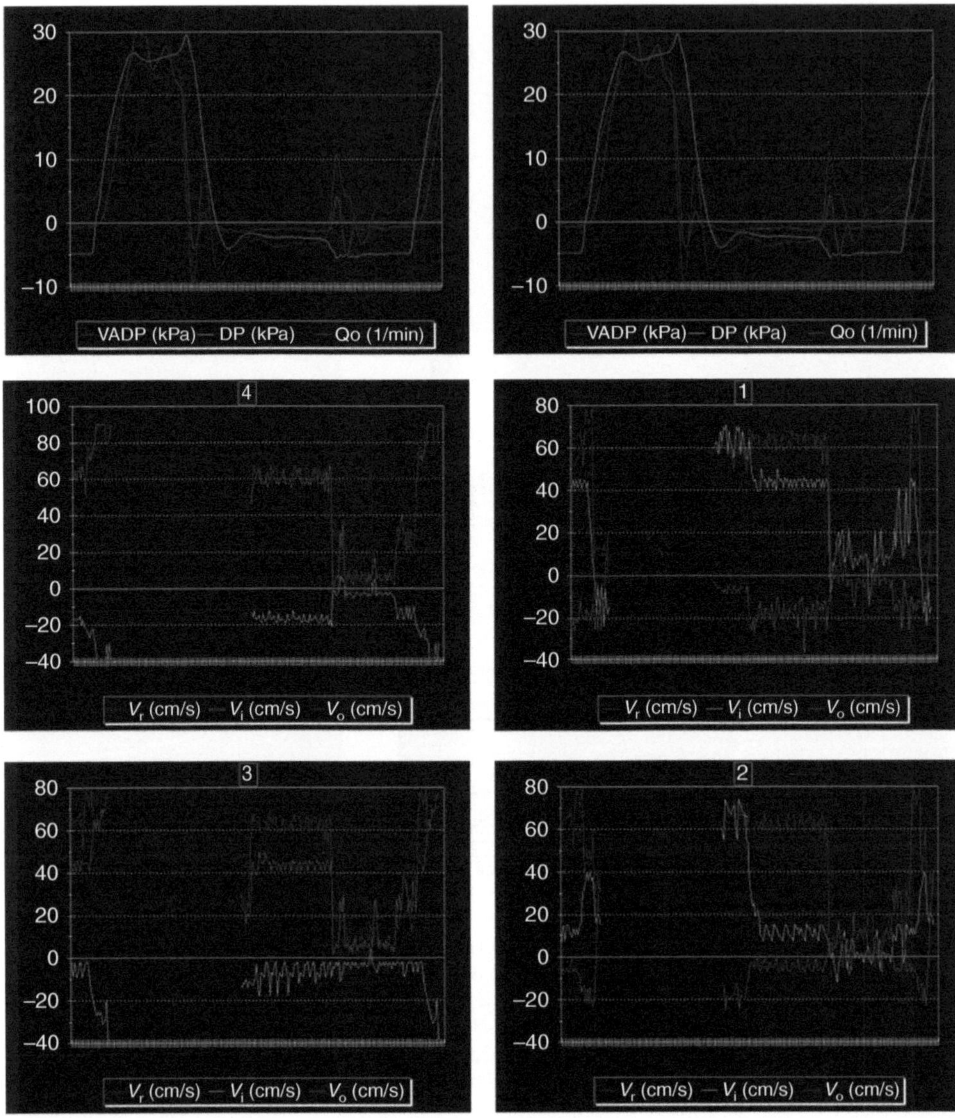

Figure 4(c) The chamber was divided into areas where, for a given section, and for every local flow velocity, vectors are assigned with its components in the direction of an outlet V_o and inlet V_i channel as well as the component V_r tangent to the chamber's wall. The results depend on the phase of work cycle. The analysis of time changes of the first two components allows to track down the dynamics of flow structure organization during an entire cycle, to detect inertial effects and passage time, while the tangent component is responsible for a good washing of wall vicinity areas. The full analysis makes possible the classification of obtained pictures (since digital picture recording is suitable for computer analysis and gives results that are easy to interpret), so that the conclusions regarding the causes of phenomena and their regulation may be drawn easily. (See color plate 2(c)).

Natural valve leaflets are constructed from layers of connective tissue lined with endothelium. The zones of stress concentration are reinforced with bands of collagen fibers. Artificial heart valves should be designed to mimic the natural function and flow dynamics that are optimized in the course of million years of evolution.

The prosthetic valve must be durable and biocompatible, should quickly change phase (close to open, or vice versa), and must allow adequate blood flow with a minimum pressure gradient (energy dissipation), turbulence, and stagnation zones (Figure 5). Optimum leaflet valve design creates minimal stress concentration in leaflets, and optimal cusp shape effects smooth washout, minimal pressure gradient, and adequate longevity. The geometry and the construction of the valve should be surgically convenient to insert in the heart. Most number of valves, including natural ones, are constructed from three leaflets. The simplest configuration (Figure 6) shows that it is due to the Π number whose value is about 3.

There are two main types of prosthetic valves: mechanical and biological. Mechanical valves can cause various complications, including thrombosis, ischemia, and

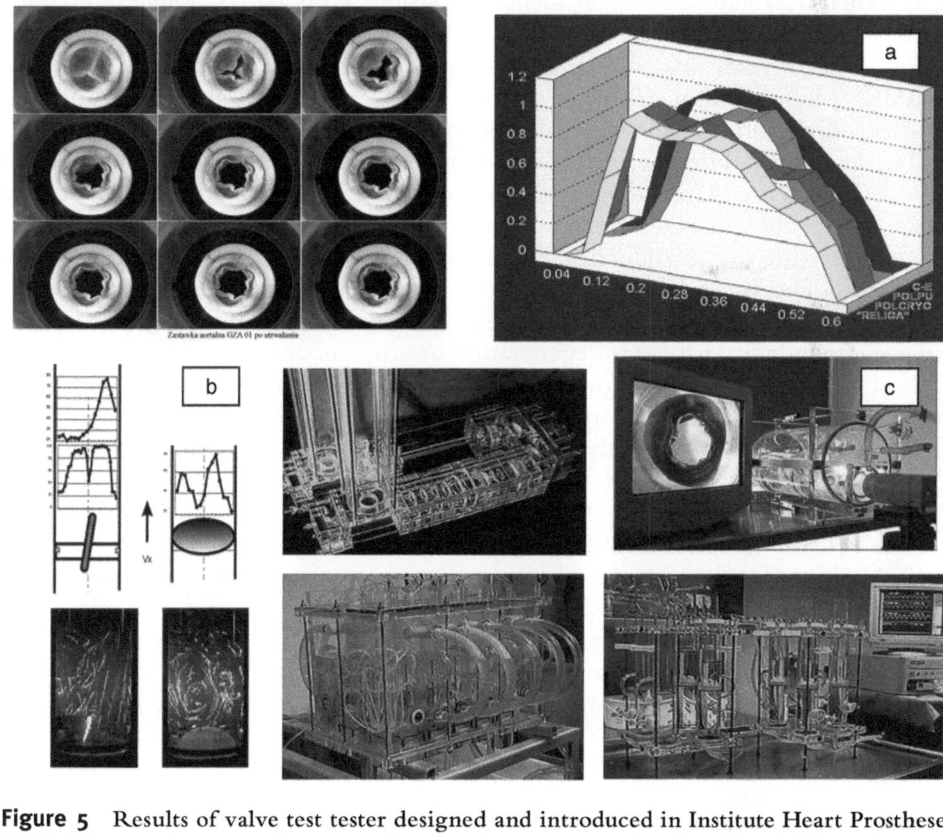

Figure 5 Results of valve test tester designed and introduced in Institute Heart Prostheses, FCSD. The opening area valve test and comparison in one cycle of working biological Religa, Polcryo, and synthetic Polpu prototypes with Carpentier–Edwards bioprotheses (a). The laser visualization and laser anemometer test results of Polish disk valve prototype (b). Several testing apparatus used in our laboratories (c). (See color plate 3).

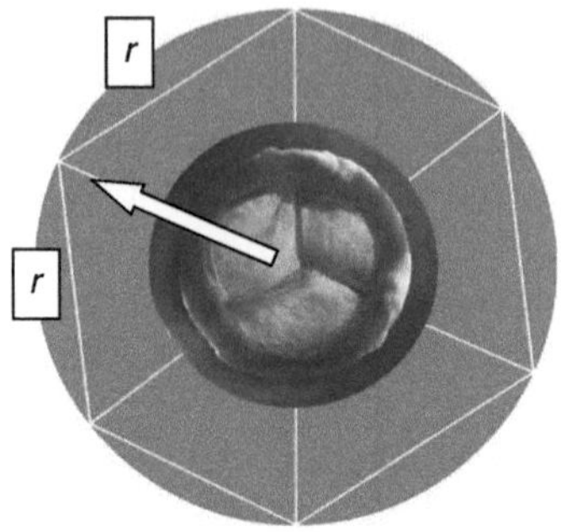

Figure 6 The simplest configuration of a valve:. An open valve ideally constructed using three leaflets lies on the perimeter of the wheel created by the aorta's cross section ($2\pi r$ is equal to about $2r + 2r + 2r$).

turbulent flow. Advantages of bioprosthetic valves (porcine and bovine xenografts) include their durability and low incidence of thromboembolism. Complications associated with bioprosthetic heart valves include hemorrhage and stiffening of the leaflets.

For over 50 years, materials and construction of cardiovascular prosthesis have been under investigation. Hufnagel was the first to implant a valvular prosthesis in the descending aorta for the treatment of aortic insufficiency (in 1952).

The first artificial heart valve prosthesis for routine prosthetic valve replacement clinical use was the Starr–Edwards mechanical valve, which was introduced in the 1960s. The problem of lifelong anticoagulation therapy in mechanical heart valves was solved with the introduction of biological valves in the 1970s: porcine or pericardial. The most popular tissue valves, the Hancock and the Carpentier–Edwards prostheses, are porcine xenografts. However, their durability was limited and age dependent with faster degeneration at younger age.

In 1977 the new generations of mechanical prostheses were introduced: the monoleaflet Medtronic-Hall and the bileaflet St Jude Medical valves and next CarboMedics valves. Currently, mechanical heart valve prostheses account for 60–70% of the prosthetic heart valves implanted worldwide, with bileaflet tilting disk prostheses accounting for the majority.

The development of flexible polymeric heart valves started as early as in 1958 (Roe implanted valves made of silicone rubber). Between 1961 and 1963, 23 patients underwent aortic valve replacement with a tricuspid polytetrafluoroethylene (PTFE) prosthesis by Braunwald et al. Again, mortality was high, and the explanted valves showed severe thickening and rupture of the leaflets. Next aortic prosthesis made of Dacron and silicone was implanted by Roe in 1966, but only four patients survived the operation and lived for 33–61 months. In 1977, Hufnagel implanted a single leaflet aortic prosthesis made of Dacron into the aortic position. Most valves failed, but some patients survived up to 15 years. Embolic complications and fatigue failure were observed frequently in silastic, collagen, and PTFE valves; the latter also exhibited shrinkage. Calcification was also observed frequently, hinting at a destruction of the chemical integrity of the polymer. Polyurethanes have demonstrated the best biocompatibility, durability, and resistance to thromboembolism among all polymers and are, therefore, also used for a variety of medical devices [7].

Currently, some polymeric valves have proven efficacious in assist devices (Abiomed, Medos, and Berlin Heart). However, so far no flexible polymeric heart valve has proven durability for long-term implantation.

All mechanical valves are made up of an orifice ring and occluders, either one or two leaflets through which blood flows through the heart in a forward direction. Because of susceptibility to thromboembolic events, mechanical devices are used in a lifelong anticoagulation therapy (using agents such as Heparin, Warfarin, or Coumadin that delay the clotting of blood, which can cause a heart attack or stroke). The abovementioned drawbacks are minimized or eliminated when bioprostheses are implanted.

Tissue valves (bioprosthetic valves) are made of human or animal tissue. There are three types of tissue valves: pig tissue (porcine), cow tissue (bovine), and human tissue (e.g., aortic or pulmonary valves obtained from human cadavers), or autografts (e.g., the patient's own pulmonary valve, pericardium, or fascia lata). Homograft aortic or pulmonary valves (and associated portions of aortic or pulmonary root) obtained from human cadavers are cryopreserved and implanted directly in place without a synthetic frame.

Heterograft tissue valves for aortic or mitral valve replacement (MVR) are made from natural, animal tissue (porcine aortic valve or bovine pericardium) fixed, usually in dilute glutaraldehyde (GA) and mounted on a synthetic stent – a semirigid frame comprising a base ring and struts.

The technically more difficult to implant, similar to the homograft valve, porcine stentless valve is used only for aortic valve replacement.

The mechanical valves are made from Stellite 21, Haynes 25, Titanium (the housing/strut), Delrin, pyrolytic carbon, carbon/Delrin composite, and an ultra-high-molecular polyethylene (UHMPE) (the leaflet).

In bioprostheses, leaflets are made from porcine aortic valve or porcine pericardial tissue fixed by stabilized GA mounted on stents (polypropylene stent covered with Dacron and Elgiloy wire and nylon support band covered with polyester and Teflon cloth).

The sewing ring is made from Dacron and soft silicone rubber insert covered with porous, seamless Teflon cloth or PTFE fabric over silicone rubber filter.

The physical properties of material, valve geometry, stress–strain distribution in leaflets, and the local pressure and flow values near the valve influence the risk of valve failure and can cause damage to the blood components. Dangerous for blood local space are associated with shear velocities and shear stress reaching critical values for morphotic blood elements. For soft (biological) valve durability, the stress on leaflets is the most important factor, which lead to calcifications and, as a result, valve damage.

The global geometry of valve is enforced by the fit criteria and is characteristic of physical nature of work. The biological valve design made from donors (humans or animals) is limited by anatomical valve tissue and heart geometry.

The use of xenograft and allograft biomaterials has typically required chemical or physical pretreatment aimed at preserving and sterilizing the materials and reducing the immunogenicity of the tissue. Multiple chemical and physical crosslinking techniques have been explored to stabilize the collagen-based structure of the tissue, maintaining its mechanical integrity and natural compliance. Bioprostheses are mainly prepared from animal structures, e.g., valves or pericardium, composed primarily of collagen. The main advantage of xenogenic bioprotheses is their structural similarity to human tissues. However, in order to improve prosthesis durability reflected by the enhanced resistance to enzymatic degradation and reduced immunogenicity, animal tissues must be chemically or physically pretreated (fixed). Fixation procedures are connected with crosslinking of tissue proteins. Among chemical methods of crosslinking, the most commonly accepted is treatment with GA. This five-carbon bifunctional aldehyde reacts with free amino groups of proteins – mainly ε-amino groups of collagen lysyl residues – forming inter- or intrachain crosslinks. Prostheses obtained after tissue treatment with GA reveal remarkable reduction in immunogenicity and sensitivity to in vivo degradation.

Patients with mechanical valves require long-term anticoagulant therapy owing to the risk of thromboembolic complications. Reports of strut failure, material erosion, and leaflet escapes as well as pitting (cavitations) and erosion of valve leaflet and housing have resulted in numerous investigations on the closing dynamics of mechanical valves and the pressure distribution on the leaflets and impact forces between the leaflets and guiding struts. This requires the flow through the clearance between the leaflet and the housing in closing position influencing the hemolysis and thrombus initiation.

The valve-related problems are as follows:

- *Mechanical valves*: thromboembolism, structural failure, red blood cell and platelet destruction, tissue overgrowth, damage to endothelial lining, para/perivalvular leakage, tearing of sutures, infection
- *Bioprosthetic valves*: tissue calcification, leaflet rupture, para/perivalvular leakage, infection

The cause of valve failure can be related to the valve components (Figure 7).

The probability of survival 5 and 10 years following heart valve replacement is approximately 70 and 50%, respectively. Prosthesis-associated complications often lead to reoperation such that replacements currently account for 15–25% of all valve operations. Thrombotic deposits may form on heart valve prostheses, which is more likely to occur in mechanical heart valves which are more thrombogenic than tissue valves. Hemorrhage rates are higher in patients with mechanical valves in the aortic site than in patients with aortic tissue valves. However, these rates are similar for mechanical and tissue valves in the mitral location. Rates of infection (1–6%) do not differ significantly between tissue and mechanical prostheses. In tissue valves, infection may be localized in the vicinity of the sewing ring. However, the cusps may also be a focus of infection [6].

From the early generation of heart valves, only the Starr–Edwards ball valve design remains in clinical use today. Currently available mechanical valves have been designed with a lower profile and a more effective orifice area, to improve hemodynamics.

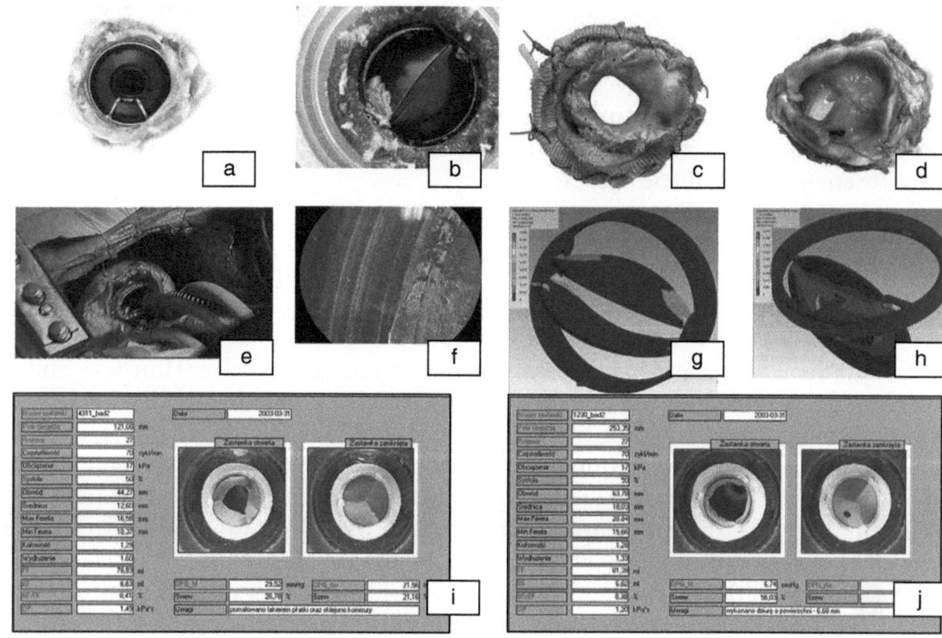

Figure 7 The samples of explanted valves collected in the Institute of Heart Prostheses, FCSD (COST Action 537 project). One-disk (a) and two-disk mechanical valve (b) and two bioprostheses (c,d). The in vivo valve test on ship (e). The microscopic evaluation of valve's disk (f) and modeling of mechanical valves – equivalent elastic strain values (g,h). The effect of biological heart valve prosthesis damage on hemodynamic efficiency has been investigated and analyzed: improper functioning due to leaflet calcification – the opening area of "calcified– (leaflets partially glued (i)) valve decreased by 44%, while the pressure gradient grew up to 22 mmHg. Perforation of a leaflet (j) – significant perforation (hole area 6.68 mm^2) of leaflet caused increase in back-leakage of 147% to 8.38 (6.8 ml). (See color plate 4).

The initial design of tilting disk valve consisting of Delrin disk exhibited good wear resistance and mechanical strength. But due to the swelling phenomenon, it was replaced by pyrolytic carbon. Structural dysfunction of the B–S (Shiley Inc., USA) 60° and 70° convexo-concave mechanical heart valve prosthesis occurred at a relatively high frequency (2.2–8.3%). The B–S tilting disk design has been withdrawn from the market, while the Duromedics bileaflet prosthesis was reintroduced as the Edwards–Tekna (Edwards Lifesciences, USA) valve following design modifications. The failure mode of the B–S prosthesis is due to failure of the welded outlet strut with resultant embolization of the disk. Fracture of the carbon component in a small number of mechanical valves, including the Edwards–Duromedics (Edwards Lifesciences, USA) and St Jude Medical (St Jude Medical Inc., USA) bileaflet tilting disk valves, has also been reported [6].

The most common cause of failure during implantation is leaflet fracture. The most common causes of mechanical valve failure are pannus formation and thrombosis. Structural dysfunction occurs more commonly with tissue valves than with contemporary mechanical valves [6].

The rate of bioprosthetic valve failure increases over time, particularly after the initial 4–5 years after implantation. At 10 years after implantation, 20–40% of porcine aortic valves implanted in either aortic or mitral sites require replacement for primary tissue failure. Up to 50% of such valves fail after 10–15 years. Calcification, cuspal tears, or both are the most common manifestations of tissue failure in bioprosthetic porcine aortic valves.

Stentless bioprosthetic porcine aortic valves have shown minimal cuspal calcification or tissue degeneration for periods up to 8 years following implantation. As with bioprosthetic porcine valves, bioprosthetic heart valves made from bovine pericardium develop both calcific and noncalcific tissue failure. The second generation of bovine pericardial prostheses, such as the Carpentier–Edwards pericardial valve, have increased durability compared with first-generation pericardial valves [Ionescu–Shiley (Shiley Inc., USA), Mitroflow (CarboMedics, Canada), and Hancock] and excellent hemodynamics. Cryopreserved human homograft (or allograft) aortic valves have equivalent or slightly better durability than contemporary bioprosthetic porcine valves with valve survival rates of approximately 50–90% at 10–15 years [6].

Technology continues to improve the durability of bioprostheses and to reduce the thrombogenic effects of mechanical prostheses.

The most significant changes in mechanical heart valves over the last decade have focused on the sewing ring and the ability to rotate the valve after implantation. The St Jude Medical Regent prosthesis and CarboMedics Top Hat improve valve hemodynamics by introducing modification that allows for the placement of the device in a supra-annular position. The St Jude Medical Silzone (silver nitrate incorporated in the sewing cuff) was withdrawn from the clinical trials due to increased incidence of paravalvular leak in the silzone cohort. The newer generation mechanical prostheses, e.g., the Edwards–MIRA (Edwards Lifesciences, USA) (Sorin Bicarbon mechanical prosthesis with a modified sewing ring), ATS (Advancing the Standard) (ATS Inc., USA), and On-X (Medical Carbon Research Institute, USA), have just completed regulatory clinical trials [6].

The most significant changes in biological heart valves over the last decade have focused on the stent construction and tissue treatment method (changing fixation pressure and chemicals). Stent mounting produces higher transvalvular gradients and also causes premature valve failure. In the brief period in which homografts were stented, the average life expectancy of the valve was less than 10 years (about 90% homografts have a 10-year freedom from valve degeneration). Furthermore, to date, in two large international trials, no Medtronic Freestyle (Medtronic Inc.) and Toronto SPV (St Jude Medical Inc.) stentless valves have been explanted because of primary structural failure [6].

The valves fixed at zero pressure retain a collagen architecture virtually identical to that of native unfixed porcine aortic valve cusps. The 10-year experience with the Medtronic Intact (Medtronic Inc.) (zero pressure fixed) valve reported no cases of primary structural degeneration in patients over 60 years of age and only one case of valve failure in individuals over 40 years of age. The exposed amine residues of the GA molecule promote tissue calcification. Surfactants, particularly sodium dodecyl sulfate (T6) (Hancock II, Medtronic Inc., USA), polysorbate 80 (Carpentier–Edwards standard and supra-annular porcine bioprostheses;

Edwards Lifesciences, USA), and toluidine blue (Medtronic Intact, Medtronic Inc.), have been incorporated in the preservation process. No-React detoxification process has been proposed as a method of preventing calcification of GA fixed tissue. Detoxification with homocysteic acid is used in Sorin products (Sorin Group Inc.) [6].

The Carpentier–Edwards PERIMOUNT Pericardial Bioprosthesis, made of bovine pericardial tissue that has been preserved in a buffered GA solution and mounted on fully extensible stents and distensible struts, decreases shearing stresses on valve leaflets and maintains physiologic aortic ring movements to reduce flow turbulence and vibrations. Both the frame and the sewing ring are covered with a knitted PTFE cloth. This material helps facilitate the healing and ingrowth of tissue around the implanted valve.

The Medtronic Company is currently manufacturing two new bioprostheses. One valve is stentless (Freestyle aortic root bioprosthesis) and the other incorporates a stent design (Mosaic bioprosthesis). Both represent new concepts in the manufacture of bioprostheses, specifically, zero-pressure fixation and α-amino oleic acid antimineralization treatment. CarboMedics is investigating a Photofix bioprosthetic valve: a patented trileaflet, central-flow prosthesis, with each leaflet mounted on a flexible support frame. The leaflets are prepared from bovine pericardium treated with a unique patented dye-mediated photo-oxidation process. This new fixation process results in collagen crosslinking without the use of GA, a suspected contributor to the calcification failures of other clinically available tissue valves.

The *choice* of prosthesis is again a decision to be made by the surgeon and the patient, with full knowledge of the advantages and disadvantages of the different types available.

From a clinical point of view, the use of bioprosthetic valves is recommended for older patients (over 65–70 years) because of the increased risk of bleeding in the elderly and the low probability of structural failure of the bioprosthesis in the remaining expected life of these patients. Mechanical prostheses are indicated for patients 70 years of age or younger, even though there is significant valve-related morbidity. The outcomes 15 years after valve replacement with a mechanical versus a bioprosthetic valve have been reported by the Veterans Affairs randomized trial. All-cause mortality was not different after MVR with mechanical prostheses versus bioprostheses. Structural valve deterioration was greater with bioprostheses for MVR in all age groups but occurred at a much higher rate in those less than 65 years of age. Thromboembolism rates were similar in the two valve prostheses, but bleeding was more common with the mechanical prostheses. In 2003, the Edinburgh randomized trial extrapolated results to 20 years. The prosthesis type did not influence survival, thromboembolism, or endocarditis. Major bleeding was more common with mechanical prosthesis [6].

Thanks to EU grant COST Action 537 – "Core Laboratories for the improvement of medical devices in clinical practice from the failure of the explanted prostheses analysis (FEPA)" – organized in Europe. The goal is to improve medical devices in clinical practice from the FEPA – systematic studies of medical evaluation of cardiovascular implants requiring an explantation and assessment of generalized (immunological, inflammatory) and/or remote biological effects in patients with cardiovascular implants and with complications.

Living tissue valve replacements would solve many of the existing problems. In the future, natural biomaterials repopulated with autologous or genetically engineered cells will be used as ideal templates for the design of living tissue grafts. To date, the knowledge about these new materials is not very wide.

Further developments toward the alternative biological valve prostheses based on biological cell-free matrices as ideal valve substitutes using tissue engineering would potentially eliminate the known disadvantages of current valve prostheses. The endothelial cells harvested from the patient may be cultured in laboratory and incorporated into the synthetic scaffolds (polyglycolic acid, polyhydroxyalkanoate, and poly-4-hydroxybutrate) or both heterograft and allograft valvular decellularized tissues.

An ideal heart valve substitute should possess the following characteristics: absolute biocompatibility, long-term durability, nature-like biomechanical properties, no thrombogenic or teratogenic effects, sufficient availability in all common sizes, and moreover, particularly for pediatric patients the growth potential is crucial. Regions of stress concentration on the leaflets during the complex motion of the leaflets have been implicated in the structural failure of the leaflets with bioprosthetic valves. Computational fluid dynamics (CFD) has emerged as a promising tool, which, alongside experimentation, can yield insights of unprecedented detail into the hemodynamics of prosthetic heart valves.

3.3. Heart Pacemaker

The pacemaker was the first electronic device ever surgically implanted inside a human. In 1950, John Hopps, an electrical engineer, built the first pacemaker. The first successful attempts at designing a totally implantable pacemaker were reported by W. Chardack, A. Gage, and W. Greatbatch (New York). The group's work was recognized by Medtronic, which in 1960 signed a contract with Chardack and Greatbatch to produce an implantable pulse generator. These pacemakers and batteries have improved and saved the lives of millions of people worldwide. The pacemaker was applied for the first time in 1952 (P. Zoll) to stimulate the heart muscle through electrodes placed on the patient's chest. On the next stage, endocavitary electrodes were introduced into the right ventricle through the vein and a new, more effective way of controlling the system was carried out. The size and the weight of pacemakers and their durability had improved after replacing mercury batteries with lithium batteries, which have to be replaced every 10 years. Currently, the average weight of the device is about 15–30 g and 2.5 mm in diameter. In order to reduce the risk of system distortion, the bipolar electrodes should be fixed to endocardium using special methods. The electrode tips are made of special metal alloys and have a porous surface.

4. The Minimally Invasive Cardiology Tools

Minimally invasive cardiology requires a separate set of tools allowing the surgeon to operate through vessels. For surgical interventions, the following devices can be used: the guideline (diameter 2–3 mm), catheters (the balloon catheter has a diameter of 0.8 mm), balloons (the balloon has to hold out an inflation up to 20 atm), and broadening vascular stents and devices (about 2 mm in diameter covered with diamond particles) for drilling out and excising the atherosclerotic plaques in coronary vessels, all under X-ray control.

The human circulatory system consists of the heart and the network of vessels transporting blood throughout the body. A pathological state of the aorta, which in more advanced stage blocks the blood flow, can be treated by vascular prosthesis implantation, replacing the damaged fragment of the vessel. Tissue implants (autogenic, allogenic, xenogenic) and synthetic vessels can be used as vascular prostheses in the surgery of vessels and cardiosurgery. If the main aim is to preserve the life of the cells, then depending on the time of tissue treatment, they are stored in antibiotic bath, nutritious liquids, or in hypothermia; or deep-freezed in liquid nitrogen. The blood cannot be used as a preservation solution, because at room temperature, unoxygenated blood becomes toxic. In some cases, vascular stents can also be used, which are constructions for vessel wall supporting to preserve its physiological diameter. They are pipes with diameter up to several centimeters, made of biocompatible materials (Dacron) fixed on a steel frame, assembled in the vessel during operation.

The narrowing or blockage in one or more coronary arteries (those supplying the blood to the heart muscle) can be a direct cause of the coronary artery disease. Currently, several treatment methods are available: e.g., using bypass techniques, vessel stents and minimally invasive, mechanical cloth, or by calcification removing. Pharmacological therapy and tissue engineering are very promising tools for contemporary and future treatment.

The catheter can be brought to the desired location via navigation, propulsion, and steering to perform several functions:

- actuation (ablation of material, balloon angioplasty, deployment of stent)
- sensing (pressure)
- transportation of material or energy (contrast fluid, embolization material, or signal from sensors for diagnostic purposes)

and that it can be retrieved again from the body [8]. The physical contact of the interventionist with the patient's tissue is established by using catheter, and the visual observation is mediated by the X-ray imaging.

Since the origin of the field of interventional cardiology in 1966 with balloon atrial septostomy, a procedure directed at treating congenital heart diseases, pediatric interventional cardiologists have developed percutaneous techniques for treating congenital valvular and great vessel stenoses and transcatheter closure of aberrant vascular channels. Balloon angioplasty [percutaneous transluminal coronary angioplasty (PTCA)] is widely used for treatment of the blockages of coronary artery. Percutaneous coronary intervention (PCI) encompasses a variety of procedures used to treat patients with diseased arteries of the heart. The catheter with balloon is inserted into the vessel (most often the femoral artery) through a small prick in artery. After transferring the balloon to the destination place, it is filled with a physiological saline mixture up to several atmospheric pressure. Due to this pressure, the balloon will expand, thus compressing the plaque, and hence, the narrowed vessel will also expand. The effectiveness of coronary angioplasty can be improved by using vessel prosthesis – a stent that is placed in vessel supports it and prevents it from closing. Typically, PCI is performed by threading a slender balloon-tipped tube – a catheter – from an artery in the groin to a trouble spot in an artery of the heart (this is referred to as the PTCA, coronary artery balloon dilation, or balloon angioplasty). The balloon is then inflated, compressing the plaque and dilating (widening) the narrowed coronary artery so that blood can flow more easily. This is often accompanied by inserting an expandable metal stent. Stents are wire mesh tubes used to prop open arteries after PTCA. The stent is collapsed to a small diameter and put over a balloon catheter. It is then moved into the area of the blockage. When the balloon is inflated, the stent expands, locks in place, and forms a scaffold. This holds the artery open. The stent remains in the artery permanently, holds it open, improves blood flow to the heart muscle, and relieves symptoms (usually chest pain). Within a few weeks of the time the stent was placed, the inner lining of the artery (the endothelium) grows over the metal surface of the stent. More than 70% of coronary angioplasty procedures also include stenting. Some blockages in the arteries are hard, calcified deposits that do not respond to balloon angioplasty or stent implantation. In this case, the rotablator technique may be introduced. This is a small device used inside of the coronary arteries to "drill" through the calcified blockage. It breaks the calcium into microscopic pieces and disintegrates the blockage.

Percutaneous balloon dilation of isolated congenital semilunar valvular stenoses has proven to be highly effective in providing long-term hemodynamic and symptomatic benefit in neonatal, pediatric, and adult patient populations. Kan and associates described the first clinical application of balloon valvuloplasty in 1982, in which drastic improvement in transvalvular gradient was achieved in five children with pulmonary valve stenosis.

Percutaneous balloon valvuloplasty has thus been proposed as a less invasive means of treating mitral stenosis (Figure 8). The valve is usually approached in an anterograde direction via an interatrial septal puncture, although a retrograde technique has been reported. One or two tubular balloons or a specialized nylon-rubber (Inoue) balloon is advanced to position across the mitral valve and repetitive inflations are performed until the balloons have fully expanded.

Figure 8 The physical percutaneous balloon valvuloplasty simulation on polyurethane and mitral valve.

Hemodynamic improvement following mitral valvuloplasty occurs immediately, with valve area increasing on average by $1.0\,cm^2$, with a prompt drop in pulmonary pressures.

A randomized trial comparing mitral balloon valvuloplasty to open surgical commissurotomy among patients with favorable valvular anatomy demonstrated comparable initial hemodynamic results and clinical outcomes among patients treated with either technique, although hemodynamic findings at 3 years were more favorable in the balloon valvuloplasty group.

Pulmonary artery stenosis or hypoplasia may be effectively treated using catheter-based technologies. Success rates following balloon dilation of these vessels have ranged from 50 to 60%, with failures due primarily to elastic recoil. There has also been a limited experience with balloon angioplasty or stent placement for management of aortic coarctation, venous obstruction, or stenoses of Fontan shunts.

Advances in the technology for percutaneous coronary revascularization have been accompanied by a dramatic increase in the number of procedures carried out in the world.

Intravascular stenting of arteries is one of the most frequent operations in cardiovascular surgery. Stents are inserted into a vessel with injured inner wall to avoid its destruction, thrombus formation, or vasoconstriction. Vascular stents should have a shape rendering it relatively elastic and durable with respect to the vessel (Figure 9). There have to be (e.g., gold) marker lines, allowing stent observation during procedure (RTG monitoring). Stents are made from metal alloys using various techniques. They can be formed from the wire or through excising from tubes using laser. More recent research uses shape memory alloys (Nitinol). Such stents in cooled stage are transported to the target place, where after warming up to physiological temperature, they reach the final shape. One of the first stents clinically applied (Wiktor-GX stent, Medtronic Interventional Vascular, Holland) is an original project of Pole, Dominick Wiktor, in the United States. This stent was created from wire with sinusoidal shape wounded on the cylinder. It

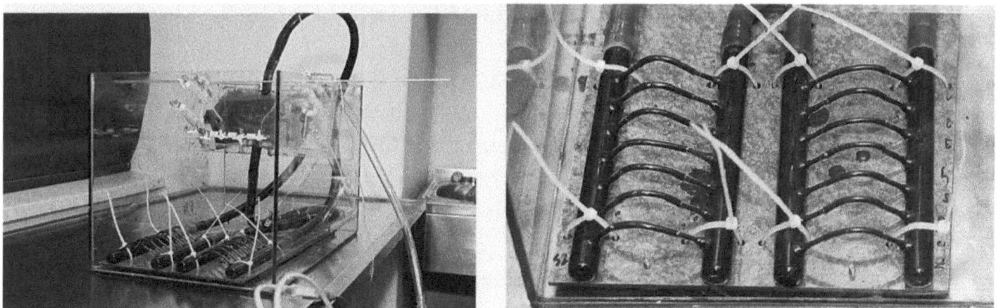

Figure 9 Fatigue tests on the nanocrystalline diamond and steel coronary stents were carried out: Tyrode's physiologic solution; frequency 180 cpm; 3 months. Coronary arteries were simulated using a tube of inner diameter 3 mm and wall thickness 1 mm. After investigations, the corrosion resistance tests of the implants have been additionally performed.

is made from tantalum, instead of steel, because of its better contrast in RTG visualization and the electronegative surface improving thromboresistance of the stent. The first clinical implantation of stent was performed in 1989. The materials used for the construction of these stents have changed dramatically over the past 10 years. Unfortunately, metal stents induce thrombus formation. Stent biocompatibility can be enhanced by either modifying the metal surface of the implant or applying biocompatible coatings. To improve its biocompatibility and to prevent the closing of renewed vessel, the surface of stents is covered with heparin. Stenting procedures can also be connected with genetic therapy by covering its surface by a layer of polymer, which when placed into the vessel wall releases the genetic material, which blocks the cell growth. In Poland, the multidisciplinary team works on original vascular stents covered with a layer of nanocrystalline diamond.

Currently, multiple research projects are focused on the design of a stent with better radiopacity, reduced stent–vessel wall contact area, and improved elasticity, and thrombogenic effects are lowered when the stent is coated with the proper antithrombotic substances. The problem that plagues about one-third of patients who receive angioplasty with or without stents is "restenosis" – the recurrence of the narrowing of the blood vessel. The new drug-coated stents release a medication that blocks this scarring process.

Novel technologies will be introduced and developed, directed at the efficient removal or modification of arterial plaques with minimal arterial trauma or at site-specific drug delivery to inhibit thrombosis and restenosis. Concurrently, intravascular ultrasound and angioscopy will be further refined and integrated into revascularization devices, allowing precise guidance and optimization of plaque ablation or remodeling while limiting associated coronary injury. Finally, understanding of the vascular biology of percutaneous revascularization, thrombosis, and restenosis will be improved, leading to pharmacological therapies designed to ameliorate adverse thrombotic, proliferative, and remodeling responses.

5. The Technology for Atrial Fibrillation

The evolving technologies for ablation of atrial fibrillation include radio frequency, cryotherapy, microwave, laser, and ultrasound. The efforts directed at isolating and ablating percutaneous catheter pulmonary vein have essentially been abandoned because of the extensive length of procedures and the high incidence of pulmonary vein stenosis. The newer technologies have been developed to create transmural lesions during cardiac surgery [6].

Radiofrequency used in ablation: Radio frequency corresponds to the frequency of unmodulated alternating current delivered in the range of 0.5–1.0 MHz between two electrodes, one located on the endocardial surface and the other on the skin. Its heating effect produces homogeneous lesions that measure a few millimeters in diameter and depth. It has been demonstrated that reliable and effective ablation is performed at 70°C for 60 s. The goal temperature should never be set at more than 95°C to avoid potential tissue disruption.

Cryoablation: Cryoablation has an excellent clinical safety record, although its use in atrial fibrillation surgery has been limited to creating spot lesions over the tricuspid and mitral valve annuli. The salient features of the procedure are rapid freezing, and slow thawing with repeated freeze–thaw cycles. The coldest temperature (the prime determinant of cell death) may range from −50°C to −150°C and the application time can vary between 0.5 and 5 min, depending on the area of application.

Laser ablation: The laser lesion formation is thermal through photon absorption at the surface, with deeper myocardial sites heated through passive conduction. The wavelength chosen for good penetration is 980 nm (using a 980-nm diode laser). This wavelength ablates tissue with absorption of actual laser energy as deep as 4 mm into the tissue and further ablation by conductive heating mechanisms. The lesion times are for 36 s utilizing 5 W/cm, but ablation cannot be longer than 5 cm.

Microwave ablation: The electromagnetic microwaves occur at 2.45 GHz to generate frictional heating by induction of dielectric ionic movements. The microwave device can provide a range of 40–45 W of power for 20–30 s, generating a consistent 3–6 mm lesion depth sufficient to produce transmural ablation.

Ultrasound ablation: This technology uses an ultrasound transducer to deliver mechanical pressure waves at high frequency. The tissue destruction is thermal and lesion depth corresponds to vibrational frequency. The ultrasound wave emitted from the transducer travels through tissue causing compression, refraction, and particle movement, resulting in kinetic energy and heat [6].

6. Minimally Invasive Surgery

Minimally invasive surgery, a keyhole surgery, is an operation technique in which access to the inside of the patient's body is achieved via small incisions. The long rigid instruments inserted via trocars can be used to move, retract, and cut tissue in abdomen (laparoscopy), chest (thoracoscopy), blood vessels (angioscopy), gastrointestinal tract (colonoscopy), uterus (hysteroscopy), and joints (artheroscopy). Insufflation of the body cavity may be carded out and maintained by seals on the trocar port, which allow tool manipulation without excessive loss of insufflation gas [8]. The trocars may be sized to provide ports of differing sizes, typically of about 5–11 mm diameter. Currently, tools such as retractors, forceps, graspers, suture needles, scissors, different types of knives, laser incision instruments, and specimen bags are available. A video camera is generally operated through a trocar port for visualization and control of the procedures in the cavity. High-definition video cameras and monitors are then attached to the camera and a surgical team can obtain a clear picture of the affected internal area without resorting to radical, disfiguring surgical incisions to physically open the patient. Another benefit of laparoscopic surgery is the significantly reduced recovery time, when compared to standard surgical procedures, due to the minuscule size of the scalpel incisions and avoidance of the massive internal traumatization known in standard surgical procedures.

Due to fixed entry points of the instruments in the abdominal wall, the DOF is reduced, and the movements are mirrored and scaled. The coupling between observation and manipulation, the hand–eye coordination is disturbed, tactile information about tissue properties partially, due to friction and the poor ergonomic design of instruments, lost. The future is open for semiautomatic tools for MIS. As an example, Medtronic, Inc., has introduced the world's first minimally invasive epicardial lead placement tool. The Medtronic Model 10626 Epicardial Lead Implant Tool allows for perpendicular alignment to the heart from different angle approaches. The tool furthers the trend moving away from invasive sternotomy procedures. Laparoscopy reduces patient trauma, but eliminates the surgeon's ability to directly view and touch the surgical environment. This is reason that the surgeon introduced the small-incision surgery when they have a direct view of the tissue.

The rapid development and growing interest in MIS methods impelled creation of modern, complicated tools: cardiosurgical robots.

6.1. The Classical Thoracoscopic Tools

Laparoscopic or thoracoscopic surgery requires that surgeons perform complex procedures using a standardized set of tools. For laparoscopic procedures, special sewing devices, ligating instruments, knot pushers, clips, and clip appliers have been developed. The clamp is an essential tool for stopping or controlling blood flow to an organ during surgery. There is a need for a secure and easy method of suturing in laparoscopic surgery. Laparoscopic suturing can be performed with a suture, using automatic sewing devices or clips. When a surgeon wants to suture laparoscopically in the traditional way by using a Roeder knot for instance, the tying of the knot requires some expertise and makes the procedure more complicated and time-consuming. The vascular clips provide a very easy (and secure) method of stopping blood flow while performing procedures on vessels. The locking forceps allows easy, precise, and secure grasping of the surgical clips. The risky point of a clip is the open end of its U-shaped design, which makes it possible for the clip to slip off. Sutures avoid this risk by totally enclosing the structure to be occluded. Knotting instruments require a two-forcep technique, whereas sewing devices require only one working channel.

There has been a tremendous interest in the development of *anastomotic devices* for coronary surgery. Coronary artery disease is a major health problem worldwide, with approximately 850,000 bypass procedures being performed every year. The current methods of coronary bypass surgery can be time-consuming and technically difficult especially when minimally invasive instruments for suturing the blood vessels (the connection is called anastomosis) are used. Alexis Carrel, who many consider the father of anastamoses, received a Nobel Prize for the suturing of blood vessels about 100 years ago. More than 5 million vascular anastomoses are performed annually in the world and more than 70% often are on the CABG market [9].

The requirements of anastomotic devices include reproducibility, ease of use, and a short anastomotic time (less than1 min). The anastomotic device ideally would work with either vein or artery and so the proximal or distal order of anastomoses should be interchangeable. The device should facilitate a wide range of surgical access points and should have patency comparable to that of hand-sewn techniques [9].

There are manual and automatic anastomotic devices. The manual proximal anastomotic devices Heartstring *Aortic Occluder*™ from Guidant (that acts as an umbrella inside an aortotomy) and the Enclose™ device from Novare Medical (require a separate insertion through the aorta) create a dry area for conventional suturing in aortotomies.

The Symmetry™ Bypass System Aortic Connector is the first in a line of sutureless anastomoses devices for CABG developed by St Jude Medical, Inc., Cardiac Surgery Division – Anastomotic Technology Group (ATG), formerly Vascular Science, Inc. (VSI). There have been more than 50,000 implants performed.

CardioVations™ has released the *Corlink*™ proximal device in Europe. This is a Nitinol-based stent device. The major difference in comparison with Symmetry is that the Corlink device sits outside of the vein graft wall and the Symmetry device lies inside the vein graft wall.

The *Passport*™ automatic proximal device consists of an integrated system that performs the aortotomy, delivers the graft to the aorta, and deploys a stainless-steel connector end-to-side the graft using a delivery device. The stainless-steel stent (more rigid than Nitinol) creates a widely patented anastomosis.

Another proximal anastomotic device is the *Magnaport* from Ventrica. This is also a one-step anastomotic device using rare earth magnets.

Coalescent Surgical™ has developed an automatic proximal anastomotic device, called the *Spyder*® based on their proprietary Nitinol (nickel–titanium) clip technology.

The first distal device to market (launched in 2000) was the manual *Coalescent U-CLIP*™. This is a self-closing, penetrating device with bailout. The U-CLIP device is designed to provide an alternative to conventional suture and surgical clips in a variety of applications including anastomosis creation and tissue approximation. The device consists of a self-closing surgical clip attached to a conventional surgical needle by a flexible member. U-CLIP device placement is easy and requires only standard surgical instrumentation. The self-closing clip eliminates the requirement for knot-tying and suture management. A fresh, sharp needle is provided with each clip. The clip is easily removed. The U-CLIP anastomotic device technology consistently facilitates the interrupted technique in both on-pump (or surgery performed with a cardiopulmonary bypass machine) and off-pump (also known as beating-heart, CABG procedures). Independent core laboratory analysis demonstrated that the left internal mammary artery to the left anterior descending coronary artery (LIMA to LADA) anastomoses created with the U-CLIP anastomotic device were 100% perfectly patented at 6 months [9].

The Spyder device provides fast, automated proximal anastomotic connection by simultaneous delivery of six U-CLIP devices without the use of a side-biting clamp or second manipulation of the aorta. The *JoMed Solem* connector consists of a stented "T"-shaped PTFE graft connected to the internal mammary artery. The *St Jude* distal connector consists of a stent that is deployed with a balloon using a side-to-side technique to create a distal coronary anastomosis. The *Converge Medical* device consists of outer and inner Nitinol couplers that are placed on the vein graft. This device allows for an angled anastomosis and creates elliptical interface. Magnetic Vascular Positioner (MVP), developed by Ventrica (received CE Mark approval in February 2002), uses magnetic attraction to form an instantaneous self-aligning, self-sealing connection between two blood vessels. The elliptical magnets are placed inside the conduit and native vessel. The MVP device connects the "bypass graft" to the coronary artery in seconds, and is designed to be used in the distal, or smallest, portion of the coronary arteries.

Anastomotic devices in OPCAB may play their most important role in the proximal anastomosis. The major meta-analyses reveal the significant benefits of performing OPCAB (off-pump coronary artery bypass) over CCABG (conventional coronary artery bypass) for certain patient populations requiring myocardial revascularization procedures. The technology has potential benefits as a replacement for sutures due to shorter procedure times, uniform and repeatable connections, and the potential to allow for significantly smaller incisions. This would allow proximal anastomoses to be performed with minimal aortic manipulation, which may result in a lower rate of intra- and postoperative embolism. An automated distal device could facilitate lateral wall grafting in OPCAB. In total endoscopic coronary artery bypass (TECAB), a distal automatic device could be critical in evolving the endoscopic procedure. Currently, the Ventrica device is being modified for deployment in a robotic environment. Finally, in multivessel TECAB procedures, distal anastomotic devices are predicted to make a major contribution. Most minimally invasive procedures in the future will be facilitated by an automated anastomotic device. The automated anastomotic device will level the playing field and allow for more surgeons to perform more perfect anastomoses. Wolf concluded that with continued evolution of these devices, it is easy to anticipate that in the near future the majority of anastomoses may be performed with a manual or automatic device as opposed to the current conventional suture technique [9]. The leaders of clinical trials of these new methods are Randall K. Wolf and Volkmar Falk, MD.

Since 1997, Medtronic has been a pioneer in minimally invasive, beating-heart surgery with the introduction of the Octopus tissue stabilizer system and its family of products, and most recently with the release of the Starfish2 heart positioner.

In traditional open-heart surgery, the heart is stopped and an artificial pump is used to maintain blood circulation. This is highly traumatic as the blood flow is rerouted through a machine. Many patients with coronary heart disease undergo open-heart operations, especially bypass surgery. These are traditionally performed with a heart–lung machine. The machine serves as a substitute for the patient's own beating heart and lungs, allowing the organs to function while surgeons operate on a "still" heart.

Endoscopic surgery in a major cavity in the body requires space for manipulation and local cardiac wall immobilization for coronary artery grafting on the working heart. One of the revolutionary instrument used in beating-heart surgery is Octopus. The name comes from its suction device of two arms and a series of "suckers", which allow to position the heart in various places with the chest open.

6.2. The Surgical Robots

About 4 million MISs are performed in the world every year. The aim is to limit the operative field and spare surrounding tissue, which would be damaged if a traditional surgical technique was used. The number of endoscopic procedures, less invasive than traditional surgery, performed through natural orifices in the patient's body, or through special openings called ports, is on the rise.

The success of the procedures largely depends on the instruments used. Unfortunately, typical endoscopic (laparoscopic) instruments reduce precision and make the surgery more difficult because they add to hand tremor and almost completely eliminate the natural sense of touch. Additionally, the surgeon does not have a direct view of the operative field – a camera inserted into the body through a third opening transmitting the image to a display. So the surgeon's task is not easy. An ideal noninvasive surgery can be compared to renovating a house through a keyhole without disturbing the household members. Across the world, physicians and engineers are working together toward developing increasingly effective instruments to enable surgery using the latest technology. But how can one enhance instrument precision and maneuverability, which are so important in the case of surgery on the beating heart, for instance? Surgical robots provide such capabilities.

Surgical robots improving precision and facilitating minimally invasive access to the area of the operation make up a great potential instrument for the surgery. Currently, these are mainly telemanipulators, where on the "master" side is the surgeon giving the instructions (movements, tasks), on the "slave" side is the end-effector as a surgery tool tip executing tasks in the definite working space, and in the middle of these "actors" is a control system, which reads, processes the input data, enriches them, and computes the output commands.

Currently, two types of medical robotic systems are used in the surgery:

Robots replacing the assistant during the operation: By using systems such as AESOP (Computer Motion – production stopped at present) or EndoAssist (Armstrong Healthcare Ltd, High Wycombe, UK), the surgeon can unaidedly control the position of endoscopic camera, serving as their "eyes" in the closed area of the operation field.

Surgical robots: Robots invented for less minimally invasive cardiac surgery are computer-controlled devices, located between surgeon's hands and the tip of a surgical instrument (Figure 10). Currently used cardiac surgery robots fulfill the role of telemanipulators, the main task of which is to detect and scale up or down the surgeon's hand movements and precisely translate them to the movements of robot's arm equipped with tools. A cardio-robot was invented in the United States in the 1990s. Two companies, currently merged, Computer Motion® (Computer Motion) and Intuitive Surgical® (Intuitive Surgical, Mountain View, CA, USA), were set up based on Star Wars project technologies developed by NASA and Pentagon. In Europe, Poland, the Robin Heart family robots are

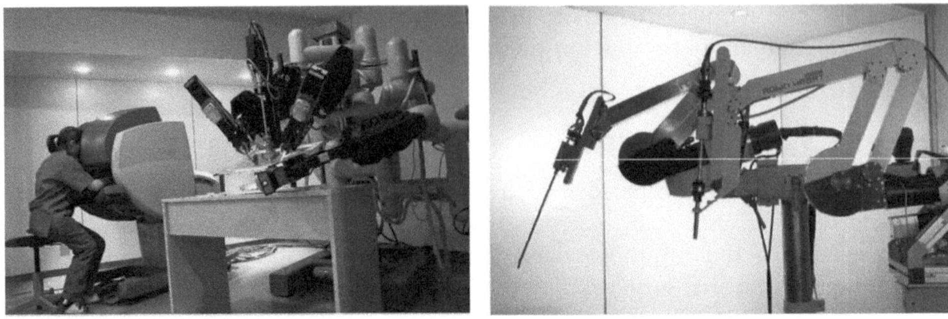

Figure 10(a) Robot daVinci (*left*) and Robin Heart cardiosurgery robot (*right*).

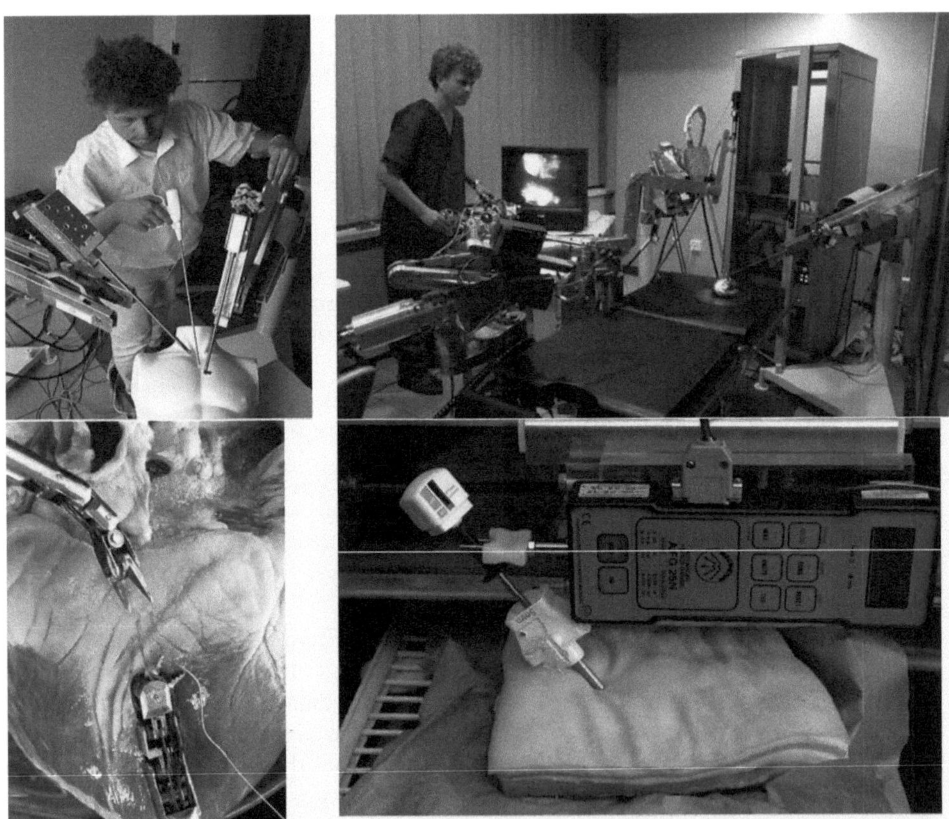

Figure 10(b) Comparison of different tools: Robin Heart 0,1 and laparoscopic tools; test of Robin Expert advisory system (presented by the author); and the test using animal tissue.

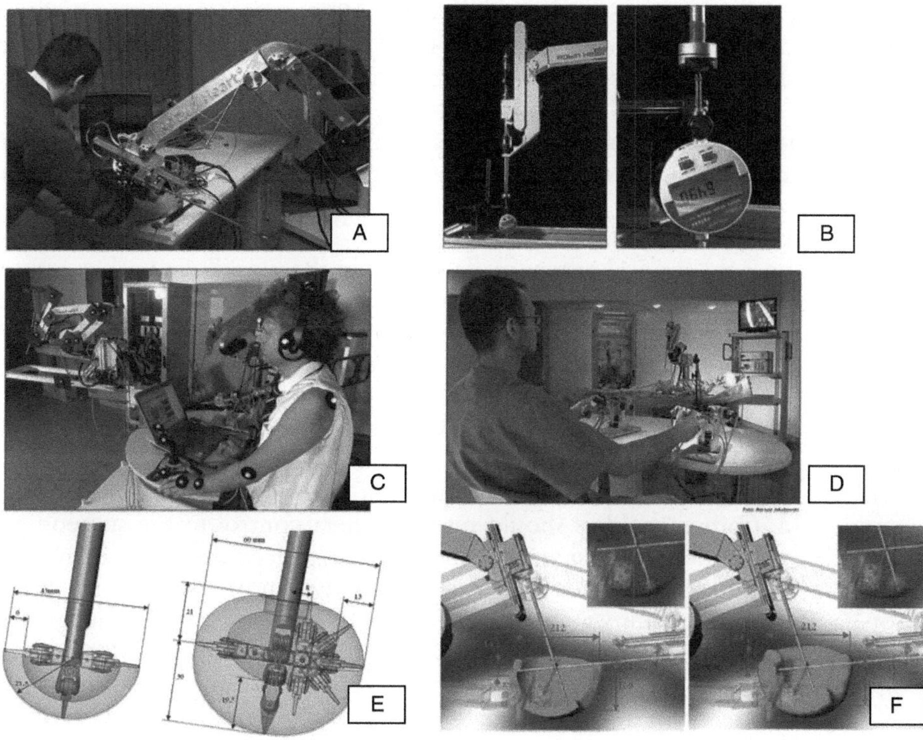

Figure 10(c) Evaluation of Robin Heart robot is carried out using the following measurements: vibration (new accelerometer sensors) [A]; linear movement (using digital micrometer) [B]; system of external trajectory recording (several digital cameras, markers, and image analysis methods) [C]; and physical [D] and computer surgery simulation (*bottom*: general range of robot mobility and workspace equipped with [E] standard laparoscopic tool and [F] Robin Heart I instrument). (See color plate 5(c)).

developed (Figure 10). The Foundation of Cardiac Surgery Development (FCSD) in Zabrze in 2000 started issuing the grant for realizing the prototype of a robot useful for cardiac surgery. The multidisciplinary team including specialists in medicine and techniques prepared until now four robot prototypes named Robin Heart [10].

An application of teleoperation allows to remove the tremor and to introduce the scaling of hand movement range via interface of manipulator on exact movements of tool inside the body, thus improving the ergonomics and precision. Supervision is held using visual observation via voice- or manually controlled endocamera (2D or 3D). American cardiosurgical robots have been produced by two, currently merged, companies, Computer Motion and Intuitive Surgical, which was firstly clinically used in Europe.

The first mechanical assistant of surgeon – voice-controlled endoscope positioner AESOP 1000 (automated endoscopic system for optimal positioning) – was introduced by firm CM in 1994. In January–May 1998, a French team in Paris and a German group from Leipzig performed using da Vinci (Figure 10a) (IS) telemanipulator the first endoscopic operation of single coronary bypass and mitral valvuloplasty. About 1000 surgical and endocamera robots were installed in clinics; to date, more than 130,000 minimally invasive procedures, across a wide range of surgical applications such as general surgery, gynecology, spinal, urology, and cardiothoracic surgery, have been performed in several counters (also in Poland, Katowice) [11].

In 1999, the Zeus surgical system made history in the world's first robot-assisted beating-heart bypass surgery, by Douglas Boyd, MD. In 2001, the first transatlantic telesurgical procedure was performed using the Zeus system. The doctors in New York removed a gallbladder of a 68-year-old patient in Strasbourg, France, and the procedure was successful with no complications.

In traditional laparoscopic surgery, the operating surgeon does not have direct visual control of the operative field due to manual camera control by an assistant. Ideally, the surgeon should have full control of operative instruments and the operative field. Camera holders should return camera control to the surgeon and stabilize the visual field during minimally invasive procedures. The active and passive camera holders offer the surgeon an alternative and better tool to control the operating surgeon's direct visual field. One of the first, active teleoperated robots introduced into clinical practice was produced by Computer Motion (US) Company. Currently, more than 1500 of Computer Motion's robotic systems (AESOP, Zeus) are in use worldwide in 900 hospitals by more than 3000 surgeons in 32 countries. The Zeus system consists of three robotic arms mounted onto an operating table. Two arms hold surgical instruments, and the third arm holds and positions the endoscope via voice control. All three arms are connected to a master console where the surgeon telemanipulates the arms. The handles used to control the movement of the surgical instruments are similar to the instruments used in conventional surgery. The surgeon's manual movements are filtered and scaled for the slave instruments to perform precise microsurgery. The Zeus system can be used in combination with an independent 3D visualization system (AESOP). More than 300,000 surgical procedures have been performed with Computer Motion's robotic systems assistance. The AESOP system is both CE marked and FDA approved and has been used in more than 100,000 laparoscopic and endoscopic procedures. The duration of several types of endoscopic surgery was reportedly faster using the AESOP; voice control was considered to be more efficient and faster than either the foot or hand control.

The disadvantages for the AESOP include the constant voice commands, which may be distracting. In addition, the voice control is slower compared to the rapid camera movements achieved by an experienced assistant. The AESOP is intended to facilitate solo-surgeon laparoscopic procedures; however, the surgeon may still need an assistant to control the fourth laparoscopic port. The Intuitive Surgical Inc. (US) da Vinci telemanipulator consists of three or four robotic arms in one set fixed on one common column, placed near the operating table.

In contrast to the AESOP system, da Vinci arm for camera holding is integrated with all robotic arms. Today approximately 400 da Vinci systems have been installed worldwide, and its applications have been described in thousands of scientific publications and presentations. It is CE marked and FDA approved and used in more than 300 hospitals in America and Europe. The da Vinci was used in at least 16,000 procedures in 2004 and sells for about $1.3 million.

In Europe, the Armstrong Healthcare Ltd (UK) produced telemanipulator EndoAssist. A robot system EndoAssist holds a conventional laparoscopic telescope and camera, coordinated by the surgeon's head movements. EndoAssist (CE marked and approved by FDA) has also been used in telesurgical applications with control via a joystick communicating with the robot over a telephone line. About 30 systems installed worldwide now has been used in several thousand clinical procedures.

The future of robotic surgery has significant potential that has been proven in many clinical applications. The procedure completed without the need for an additional assistant is called a "solo surgery". Robotic assistance has enabled a solo surgery approach. The basic principle of manipulator construction is that of a serial architecture of joints and links with a fixed remote center in the corporeal wall. For instance, the da Vinci robot have seven DOFs: three (yaw, pitch, and insertion) have surgical manipulators provide motion coupling to the end-effector, an exchangeable instrument that adds four DOFs (roll, pitch, yaw, and grip) by means of a cable-driven mechanical wrist. The surgeon controls the movements of tools via a human machine. The slave is capable of detecting force feedback (tissue contact, collision of manipulator) greater than 2.2 N. Scaling is useful in the range from 1:1 to 10:1, in combination with tremor filtering greater than 6 Hz [12].

The computer and laboratory tests are required to design a new, smart surgery robot and expanding its applications to the field of medicine. The 3D computer simulation and physical modeling will continue to be supported on Polish grants.

The use of conventional endoscopic instruments allows a limited range of motion by the trocars through which the instruments are introduced and by reduction to only four DOFs for tools. Long and rigid instruments are harder to control and amplify tremor.

Six DOFs are required to perform a free motion. For instance, suturing perpendicular to the four-DOF tool's shape becomes impossible. The study shows that using four-DOF robot (Zeus system, Computer Motion) the 1.5 mm tube anastomosis was performed in 46 min compared to only 12 min performed by six-DOF robots (da Vinci). Endoscopic tissue manipulation and suturing using four-DOF instrumentation requires skill and training.

Cardiac surgery is carried out on soft tissues. Results obtained from surgical action analysis allowed to determine the maximum values of forces needed for typical procedures performed in heart area, using different types of tools. This can be the basis for cardiac surgery robot design assumption, in the field connected with controlling of robot tool movements. The penetration of soft tissue involves actions such as cutting, slicing, inserting a needle, and knotting. The difficulty with soft tissue is that it deforms and changes shape. The map of force resistance during pricking for left and right heart chamber was obtained (i.e., the basic mechanical properties of typical surgical actions; from the FCSD). For example, the maximal force value for surgical needle Prolene 3/0 during pricking through papillary muscle reached the 150 G, the measured load value equaled 200 G (2 cm depth) during scalpel cutting procedure for the left ventricle (the mitral valve ring), and up to 200 G with 0.1 mm/s test speed for sewing tests – the knot-tying using Prolene needle.

One of the limitations of current surgical robots used in surgery is the lack of haptic feedback. Although current surgical robots improve surgeon dexterity, decrease tremor, and improve visualization, they lack the necessary fidelity to help a surgeon characterize tissue properties for improving diagnostic capabilities. Many research groups focus on the development of tools and software that will allow haptic feedback to be integrated in a robot-assisted surgical procedure.

The surgeon needs assistants of next robot arm during operation mainly for stabilization of the heart (beating-heart bypass grafting) or another tissue. Several trials using two da Vinci consoles delivered good results. In response, the IS prepared the da Vinci S model with four arms.

The devices providing an "inside equipment store" for the surgeon are also needed. A self-sufficient Cargo Module was developed as a transportation and depot device by Dresden group. With the Assist Module, the surgical equipment, tissue, and vessels can be positioned on a desired place in the operating field. This module provides the surgeon an "assistant" inside the closed chest.

Potential applications of robotics in cardiac surgery includes aortic valve replacement (standard or percutaneous technology); tricuspid valve repair; descending thoracic aortic surgery; transmyocardial laser revascularization; ventricular septal defect, patent ductus arteriosus, coarctation; and intramyocardial delivery platform for biological agents such as stem cells, molecular therapeutics, genetic vectors, and AORobAS (Artificial Organs Robotically Assisted Surgery).

In the Institute of Heart Prostheses, FCSD, surgeries related to artificial organ (AO) implantation are performed. Our future plans, AORobAS, include carrying out a robotically assisted less invasive procedures to implant pumps and valves.

The da Vinci system is therefore the only surgical telemanipulator that is currently clinically being used. Due to the actual range of the system acceptance, the number of endoscopic cardiac procedures performed in 2005 worldwide (reported in a company-based registry) was 2984 and it increased steadily. This includes TECAB procedures or small access single- or multivessel coronary artery bypass procedures with endoscopic uni- or bilateral ITA harvest (1784 procedures). The number of MVR increased to about 450 cases in 2004, to more than 600 in 2005, to 850 in the first 10 months of 2006 (reported in a company-based registry) [13].

The number of procedures performed may increase due to development of and improvement in facilitating anastomotic devices (MVP magnetic coupling device) and augmented reality with preoperative planning and intraoperative navigation. Based on the preoperation cardiac surgery simulation results, the optimization of cardiac surgery procedures can be achieved.

The *effectiveness* and *expansion* of robot's application fields requires searching the most effective cardiac robots in the wide application range, building the strategy of its usage, simulating the operation results, and creating the knowledge base supporting the robot's arm navigation and cardiac surgeon decision making, and studying image processing methods for optimal robot's arm navigation [13].

6.3. Blood Pumps – MIS Application Study

The use of the mechanical circulatory support systems (TAHs and VADs) has evolved significantly over the last 30 years, with more than 10,000 patients being supported by these devices (TAH 2%). The devices have evolved from extracorporeal devices (such as Jarvik 7, Thoratec, Abiomed systems), driven by large pneumatic consoles, to electrically driven partially implantable devices with portable controllers carried by the patient (such as Novacor, HeartMate).

The future prospects of the mechanical circulatory support clearly lie in three major areas:

1. active left ventricular pump system, which supports and stabilizes the heart in extremely critical situations (e.g., acute heart attack), and can be implanted for a period of up to 5 days (e.g., Impella® Acute)
2. devices as a bridge to recovery (e.g., pulsatile pumps: Novacor, HeartMate; rotary pumps: MicroMed, Jarvik-Heart, Nimbus-HeartMate II)
3. devices as an alternative to the cardiac transplantation and conventional therapy (e.g., destination therapy : LionHeart, AbioCor TAH)

6.3.1. Minimally invasive VAD implantation

Dr J. Donald Hill of the California Pacific Medical Center (San Francisco, CA, USA) has successfully performed the less invasive Thoratec LVAD surgery on several patients.

The patient is placed on the cardiopulmonary bypass using the right or left femoral artery or vein. The VAD cannulation system is placed in the left ventricular apex (VAD inflow) and the ascending aorta (VAD outflow), using two separate incisions. The procedure offers the following advantages: no sternotomy, easier reoperation, less blood use, reduced risk of sensitization, and it is psychologically more acceptable (according to Dr Hill, the Thoratec Laboratories).

General steps in VAD implantation are as follows: anesthesia is administered to the patient; preparing the operating area – cardiopulmonary bypass (if required) and lung deflation; choosing the exit site for LVAD – depending on the need of LVAD alone or of BiVAD, one or two outflow and inflow grafts will be necessary; aortic and pulmonary artery anastomoses performed; left and right ventricular apical or atrial cannulation performed; connecting the VAD to the cannulas, deaerating the VAD; gradually discontinue cardiopulmonary bypass and allow the VAD to run in "fill-to-empty" mode; completing surgery.

Our plans include carrying out robotically assisted less invasive procedures to implant VADs (Figure 11), TAHs, valves, and vessel prostheses. The rapid evolution of MIS techniques for heart prostheses will have implications for designing and construction of heart prostheses. Our team works on the designing of heart pumps and valves particularly for robot and MIS applications and special tools of robot are being constructed. I am sure that this is the future of this domain of biomedical engineering.

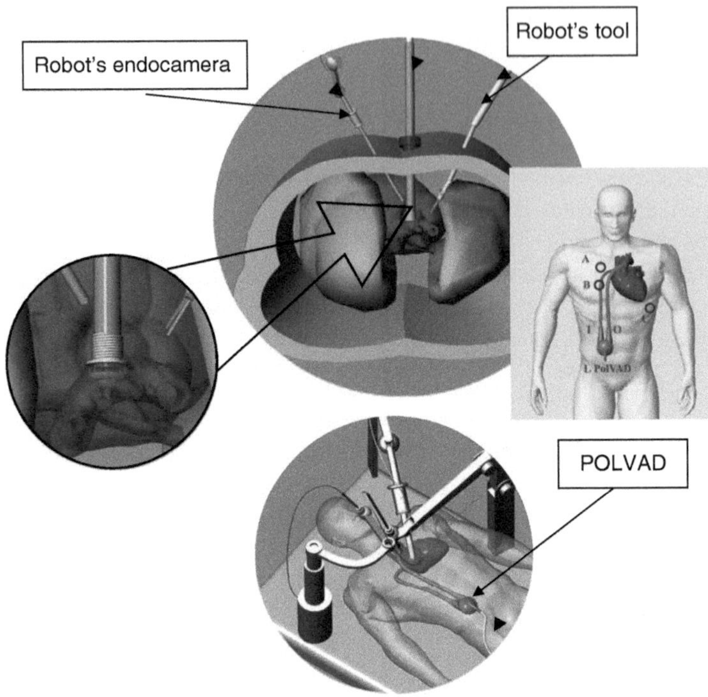

Figure 11 Semiautomatic tools for robotically assisted POLVAD implantation. (See color plate 6).

6.3.2. The AORobAS idea

Currently used VADs, pumps, valves, and pacemakers require to be replaced and repaired throughout the patient's life. A better solution seems to be application of MIS at once by designing AOs with construction ready to easy assembly and disassembly as well as to ensure an access to replace the used parts. For external pumps, an MIS technique of cannula implantation should be performed.

In 2000, a new type of blood pump was introduced into clinical practice. In robot-assisted surgery, the use of the small, axial heart assist pump is very promising. The Jarvik 2000 VAD and the DeBakey VAD, which have been approved for evaluation as a bridge to transplantation, are valveless devices that are electrically powered miniature axial flow pumps. These fit directly into the left ventricle, which may eliminate problems with clotting. The outflow graft connects to the descending aorta. The DeBakey VAD is 30 mm × 76 mm, weighs 93 g, and is approximately 1/10th the size of pulsatile products on the market. The development of this idea causes changes in requirements for blood pump. It should be folding device, with dimension not exceeding the port hole incision diameter. It means that this part must be a cylinder with about 1 cm diameter and the technology for quick mounting of this part on the whole pump inside the chest must be developed. Robots also require special constructions of semiautomatic tools. The surgical procedure requires faster and efficient connection to pulmonary bypass apparatus, because the surgery is performed on a stopped heart. Of course, there are many problems to solve, but this type of surgery has wide perspectives. In our laboratory, the design and simulation study has been conducted.

I work on the realization of the following procedure:

- *Stages of AO implantation and assembly*: Parts of the AO, put into capsule through mechanically or pneumatically controlled channel, run into pumped balloon, where they are assembled. The balloon has valves that allows for insertion of additional tools into the workplace. The balloon is supported mechanically or by compressed CO_2. In the next stage, the balloon is removed and the AO is implanted. Pumps that fulfill the role of a vessel bypass seem to be most suited to this procedure because their implantation does not require stopping (and hence external circulation) of the heart.
- *Stages of repairing of replacing of the AO*: Laparoscopic tool inserts a tight balloon around the AO, which is removed after repairing. The parts for replace are inserted in special capsules.

Great progress in MIS technique will influence the design and construction of implantable AOs. The goal of this work expresses the evolution of telemanipulator and AOs dedicated to the MIS. The system is named AORobAS (Figure 12). The future plans regarding development of Polish robot Robin Heart include carrying out of a robotically assisted MIS to implant AOs,VADs, TAHs, valves, and vessel prostheses. Currently used blood pumps, valves, and pacemakers require to be replaced and repaired throughout the patient's life. This is only a temporary solution. Probably, in the near future, the progress in the subject of biocompatible, long-term

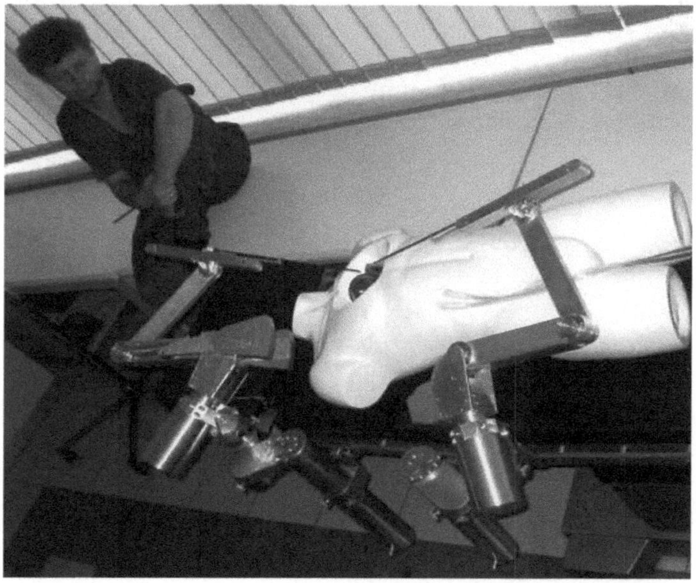

Figure 12 Artificial Organs RoboticallyAssisted Surgery (AORobAS) system.

durable materials and mechanical construction will not solve this problems. The best for patient will be realization of conception minimally invasive service of AO. Robots may be ideal for this task. The solution seems to be application of robotically assisted MIS at once by designing AOs with construction ready to easy assembly and disassembly and to ensure an access to replace used parts. The development of this idea causes changes in requirements for blood pump. It should be folding device, with dimension not exceeding the port hole incision diameter. It means that this part must be a cylinder about 1 cm diameter and the technology for quick mounting of this part on the whole pump inside the chest must be developed. Robots also require special constructions of semiautomatic tools.

6.3.3. The blood pump support in the palliative (pediatric) surgery study

The aim of the palliative procedure is to increase pulmonary blood flow in children with a congenital heart defect. As a result of a shunt operation [Blalock–Taussig (B–T) shunt or Glenn procedure], the oxygen level in the child's blood is improved. The main problem of such surgery is the small shunt effectiveness and lack of possibility of flow regulation. Our idea is to apply advanced computer simulation methods to give the information to the cardiac surgeon – which procedure will be optimal before realization of the operation. A small axial blood pump can be introduced into B–T shunt or Glenn in order to control the blood flow and prevent the growth of graft stenosis. Based on the results obtained using physical and computer 3D simulation based on finite element method (FEM) (Figure 13), it can be concluded that the module graft with axial pump makes possible the exact regulation of blood flow and blood pressures in the pulmonary artery. Its use allowed to decrease the afterload of left heart ventricle. Palliative procedures are only a temporary solution. As a child grows, the graft does not grow, and the size of graft (flow through graft) may be not sufficient. The use of axial pump across regulation blood flow allows to extend the time between necessary operations.

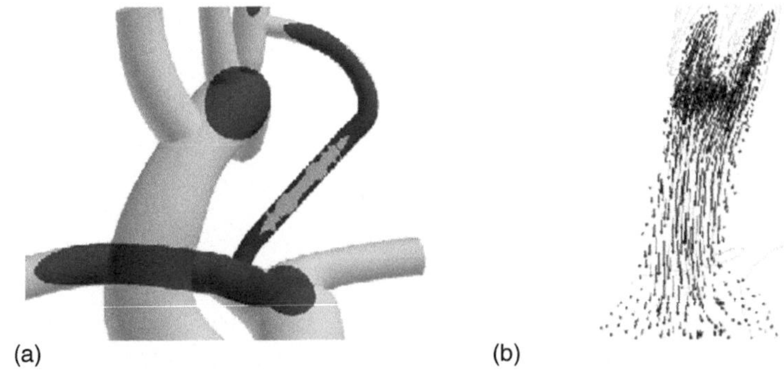

Figure 13 Pulmonary artery axial blood pump used to control blood flow in Blalock–Taussig (B–T): (a) the axial pump in B–T graft and (b) velocity vectors on inlet from pump in B–T graft. The simulations have been performed by Z. Małota.

7. The Minimally Invasive Valve Implantation

Percutaneous (pulmonary and aortic) valve replacement has recently opened new perspectives on transcatheter replacement of cardiac valves. Typically, a bioprosthetic valve is dissected and sutured into an autoexpandable Nitinol stent (which has an initial aortic diameter, for instance, 25 mm) or mounted within a balloon-expandable stent. It has been used in patients who are at high risk for valve replacement surgery. Percutaneous valve repair is also being developed for mitral regurgitation. Direct leaflet repair and percutaneous annuloplasty are being employed in clinical trials. All the percutaneous approaches are based on existing surgical techniques and offer less invasive alternatives. This marked the beginning of the era of percutaneous valve therapy, and ongoing trials will define the clinical role for these new therapeutic modalities. Percutaneous catheter-based systems for the treatment of valvular heart disease have been designed and studied in animal models for several years. Bonhoeffer et al. were the first to perform percutaneous implantations of artificial valves, using a bovine jugular vein valve mounted within a stent, in children with right ventricle to pulmonary prosthetic conduits. The first human case report of percutaneous transcatheter implantation of an aortic valve prosthesis for calcific aortic stenosis appeared in 2002 [14].

A percutaneously implanted heart valve (PHV) composed of three bovine pericardial leaflets mounted within a balloon-expandable stent was developed (Percutaneous Valve Technologies, Inc.). Using an antegrade transseptal approach, the PHV was successfully implanted within the diseased native aortic valve, with accurate and stable PHV positioning, without impairing the coronary artery blood flow or the mitral valve function, and a with no mild paravalvular aortic regurgitation. Immediately and at 48 h after implantation, valve functioning was excellent, resulting in a marked hemodynamic improvement. Nonsurgical implantation of a prosthetic heart valve can be successfully achieved with immediate and midterm hemodynamic and clinical improvement.

8. Support Technology for Surgery Planning

Modern medical imaging techniques such as CT, NMR, and ultrasonic imaging enable the surgeon to have a very precise representation of internal anatomy as preoperative scans.

For surgery robots (telemanipulators), the following distinct phases can be realized:

- *Preoperative planning*: The optimal strategy is defined using the 3D computer model.
- *Robot-assisted intervention*: A calibration routine brings robot, patient, and image system to a common frame of reference, for example, using anatomical (or artificial) landmarks.

– *Feedback and replanning*: The robot starts to operate under the supervision of surgeon. Sensor information ensures that the anatomy is as expected and stored by a model in computer. If deviations occur, the surgeon asks for a revised strategy, or for permission to continue.

Computer flow simulation method is important for diagnosing the heart disease development, for optimizing the surgery for a particular patient, and for long-term functioning of graft prognosis. During simulation, the input data (biochemical data, diagnostic data, geometrical data, biophysical data) are transformed into test output data (the hemodynamic pressure and flows or another characterization of biological object modified by surgery).

Remote-control manipulators are proposed for cardiac surgery using a computer-based advisory system. Information in a prepared database (an online expert system) may be of help to a surgeon in decision making. The first step is to simulate the robot-assisted surgery using both computer and physical models of a particular operation type. Based on the preoperation cardiac surgery simulation results, the optimization of the procedures can be achieved.

The main issues of computer simulation support to surgery robots are as follows:

1. *The operation planning*: Based on diagnostic data (images, pressure and flow signals, etc.), computer and physical models can be created. In vitro simulations performed on them may be used to find the optimal way of operation (the joint point localization, the graft selection). A prepared report can be presented to the surgeon as a hint for robot choreography planning. This stage should also include input port localization on patient skin, the type of tools, and the way of removing and preparing the graft branch.
2. *Advisory and control system*: During the operation, diagnostic images or simulation results from various sources can be called by surgeon and superimposed on the real operating image.

The introduction of robots to cardiac surgery created the possibility of direct and practical use of simulation results of surgical procedures in the robot information system.

Surgical planning and augmented reality are likely to enhance robotic surgery in the future. The interesting case of applying preoperatively in robotically assisted cardiac surgery planning, intraoperative registration, and augmented reality was performed by Falk et al. [15]. The regions of interest (i.e., the heart, ribs, coronaries, ITA) were segmented semiautomatically to create a virtual model of the animal. In this model, the target regions of the total endoscopic bypass procedure along with the ITA and anastomotic area were defined. Algorithms for assessing visibility, dexterity, and *collision avoidance* were developed after defining nonadmissible areas using the virtual model of the manipulator. Intraoperatively, registration of the animal and the telemanipulator was performed using encoder data of the telemanipulator by pointing to the fiducial points. After pericardiotomy, the reconstructed coronary tree was projected onto the videoscopic image using a semiautomatic alignment procedure. In dogs, the total endoscopic bypass procedure was successful on the beating heart.

For medical applications, matching procedures between diagnostic images and off-line intervention planning and real execution are very important. Many problems still remain for soft tissue surgery where deformations may occur. The navigation and guidance of the instruments highly depends on the surgeon's skill who has to combine their intraoperative views with the information extracted from the preoperative images.

Several scientific groups work on *computer method for preparing quasi-stationary view*. The da Vinci robot uses a two-camera endoscope that feeds images to the surgeon's viewer. Images from the left and right cameras are fed to each of the surgeon's eyes separately, providing a 3D view of the tissue being operated on. The Mylonas system exploits this principle by fitting an infrared eye tracker to the viewer. This ensures that each eye detects precisely where the surgeon is looking, and then calculates the distance to the point their gaze is fixed upon using triangulation. The software first constructs a 3D model of the heart by tracking the surgeon's eyes as they move over the organ. Then it creates a real-time moving image by recording the changes in the surgeon's focal point as the heart beats. The endoscope is calibrated to move forward and backward in time with this image, thus making the heart to appear stationary to the surgeon viewing it through the two cameras. The surgical instruments are also calibrated to move in synchrony with the beating heart, removing the need to constantly move them back and forward, and allowing the surgeon to concentrate on performing the operation. The software has so far been tested only on an artificial silicone heart using a robotic arm.

According to surgeons and device executives, surgeon training is the key to the future success of the robotics industry. Because modern surgery is very challenging, the surgeon training will become comparable to fighter pilot training while the robotic systems are having a huge impact on surgical education, and, without doubt, will be integrated into surgeons' future education process. Currently for clinical use of the da Vinci surgical system, the FDA requires a 2-day training course to understand the setup, maintenance, and applications of the surgical system, in addition to animated laboratory training. The device manufacturers have training programs in place to advance the skill sets of both new and experienced surgeons.

Both the *virtual* (Figure 14) and the *real* station are used for MIS education. As examples, commercial products include LapTrainer z SimuVision (Simulab Inc., Seattle, WA, USA; www.simulab.com), ProMIS (Haptica Inc., Boston, MA, USA; www.haptica.com), Laparoscopic Surgical Workstation and Virtual Laparoscopic Interface (Immersion Inc., Gaithersburg, MD, USA; www.immersion.com), Phantom devices (SensAble Technologies Inc., Woburn, MA, USA; www.sensable.com), Xitact LS500 Laparoscopy Simulator (Xitact S.A., Lausanne, Switzerland; www.entice.com), Lapmentor (Simbionix Inc., Cleveland, OH, USA; www.simbionix.com); Surgical Education Platform (SEP) (SimSurgery, Oslo, Norway; www.simsurgery.no;www.meti.com), LapSim (Surgical Science Ltd, Gothenburg, Sweden; www.surgical-science.com), Procedus MIST (Mentice AB, Gothenburg, Sweden; www.mentice.com), EndoTower (Verefi Technologies Inc., Elizabethtown, PA, USA; www.verefi.com), Reachi Laparoscopic Trainer (Reachin

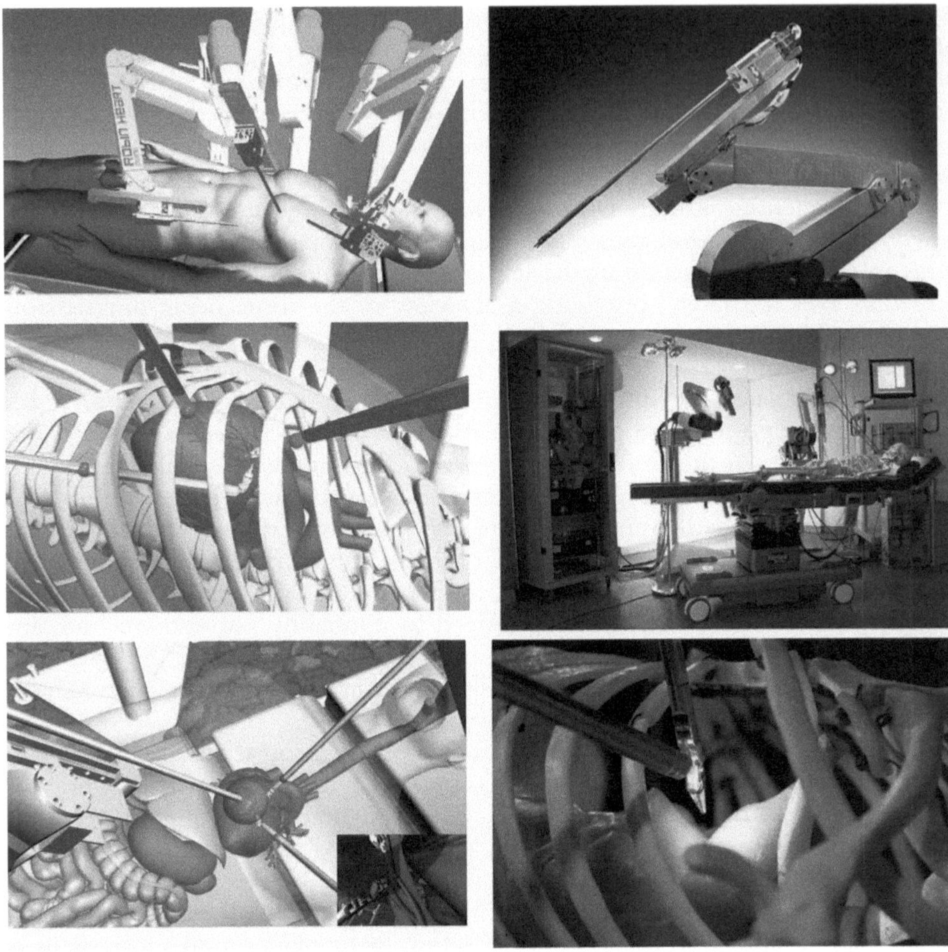

Figure 14 The virtual and the real condition for testing the Robin Heart robot. Using a virtual reality (VR) technology, an interactive model of surgery room equipped with a Robin Heart system was created using EON Professional software. This computer modeling method allows for an advanced procedure training and will be used as a low-cost training station for surgeons in the future. The model allows for a better understanding process of less invasive surgery treatment and a robot behavior. This type of modeling and a computer-aided design (CAD) technique use an accurate CAD robot model in a VR software together with a precise reflection of workspace geometry. This approach gives a surgeon easy and intuitive way to understand the technical details and use it to optimize and plan medical process. The next step in FCSD research work will be establishing the surgery workshops based on the newest technology, and some new projects using affordable semiautomatic robotic tools. Presented model of operating room in VR environment has been successfully used during Surgery Workshop in FCSD (May 2006). This system is intuitive for a user and gives them a very realistic 3D visualization. (See color plate 7).

Technologies AB, Stockholm, Sweden; www.reachin.se), Vest System (Virtual Endoscopic Surgical Trainer; Select-IT VEST Systems AG, Bremen, Germany; www.select-it.de), and Simendo (Simulator for endoscopy; DeltaTech, Delft, Netherlands; www.simendo.nl).

9. Conclusions

Notable achievements in cardiovascular research and devices that are currently under investigation and expected in the near future are listed below:

1. New pediatric devices, such as the axial-flow pumps and small-diameter valve prostheses, have been introduced, and some are under construction. As the devices become more and more reliable, mechanical circulatory support will play an increasingly important role, not only for rescue therapy but also for safe treatment of the most complex congenital heart diseases, not only with the aim of bridging to cardiac recovery or transplantation but, eventually, as a permanent solution.
2. The effective miniature blood pumps have been commonly used in minimally invasive cardiology. Transported to the destination place through the arteries, the smart pump mainly for short-term heart support can play an important role in an emergency.
3. New types of devices create possibilities of pump and blood oxygenation introduction into clinical practice. The idea is not new, but thanks to new materials (durable, semiconductive silicon membranes) that are being rapidly developed.
4. New "biomechanical" valves and vessels completely synthetic/artificial, but flexible and durable, will be introduced. The development of small-caliber vascular grafts is very important for bypass and graft surgery.
5. A significant improvement in technical support for preplanning and control surgical interventions, including telemedicine technology, will be observed.
6. Bioartificial myocardial grafts will be introduced in which perfusion by a macroscopic core vessel will be applicable.
7. Improved cell-culture techniques may render human aortic myofibroblasts a native tissue-like structure.
8. Tissue-engineered bioprosthethic valves will be commonly used in clinics.

As the field of surgery robots controlled by surgeons expands, the economical cost is expected to reduce, which will allow surgeons to more commonly use teleoperation in situations that warrants professional staff assistance such as wars, epidemics, and space trips. The situation of "no contact" between the medical personnel and the patient will reduce the risk of loss of health because of the infection(s) from the side of the operating staff.

It is most probable that contemporary telemanipulators will be replaced by adaptive robots in the near future. Currently, the trials of shifting from passive to active systems can be seen. To be clear, let's define the basic concepts:

Teleoperators are remotely controlled by operator robot transferring on distance motoric and sensoric functions, whereas adaptive robots have more advanced control system with sensoric and learning abilities. However, creating the next-generation "intelligent" robots is a true challenge. These robots must be able to work unaidedly in various environments, gathering information from their senses. Optimization of their behavior and the effectiveness of given task realization will depend on "self-learning control algorithm", which will more resemble systems based on "instincts" in relation to living creatures and contact with surroundings by means of "senses", zoom, touch sensors (inductive, supersonic, optical, pneumatic, and microwave). I am convinced that they will not be similar to contemporary cardiosurgical robots, but they will be able to replace them considerably. Probably, they will be microrobots that are able to reach, e.g., a given human internal diseased organ (e.g., the heart).

ACKNOWLEDGMENTS

This chapter has been illustrated with the experimental data obtained in the Institute of Heart Prostheses, FCSD, Zabrze. The research was supported by the Polish State Committee for Scientific Research and Foundation. The author thanks his numerous collaborators, including R. Kustosz, M. Koźlak, P. Kostka, Z. Małota, L. Podsędkowski, Z. Religa, and M. Jakubowski (special thanks for excellent photography), who prepared the artwork of this chapter.

REFERENCES

[1] Morris D., Couves C. Experiences with a sac-type artificial heart. *Can Med Assoc J* 1971;105:483–487.

[2] Copeland J.G., Smith R.G., Arabia F.A., et al. For the CardioWest Total Artifical Heart Investigators. Cardiac replacement with a total artificial heart as a bridge to transplantation. *N Engl J Med* 2004;351:859–867.

[3] Hetzer R., Stiller B. Technology insight: Use of ventricular assist devices in children. *Nat Clin Pract Cardiovasc Med* 2006;3:377–386.

[4] Short-term circulatory support with left ventricular assist devices as a bridge to cardiac transplantation or recovery. National Institute for Health and Clinical Excellence, June 2006; ISBN 1-84629-233-6. Available from: www.nice.org.uk/ip059overview.

[5] Rose E., Gelijns A., Moskowitz A., et al.; from the Randomized Evaluation of Mechanical Assistance for the Treatment of Congestive Heart Failure (REMATCH) Study Group. Long-term use of a left ventricular assist device for end-stage heart failure. *N Engl J Med* 2001;345(20):1435–1443.

[6] Jamieson E. Cartier P., Allard M., et al.; Primary Panel Members. Surgical management of valvular heart disease 2004. *Can J Cardiol* 2004;20(Suppl. E):7E–120E.

[7] Daebritz S., Sachweh J., Hermanns B., et al. Introduction of a flexible polymeric heart valve prosthesis with special design for mitral position. *Circulation* 2003;108:II–134.

[8] Dankleman J., Grimbergen C.A., Stassen H.G. *Engineering for Patient Safety. Issues in Minimally Invasive Procedures*. 2005; Lawrence Erlbaum Associates: London.

[9] Wolf R. Anastomotic devices for coronary surgery. *CTSNet*, 14 January 2005. Available from: www.ctsnet.org/sections/innovation/minimallyinvasive/articles/article-9.html.

[10] Nawrat Z. Perspectives of computer and robot assisted surgery for heart assist pump implantation. Planning for robotically assisted surgery. *Lecture Notes of the ICB Seminars. Assessment and Mechanical Support of Heart and Lungs* (eds M. Darowski, G. Ferrari), Warsaw 2001:130–150.

[11] Nawrat Z., Kostka P. The Robin Heart Vision, telemanipulator for camera holding preliminary test results. *J Autom Mobile Rob Intell Syst* 2007;1(1):str.48–53.

[12] Falk V., McLoughin J., Guthart G., Salisbury JK., Walther T., Mohr FW. Dexterity enhancement in endoscopic surgery by a computer-controlled mechanical wrist. *Minim Invasive Therapy Allied Tech* 1999;8:235–242.

[13] Jacobs S., Falk V., Holzhey D., Mohr F.W. Perspectives in endoscopic cardiac surgery. *Comp Biol Med* 2007;37(10):1374–1376 (Epub 20 December 2006; doi: 10.1016/j.compbiomed.2006.11.007).

[14] Bauer F., Derumeaux G., Anselme F., et al. Percutaneous transcatheter implantation of an aortic valve prosthesis for calcific aortic stenosis: First human case description. *Circulation* 2002;106:3006–3008.

[15] Falk V.D., Mourgues F., Adhami L., et al. Cardio navigation: Planning, simulation, and augmented reality in robotic assisted endoscopic bypass grafting. *Ann Thorac Surg* 2005;79:2040–2047.

CHAPTER 2

Biomechanical Modeling of Stents: Survey 1997–2007

Matthieu De Beule

Contents

1. Introduction 62
2. Finite Element Modeling of Stents 63
   2.1. Finite element basics 63
   2.2. Geometrical design and approximation 64
   2.3. Material properties 65
   2.4. Loading and boundary conditions 66
   2.5. Finite element stent design 66
   2.6. Effective use of FEA 68
3. Survey of the State of the Art in Stent Modeling: 1997–2007 68
   3.1. Neglect of the balloon 69
   3.2. Cylindrical balloon 74
   3.3. Folded balloon 78
   3.4. Summary 81
4. Alternative methods for biomechanical modeling of stents 84
   4.1. FEM – Prolapse, flexibility, and strut micromechanics 84
   4.2. FEM – Self-expandable stents 85
   4.3. CFD–drug elution and immersed FEM 87
5. Future Prospects 88
6. Conclusion 88
Acknowledgments 89
References 89

Abstract

The aim of this chapter is to provide an educational review on finite element modeling of stents. First, the basic principles of stent modeling are briefly described, followed by a survey regarding state of the art solid mechanical computational models of balloon-expandable stents. This review is organized chronologically according to the applied stent expansion methodology: (1) neglecting the presence of the balloon, (2) assuming it to have a cylindrical shape, or (3) taking the actual folded shape of the balloon into account. Subsequently, alternative methodologies to study the (bio)mechanical behavior of stents are described, and to conclude, future prospects for stent modeling are briefly addressed.

Keywords: Finite Element Method, stent, expansion methodology, validation

Advances in Biomedical Engineering
ISBN 978-0-444-53075-2

1. Introduction

Annually, over one million percutaneous (i.e., via needle-puncture of the skin) coronary interventions are performed world-wide to overcome the deficiency of oxygen-rich blood due to an obstruction of the coronary arteries (i.e., stenosis) [1]. Most of them involve the implantation of a stent to restore the perfusion of downstream heart tissue. The majority of stents are balloon expandable and crimped on a folded (angioplasty) balloon to obtain a low crossing profile, which allows for good deliverability and access through the coronary stenosis. After minimally invasive insertion into the stenosed artery, the stent is deployed by gradual inflation of the balloon. Upon reaching the targeted deployment diameter, the stent delivery procedure is completed by deflating and extracting the balloon catheter. The plastically deformed stent is left in situ to keep the artery open and to restore (normal) blood flow. In recent years, a variety of advances in stent technology, in particular improved deliverability and flexibility, have expanded the clinical application of stenting to complex coronary lesions [2]. However, in 20–50% of stenoses treated with Bare-Metal Stents (BMS), in-stent restenosis (i.e., re-narrowing) occurs and this phenomenon is partially linked to the stent design [3,4] and balloon–artery interaction [5], partly to an inflammatory response of the vessel wall against the stent struts and in part to local hemodynamic factors such as wall shear stress [6]. In order to reduce in-stent restenosis rates, Drug-Eluting Stents (DES) – coated with specific anti-proliferative agents – have emerged as a clinically better alternative for BMS. Despite the early promising results, reducing the restenosis rates to less than 10% [7–9], it is important to highlight that the mechanical vascular injury by stent insertion is not eliminated with these recently developed pharmacological coatings. Furthermore, both the cost-effectiveness of DES compared with new-generation BMS [10] and 1 year and beyond in-stent thrombosis (i.e., blood clot formation), risks after DE stent implantations [11] are nowadays important concerns. Finally, as the healing process of the vessel is considered to be temporary, the permanent metallic stents might rather be replaced by biodegradable stents. Therefore, further optimization of stent design, material, and (drug) coatings is warranted to improve coronary revascularization procedures.

Complementary to experimental studies, computational models provide an excellent research tool to optimize the mechanical properties of a stent, especially when physical test methods are difficult (or even impossible) to implement. Numerical simulations may sometimes be the only alternative. The main advantage of numerical modeling is that numerous 'What if?' scenarios addressing different materials, geometries, and loading conditions can easily be tested and evaluated before devices are actually manufactured [12]. Therefore, computational models are a valuable part of the stent design and development process. As numerical models are always approximations of the physical reality, it is of utmost importance to validate the numerical results by (targeted) experiments.

In essence, biomechanical modeling of stents can be categorized in three distinct domains. A first domain consists of numerical studies regarding the solid mechanical aspects of stenting (e.g., evaluation of the stresses in the vascular wall caused by stent deployment) based on the Finite Element Method (FEM). A second field studies the impact of the stent design (e.g., strut shape, interstrut distance) on the blood flow patterns using Computational Fluid Dynamics (CFD). Finally, the kinetics of the drug release – in the case of DES – can also be examined numerically. Furthermore, the above described techniques can also be coupled [e.g., Fluid–Structure Interaction (FSI)], where the interaction between the fluid flow (CFD) and the structural response to this flow (FEM) is investigated. This review focuses on FEM-based studies dealing with the mechanical behavior of stents. In particular, the possible methods to simulate the stent expansion process and the validation of the numerical results are discussed in this work. For a more general bibliographical review of the finite element modeling and simulations in cardiovascular mechanics and cardiology, the paper of Mackerle [13] is recommended. The chapter organization is as follows: in Section 2, the basics of finite element stent modeling are briefly described. Next, Section 3 offers an overview of the state of the art regarding solid mechanical computational models of balloon expandable stents presented in literature to date (i.e. summer 2007). Subsequently, alternative methodologies for biomechanical modeling of stents are described in Section 4, and to conclude, future prospects for stent modeling are briefly discussed in Section 5.

2. FINITE ELEMENT MODELING OF STENTS

The basics of finite element stent analysis are well described by Perry et al. [12] and partly summarized hereafter. In general, a finite element model is defined by its geometry, material properties, and some appropriate loading and boundary conditions. To simulate the (free) expansion of balloon-expandable stents, these prerequisites are described below. The accuracy of these prerequisites can, in many aspects, ensure the success of a simulation, or guarantee its failure. Finite element stent models are presently an important component of the design process. The actual practice has even evolved to the point that a detailed stress analysis is required by the regulatory agencies – such as the U.S. Food and Drug Administration – prior to approval of a new stent design [14].

2.1. Finite element basics

Briefly stated, a finite element analysis (FEA) is the investigation, by numerical means, of the mechanics of physical systems. The continuum (e.g., a stent) is divided into a finite number of discrete regions, named elements (Figure 1), whose behavior can be described mathematically. Partitioning the continuum into elements is generally automated through a special-purpose graphical user interface (GUI). An approximate solution of the entire continuum is solved from the assembly of the individual elements. The mechanical behavior (displacement, strain, stress, ...) in any point of an element is described in function of the behavior at a small number of control points (nodes) in the element. Usually, the displacements of the nodes are taken as the fundamental unknown quantities. At any other

Figure 1 Approximate CAD model of *Cypher* stent and finite element discretization.

point in the element, the displacements are obtained by interpolating from the nodal displacements. The interpolation order is dependent upon the number of nodes in the element. From the displacements, the strains are evaluated by taking the appropriate derivatives. The material constitutive behavior provides the necessary basis for computing stress levels from these strains. Application of the principle of virtual work to an element yields the forces exerted by the nodes on the element, which are statically equivalent with the built-up stresses, and by Newton's third law, the actions of the element on a particular node are easily found. Force contributions from all elements connected to a particular node are summed up and must be in equilibrium with any externally applied loading or force applied to the continuum.

Thus, the Finite Element Method essentially transforms the unknowns from the various continuous fields into equations of discrete nodal quantities. Assuming that certain basic numerical requirements and standards of practice are satisfied, the solution obtained from the FEA estimates the exact physical solution.

2.2. Geometrical design and approximation

Commercially available stent geometries are often (if not always) subjected to very strict patent claims. For this reason, manufacturer's specific and detailed information regarding the stent geometry is usually not available in the public domain. As typical dimensions of (coronary) stent struts are in the order of magnitude of 100 μm, an accurate geometrical representation of the stent can (only) be acquired using a microscope or micro-Computer Tomography [15]. The advantage of the micro-CT strategy is the possibility to build a precise three-dimensional (3D) reconstruction of both the stent (Figure 2) and balloon (Figure 3) directly from the CT-scans. The main stent dimensions can be measured and used to build a 3D CAD model of the (crimped) shape of the stent (Figure 1).

Figure 2 CT-Image reconstruction of *Cypher* stent.

Figure 3 CT-Image reconstruction of tri-folded angioplasty balloon.

2.3. Material properties

Angioplasty balloons are generally manufactured from either PolyEthylene Terephthalate (PET) or Polyamide (nylon) [16]. PET balloons are usually stronger and have a lower compliance than nylon balloons. Based on *in vitro* testing, the change in balloon diameter as a function of inflation pressure is determined by the manufacturer and summarized in a compliance chart in the directions for use. If neither the balloon constitutive material behavior is known, nor experimental tensile tests can be performed; the constitutive law can be derived from the compliance data based on thin shell membrane theory [17].

For balloon-expandable stents that undergo plastic deformation during stent deployment, metallic materials [e.g. (coated) medical grade stainless steel and cobalt chromium] are required. By far, the most widely used material for such balloon-expanding stents is the low carbon 316L stainless steel. As the constitutive behavior of a material is dependent on its pre-treatment (e.g., hot rolling, cold finishing, annealing, the material properties should (ideally) be obtained from tensile tests on specimens extracted from batches that are eventually used to create the stent structure. Such experimentally obtained stress and strain values (e.g., reported by Murphy et al. [18]) can be implemented in the finite element model.

2.4. Loading and boundary conditions

A finite element model should be subjected to loading and boundary conditions that represent the actual loading history of the stent. Depending on the aim of the study, the model should incorporate conditions that simulate the real-life loading on the stent during manufacturing, catheter loading (crimping and recoil after crimping), and deployment (tortuous path, expansion, interaction with the blood vessel tissue, and recoil after expansion). In general, the stent expansion can be modeled as a pressure-driven process (e.g., applying an increasing pressure on the stent or balloon) or a displacement-driven process (e.g., enforcing radial displacements on the stent or balloon).

2.5. Finite element stent design

The "ideal" stent is inexpensive to manufacture, easy to deliver and deploy, sufficiently rigid to provide support, and able to deliver therapeutic agents. Some of these design requirements (described by Lally et al. [19]) to optimize the mechanical properties of a stent can be examined in a virtual design space. This design space can be used to improve the stent characteristics in its crimped state and its expanding (and expanded) state.

2.5.1. Crimped stent requirements

2.5.1.1. *Good flexibility*

The stent, crimped on the delivery catheter, must be advanced through the vascular tree, which can include narrow, tortuous passages, to reach the targeted site. Therefore, the crimped stent (on the delivery catheter) must be flexible, and this flexibility can be examined numerically (Figure 4).[1]

[1]Furthermore, a high dislodgment resistance is required to avoid stent loss from the unexpanded balloon in clinical use.

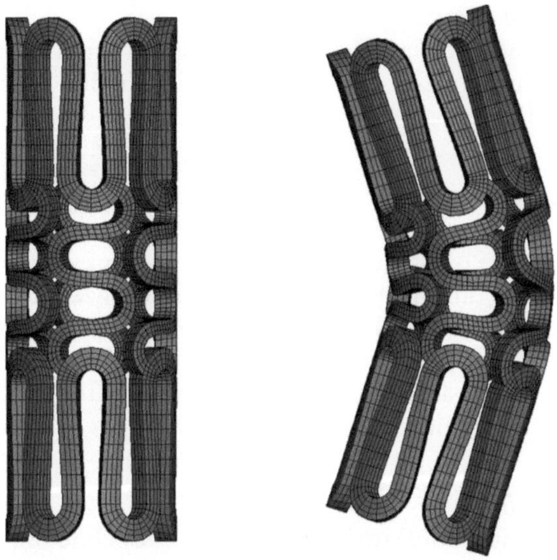

Figure 4 Virtual flexibility test on *Cypher* stent segment.

2.5.2. Expanding (Expanded) stent requirements

2.5.2.1. *Uniformity*

The implantation of a stent, in particular during and post deployment, may induce vascular injuries. The manner in which stents are implanted is a critical determinant of the degree of injury they cause [20]. The stent (transitory) non-uniform expansion in vivo, frequently in end-first manner (i.e., dogboning), is one of the (possible) reasons to induce the acute vascular injuries observed surrounding the edges of the stent. This dogboning phenomenon is directly dependent on both the balloon length and the stent design (Figure 5). Furthermore, depending on the stent design, the length of the stent may shorten during radial expansion. This foreshortening affects the precise stent positioning, and the friction between the stent and the arterial wall can injure the endothelium (i.e., vascular inner lining). Both the dogboning and the foreshortening phenomenon should be reduced to a minimum.

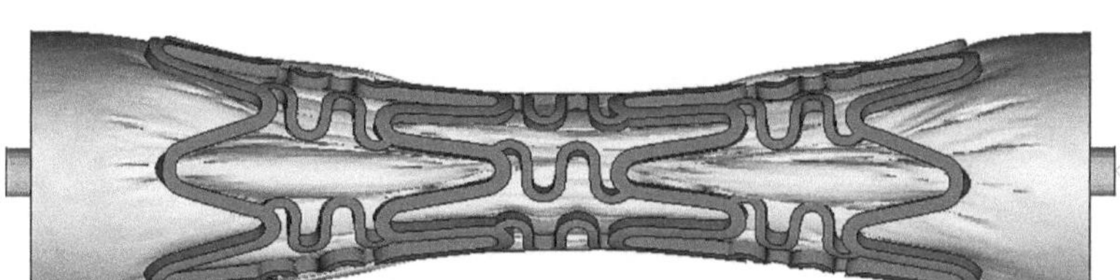

Figure 5 Non-uniform transitory *Cypher* stent expansion.

2.5.2.2. *High radial strength*

A stent should be designed as a mechanical scaffold to support the vessel lumen by minimizing post-stenting vessel recoil. The required radial support is both design and material dependent.

2.5.2.3. *Low elastic recoil*

To achieve a final lumen diameter consistent with the targeted vessel diameter, knowledge of the degree of elastic radial recoil of the stent must be factored into the expansion of the stent during deployment. To minimize the "inevitable" over-expansion, the material and design-dependent radial recoil should be reduced to the absolute minimum. In addition to foreshortening, undesirable shearing along the arterial walls may also be caused by longitudinal recoil after the balloon is deflated, and this phenomenon should be minimized.

2.5.2.4. *Good Flexibility*

The expanded stent must conform to the tortuous vessel geometry and should not straighten the vessel, as that would induce significant injuries to the vessel wall.

2.5.2.5. *Optimal Scaffolding*

To ensure vessel tissue not to prolapse between the stent struts, the stent should provide optimum (uniform) vessel coverage. In addition, a uniform strut distribution is beneficial to avoid local drug concentration and gradients in DES [21]. However, a low artery–stent contact surface area should also be maintained, because the body-foreign material of the stent can initiate an aggressive thrombotic response.

2.6. Effective use of FEA

The correlation between the finite element approximations and the physical reality is directly related to the ability of the scientist to accurately define the geometry, material behavior and loading environment of the device. While simplifying assumptions are often (if not always) inevitable, there is a risk that significant simplifications can considerably alter the problem. Therefore, the influence of the simplifications on the solution of the original problem must always be evaluated. Consequently, the numerical results should be validated by (targeted) experiments under similar conditions.

3. Survey of the State of the Art in Stent Modeling: 1997–2007

When reviewing literature regarding finite element models of the (free) expansion of balloon-expandable stents, there are three possible ways reported to simulate the inflation of the balloon: neglecting the presence of the balloon, assuming it to have a cylindrical shape or taking the actual folded shape of the

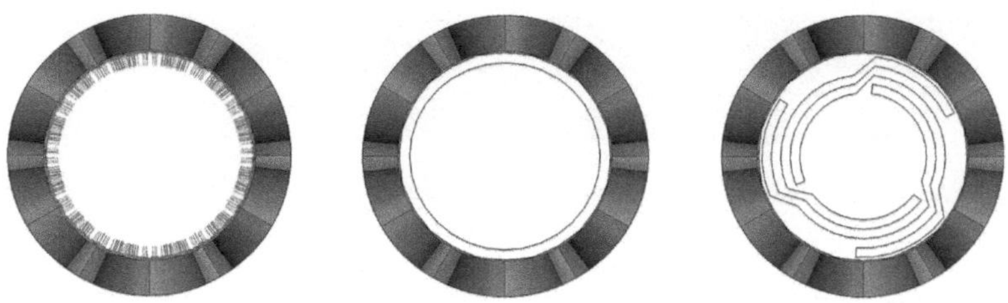

Figure 6 Stent expansion strategies: no balloon (left); cylindrical balloon (middle); and tri-folded balloon (right).

balloon into account (Figure 6). When modeling the vessel reaction to stenting, the geometry of the stenosis can be simplified (symmetric and asymmetric stenosis) or patient specific. The reviewed manuscripts are categorized chronologically (up to the summer of 2007) in the subsequent sections according to the applied stent expansion methodology. This chronological order clearly illustrates the rapid (r)evolution which this research domain has undergone since the reported results of the pioneers.

3.1. Neglect of the balloon

Dumoulin et al. [22] evaluated and characterized some mechanical properties (shortening percentage on expansion, radial and longitudinal recoil, resistance to crushing under external pressure, and fatigue) of balloon-expandable *P308 Palmaz* stents (Johnson&Johnson, Warren, New Jersey, USA) using the finite element method. The balloon is discarded from the numerical simulations and (a generic part of) the stent is expanded by applying a uniform pressure on the inner surface of the stent. This assumption is well motivated by the authors and is stated as valid for the central part of the stent. The assumption is based on a careful examination of an in vitro experiment of the free expansion of the investigated stent. Furthermore, the mechanical properties are described in terms of diameters instead of internal pressures. The numerical model is validated by comparing the expansion size (i.e., stent length) with data provided by the manufacturer, and a good correspondence is found. Dumoulin et al. clearly demonstrate that under certain specific conditions it is possible to investigate certain mechanical properties of balloon-expandable stents with the presence of the balloon discarded in the numerical models.

In the finite element stent study of Auricchio and his colleagues [23], the expansion behavior of a stent with a rectangular slotted tube design, i.e. *Palmaz-Schatz* (Johnson&Johnson) like, is investigated. Both the stent-free expansion and the vessel reaction to the stent expansion are studied and the stent design is modified in order to reach a more uniform expansion. Some typical stenting parameters (i.e. elastic recoil, foreshortening, . . .) are computed and presented. The authors stated that it is reasonable to neglect the balloon, because of its lower stiffness compared to the stiffness of the stent. This hypothesis permits the possibility of discarding the presence of the balloon in the analysis by applying the

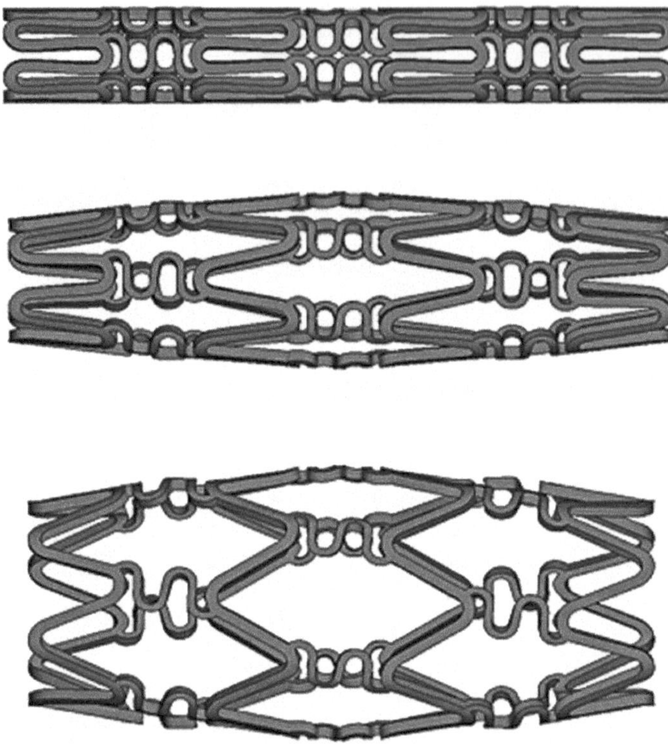

Figure 7 Pressure-driven *Cypher* stent expansion (neglect of the balloon). Top: no pressure; Middle: low pressure range; and Bottom: high pressure range.

internal pressure, which normally acts on the balloon, directly to the inner surface of the stent. Hence, the stent is loaded by an internal uniform radial pressure varying linearly from 0 to 1.3 MPa (inflation) and then again linearly back to 0 (deflation). The numerical results for some mechanical properties (elastic recoil, foreshortening, metal-artery ratio) correspond with values available in literature for the non-modified stent design. Though, the validation of the stent expansion behavior is rather limited. To investigate the revascularization, a symmetric 3D model of the plaque and artery is developed. The idealized "symmetric stenosis" is characterized by a straight artery segment and a plaque with a parabolic longitudinal profile and the constitutive material behavior is adopted from literature. This pioneering work clearly shows the enormous potential of the finite element method in this research domain, both on stent design and on procedural level.

In the same period, Etave et al. [24] used a finite element analysis to determine the "exact" mechanical characteristics of tubular (*Palmaz-Schatz*, Johnson&Johnson) and coil (*Freedom*, Global Therapeutics Inc.) stents. Seven mechanical properties were studied numerically with determination of (1) stent deployment pressure, (2) intrinsic elastic recoil of the material used, (3) resistance of the stent to external compressive forces, (4) stent foreshortening, (5) stent coverage area, (6) stent flexibility, and (7) the stress maps. The balloon is ignored in the numerical simulations and the stents were expanded by a radial displacement-driven process, forcing the stent to expand as a cylinder during the complete expansion (i.e., dogboning is excluded from the analysis). This numerical study shows the possibility of studying several important mechanical characteristics of different stent designs, but no experimental validation of the numerical results is reported.

Migliavacca et al. [25] investigated the mechanical behavior of coronary stents by means of the finite element method. Finite element analyses were applied (1) to understand the effects of different geometrical parameters (thickness, metal-to-artery surface ratio, longitudinal and radial cut lengths of a *Palmaz-Schatz*-like stent) of a typical diamond-shaped coronary stent on the device mechanical performance, (2) to compare the response of different actual stent models (*Multi-Link Tetra*, Guidant and *Carbostent*, Sorin Biomedica) when loaded by internal pressure, and (3) to collect suggestions for optimizing the device shape and performance. The presence of the balloon is discarded in the analysis, and the internal pressure on the balloon is directly applied to the inner surface of the stent corresponding with the expansion strategy proposed by Auricchio et al. [23]. The introduction of new-generation stent geometries opens the door to study stent designs currently used in the clinical practice. Experimental evidence of the numerical results would certainly create a more solid base for the proposed design optimization.

More recently, McGarry et al. [26] examined the mechanical behavior of a stainless steel balloon-expandable stent design using computational micromechanics in the context of the finite element method. Stent deployment and cardiac pulsing loading conditions were considered. As the typical dimensions of stent struts (e.g., 100 μm for coronary stents) are of a similar order of magnitude as the average grain size in stainless steel (i.e., 25 μm), continuum approaches relying on macroscopic material properties may be regarded as somewhat questionable, and alternative numerical analysis, based on crystal plasticity for example, may need to be used. McGarry and colleagues used classical phenomenological plasticity theory (J_2 flow theory) and physically based crystal plasticity theory to describe the stent material behavior. Important stent deployment characteristics such as recoil and foreshortening were determined using a two-dimensional (2D) model of a unit cell (i.e., characteristic pattern, which is repeated both circumferentially and longitudinally) of the *NIR* stent (Medinol/Boston Scientific). The deployment of the 2D unit stent cell was modeled using applied displacements. Furthermore, elementary fatigue factors of safety were determined by plotting Goodman diagrams to compare both material models, leading to a lower (and thus conservative) factor of safety for the classical plasticity theory in comparison with the crystal plasticity model. The numerical results in terms of average hydrostatic stresses are in the correct order of magnitude relative to experimental data reported in literature. McGarry et al. clearly demonstrate that under specific conditions it is possible to investigate certain mechanical properties of balloon-expandable stents (e.g., recoil, foreshortening, etcetera) with the presence of the balloon discarded in the numerical models. Nevertheless (experimentally validated), full 3D stent modeling should be performed in order to improve the accuracy according to McGarry and colleagues.

Gu and his colleagues [27] performed finite element analyses of covered microstents (i.e., bare metal stents with ultra-thin tubular coverings to treat for example wide-necked intracranial aneurysms) to assist in the design of stent coverings and to select materials for the cover in 2005. The primary objective of this study was to determine the mechanical properties of the covered microstent (e.g., longitudinal shortening, elastic recoil.) and investigate the effects of the covering on the mechanical behavior of the stent (e.g., deployment pressure vs. covering thickness). Careful observation of in vivo stenting experiments showed an almost uniformly inflated balloon except at its two ends, and the central portion of the stent is expanded by the uniformly inflated part of the balloon. Therefore, the authors decided to expand the stents in the model by a uniform internal pressure that was applied in small increments on the internal surface of the covered microstent. During loading, pressure was increased until the final diameter of the stent reached a certain value and then the pressure was decreased to study recoil. Although a simple non-specified in vitro experiment was performed to study the expansion of the stent, surprisingly no quantitative validation of the numerical results was presented in terms of the recorded deployment pressure-diameter data. Consequently, the accuracy of the proposed methodology has not been confirmed quantitatively.

A computational methodology that allows a set of stent parameters to be varied, with the aim of evaluating the difference in the mechanical environment within the arterial wall before and after angioplasty with stenting is proposed in the same period by Holzapfel et al. [28]. This methodology is applied to an image-based 3D geometrical model of a postmortem specimen of a human iliac artery with a stenosis. 3D stent models are parameterized in order to enable new designs to be generated simply with regard to variations in their geometric structure. The following parameters are altered to analyze their inter-relationship: (1) the type of stent cells, similar to three commercial products (*Multi-Link Tetra*, Guidant; *Niroyal-Elite*, Boston Scientific and *Inflow-Gold-Flex*, InFlow Dynamics), (2) the geometry of stent struts, and (3) the radial mismatch between the smallest lumen diameter in the stenosis and the expanded stent diameter (nominal stent diameter). Furthermore, the numerical simulations allowed to characterize the following indicators: (1) the change of the intimal pressure caused by the struts of the stent, (2) the stress change within the arterial wall caused by the stent, and (3) the luminal change due to angioplasty. The obtained numerical results are interpreted in terms of restenosis rates reported in clinical studies. This approach allows the analysis of the 3D interaction between the stent models and patient specific wall models, the quantification of change in the mechanical environment which occurs during stent placement, and the comparison of effects of different parameters to identify optimal stent designs. In the numerical simulations, the balloon is ignored and the (deformation dependent) pressure load is applied directly on the stent struts. This ground breaking work combines parametric modeling of new stent designs with the possibility to study the interaction of these designs with a patient-specific stenosis

In the same year, Lally et al. [29] analyzed the stress levels within a symmetric atherosclerotic artery provoked by two different stent designs (*S7*, Medtronic AVE and *NIR*, Boston Scientific). These stress levels are linked to possible vascular injury and thus, to the degree of restenosis reported in clinical studies. The developed methodology consists of two steps: firstly, the vessel was expanded to a diameter greater than that of the "idealized" expanded stent by applying a sufficient internal pressure to the vessel, and secondly, the pressure on the inner lumen of the artery was gradually reduced. The arterial tissue material model was determined by data from uniaxial and equibiaxial tension tests of human femoral arterial tissue. The hyperelastic constitutive model used to characterize the plaque tissue was determined by fitting to published data for human calcified plaques. Due to the elastic nature of the hyperelastic arterial tissue the vessel contracted around the scaffolding stent. The major limitation of the proposed methodology is the absence of the stent expansion process. Therefore, the influence of the stenosis on the stent expansion and the influence of the shear force during the expansion remain unstudied.

Migliavacca et al. [30] applied the finite element method to study the mechanical behavior of a new generation coronary stent (*BX Velocity*, Cordis). The results from the computations were compared with those from a laboratory experiment in terms of radial expansion and elastic recoil. Furthermore, Migliavacca and his colleagues were the first to compare the compliance of the stent in their simulations with the compliance chart of the manufacturer. The presence of the balloon is discarded in the analysis and a uniform linearly increasing radial pressure was applied to the internal surface of the stent to mimic the free expansion of the stent. In addition, to investigate the mechanical properties of the stent after the load removal, the stent was unloaded decreasing the internal pressure back to zero. The authors state that the absence of any balloon model is the real decisive element in interpreting the discrepancies between the experimental and the computational tests. Therefore, the development of more realistic computational models accounting for the presence of the balloon (in its actual folded shape) is required.

In order to reduce the dogboning to a minimum, De Beule and his colleagues [31] exploited the finite element method to investigate and compare the expansion of different designs of a first-generation *Palmaz-Schatz* stent. The stent expansion is studied by applying a uniformly distributed internal pressure directly to the inner surface of the stent as proposed by Auricchio et al. [23]. The dogboning effect even vanished when altering the original symmetric stent design to an asymmetric design. Thus taking asymmetry into account in the design seems very promising, at least from a dogboning point of view. However, no experimental validation of the numerical results has been performed and the authors acknowledge that more realistic balloon–stent interactions should be taken into account in future studies as the stent expansion behavior might be governed by the unfolding and expansion of the balloon [32,33].

Finally, Bedoya and co-workers [34] numerically studied the effects of stent design parameters on the stress field induced in the normal artery wall and the radial displacement achieved by the stent during systole and diastole in the same year. Each stent was positioned in a straight homogeneous cylindrical artery model with non-linear hyperelastic mechanical properties. These mechanical properties were obtained from pressure-diameter and force-elongation tests on a porcine artery specimen. The study indicated that stent designs (consisting of concentric rings of sinusoid like curves linked by straight bars of various lengths) incorporating large axial strut spacing, blunted corners at bends, and high axial ring segment amplitudes exposed smaller regions of the artery to high stresses, while maintaining an adequate radial displacement. The results of this interesting study may have limited applicability to the considered stent design. Timmins et al. [35] extended the work of Bedoya et al. [34] by developing an algorithm to optimize stent design, allowing for consideration of wall stress, lumen gain and cyclic deflection. The stent strut configuration was refined, via varying specific design parameters, to optimize stent performance. This interesting study demonstrates the feasibility of medical device design optimization under complex circumstances. Analogous to the methodology developed by Lally and her colleagues [29], the stent expansion process is not taken into account in both studies and the stent was initially a straight cylinder with a constant diameter (larger than the systolic arterial diameter). Therefore, the proposed design guidelines might alter when studying the real expansion in a patient specific stenosis, a logical topic for further investigation.

3.2. Cylindrical balloon

Prior to the millennium, in 1999, Rogers et al. [5] were the first to report an innovative experimental and numerical study on the balloon-artery interactions during stent placement, hypothesizing that balloon-artery interaction is a mechanism of vascular injury during stent deployment. Therefore, a two-dimensional finite element model was made to investigate how balloon-artery contact stress and area depend on stent-strut geometry, balloon compliance, and inflation pressure. The model assumed a linear elastic balloon membrane under pressure loading conditions to study balloon extrusion between struts.

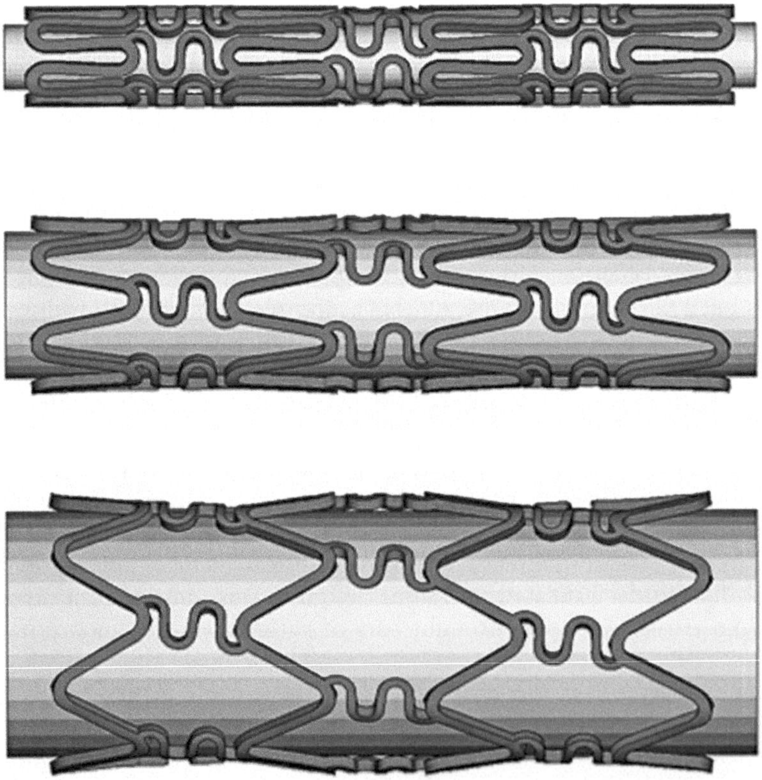

Figure 8 Displacement driven *Cypher* stent expansion (cylindrical balloon).

The constitutive behavior of the arterial wall was also considered to be linear elastic, based on previously published studies. The study demonstrates the influence of balloon compliance, stent geometry, and pressure load on contact pressure and surface.

One year later, Holzapfel et al. [36] studied the solid mechanics of angioplasty with and without stenting to provide essential insight into the mechanisms of angioplasty such as overstretching the disease-free tissue, plaque disruption or dissection, redistribution inside the wall, and lipid extrusion. Magnetic Resonance Imaging (MRI) was used to obtain accurate geometrical data for the vessel wall and plaque architecture and to identify different types of soft (biological) tissues and calcifications. New experimental results showing strong non-linearity and anisotropy were presented based on a sample of a human iliac artery with eccentric stenosis. The numerical model – to study both angioplasty procedures with and without the *Palmaz-Schatz* stent (Johnson&-Johnson) – assumed a cylindrical balloon with pressure loading in the balloon. Their non-linear numerical analysis showed that for this type of stenosis, with the presence of a complete collagenous cap, the disease-free segments of the vessel wall were overstretched, i.e., loaded beyond the elastic limit, while the plaque region was relatively unstretched. This (over)stretch leads to gains of the cross-sectional lumen area, which turns out to be a major mechanism of angioplasty for this type of stenosis as also suggested by other scientists. Subsequently, Holzapfel and his colleagues [37] analyzed the 3D stress states of the investigated artery during balloon expansion and stent deployment. Changes of 3D stress state due to model simplifications, which are characterized by neglecting axial in situ prestretch, assuming plane strain states and isotropic material responses, were studied. These simplifications lead to maximum stress deviations of up to 600% and are thus in general inappropriate. The (fully inflated) balloon is modeled as a rigid cylinder-shaped structure, and both the balloon inflation and the stent expansion are modeled as displacement-driven processes. The authors state that the assumption of a rigid cylinder is justified by the fact that fully inflated angioplasty balloons behave as non-compliant tubular structures, though this assumption has not been validated. The work by Holzapfel and his colleagues is groundbreaking on the basis of the simulation of the stenosed artery, taking into account realistic non-linear, anisotropic, and eccentric characteristics.

Recently, Liang and co-workers [38] investigated the biomechanical characteristics of intracoronary stent implantation by developing a 3D model of the complete stenting system and self-defined constitutive models for the plaque, the arterial wall, and the balloon. The stress concentrations in the contacting areas between stent and plaque, and the recoil ratios were examined. The artery and plaque were modeled as two symmetrical concentric cylindrical tubes. Due to the model symmetry, only a 120° segment was considered in this work. Uniaxial and biaxial experiments on human arteries were used to develop a hyperelastic constitutive model for the artery. In order to characterize the non-recoverable deformations of the interlining between plaques and arteries, a visco-plastic material model was established for the plaque. The authors state that this is a clearly limiting and somewhat unrealistic assumption due to the lack of experimental data. The balloon was modeled as a cylindrical plane with a fictitiously adopted hyperelastic material model in order to try to capture the unfolding of the balloon. This is an interesting approximative methodology, but unfortunately not validated.

Even more recently, Ballyk [39] evaluated the impact of stent oversizing on resultant arterial wall stress concentrations and examined the concept of a "stress threshold" for neointimal hyperplasia development in 2006. A 3D large-strain hyperelastic numerical model was used to examine the non-linear isotropic behavior of a 6 mm diameter straight cylindrical artery during stent deployment. In a fashion similar to that used by Holzapfel et al. [36], a *Palmaz-Schatz* stent was deployed to a diameter 30% greater than that of the native artery by a displacement-driven process. Ballyk found that the order in which location-specific peak stresses exceeded a predetermined stress threshold was constant: the threshold was first surpassed at the stent ends, followed by the stent cross-links, then the stent struts, and finally the bare (non-stent-covered) area between the stent struts. According to several (experimental) studies, these locations corresponded to the most common locations of intimal proliferation after stent deployment. Furthermore, an exponential relationship between peak stress concentration and percent stent inflation was formulated. These considerations provide a rationale for a stent design strategy aimed at minimizing vascular injury by optimizing rather than maximizing balloon inflation pressure and diameter. This interesting study is mainly limited by the absence of a realistic balloon, which will induce additional vessel stresses when protruding the bare (non-stent-covered) area between the stent struts (as shown by Rogers et al. [5]) and might also have an impact on the order in which the threshold is exceeded by the peak stresses. These findings could be supported by targeted (e.g., in vitro) experiments.

Also in 2006, a new methodology for quantitatively predicting the fatigue life of cobalt-chromium stents was presented by Marrey et al. [40]. In addition, fracture mechanics analysis is used to evaluate the role of pre-existing microstructural flaws. Therefore, the severity of such flaws observed on expanded stents are assessed quantitatively in terms of the propensity of the flaw to propagate and lead to in vivo failure of the stent when subjected to cyclic systolic/diastolic pressure loads within the implanted artery. The loading on the stents was modeled in terms of four discrete phases that are experienced in service, namely (1) assembly onto the balloon catheter (i.e., crimping), (2) balloon-inflation, (3) recoil, and (4) physiological loading within the artery. At the end of the crimping phase, a semi-rigid expansion cylinder was used to expand the stent into a hyperelastic tube, simulating the controlled inflation of the balloon. The symmetric hyperelastic tube represented the section of the coronary artery into which the stent is implanted. The material constants of the hyperelastic tube were numerically calibrated to exhibit a physiologically relevant coronary arterial distension. Marrey et al. clearly demonstrate that under certain specific conditions it is possible to investigate the mechanical properties of balloon-expandable stents with the presence of the actual folded balloon shape discarded in the numerical models.

In the same period, an interesting comparison of element technologies for modeling stent expansion was published by Hall and Kasper [41]. The computational efficiency and results achieved using continuum solid, shell, and beam elements are compared, and the study shows that all of the element formulations provide (asymptotically) a similar response in the context of stent deployment simulations. Among the element choices, the beam formulation is computationally most efficient by a significant factor with regard to problem size, time, and memory requirements. Furthermore, the beam simulations predict nearly the same extreme values of stress, strain, and equivalent plastic strain at identical locations on the stent. Hall and Kasper clearly show that beam discretization of stents might be very useful to speed up some aspects of the design phase; however, modeling the vascular reaction to stenting and the consequent flow disturbance might be more challenging using beam elements.

Next, Wang et al. [42] applied the finite element method to simulate the transient expansion phase of a stent/balloon system with different stent structures and balloon lengths under internal pressure-loading conditions. Two types of stents (with length equal to 9.8 mm) and six collocations of stent and balloon were modeled to investigate the reduction (elimination) of the dogboning phenomenon by decreasing the balloon (over)length and increasing the stent-end stiffness (i.e., increased strut width). The authors state that a "virtual" linear elastic material – designed to imitate the behavior of the real balloon with a simple cylindrical model – does not notably affect the transient deformation of a stent. The finite element results were validated *qualitatively* by recording the transient stent (with length = 17 mm!) expansion process and by measuring the outer diameter and stent length at different inflating pressures (providing pressure–diameter curves). Although this methodology has several interesting features, certain limitations in the approach need to be addressed. Firstly, the discrepancy between the stent length in the numerical and experimental models is not motivated, and secondly, the numerical results are not validated quantitatively, whereas the proposed methodology provides both numerical and experimental pressure–diameter information. Therefore, the accuracy of the "virtual" cylindrical balloon hypothesis is not (yet) demonstrated.

Takashima et al. [43] characterized the stress distribution at the contacts between the stent and the artery using mathematical and experimental modeling. Comparison of the experimental with the finite element results revealed that the contact area between the stent (with a high number of cells and links) and the artery model was distributed over the total surface of the stent. The stents are expanded "in numero" by the inflation of a very stiff cylindrical balloon (Young's modulus of 400 GPa) controlling the displacement of the stent. The finite element results (in terms of contact area ratio) were validated both qualitatively and quantitatively, and the contact areas of the stents corresponded to the areas of high radial compressive stress. The contact area ratios were (much) larger in the experimental model than in the finite element analysis. The authors suggest that this discrepancy may be the result of several factors (contact surface area calculation, problem non-linearity, material parameters, mesh size), which are all variables that can (easily) be examined by extending the numerical model. However, the main limitations seem to be the discrepancy between the geometry and the loading conditions of both the numerical model (length of vessel and plaque: 18 and 7 mm; vessel outer diameter: 3.6 mm; displacement-driven process) and the experiment (length of vessel and plaque: 50 and 30 mm; vessel outer diameter: 4 mm; pressure-driven process). Consequently, simulating the exact experiment could surely reduce the variation in the contact area ratios.

Finally, Wu and his colleagues [44] proposed a finite element-based methodology to study the stent expansion in both a straight and a curved vessel. The stent expansion in the curved vessel was simplified as follows: firstly, the vessel was expanded by applying a pressure on the inner surface of the tissue. Subsequently, the stent was expanded with a straight semi-rigid balloon and the pressure on the vessel was reduced. Lastly, the balloon diameter was reduced to its original shape. Their results show a hinge effect at the stent extremities and a straightening of the curved vessel by stenting. Furthermore, the maximum tissue prolapse was more severe, the minimum lumen area decreased and the tissue stress levels were higher for the curved vessel model when compared to the straight vessel model. The proposed methodology is one of the first to investigate the stenting procedure of curved vessels, but (some) of the results and conclusions should be interpreted skeptically as the validation is limited. The main limitation seems not to be the absence of the fluid dynamical aspect (i.e., study of the flow disruption in the stented vessel), but the applied stent expansion methodology which might influence tissue prolapse, minimum lumen area, and tissue stress levels. Therefore, experimental evidence seems necessary to show whether the actual stent expansion might be simplified as proposed by the authors. Furthermore, the real non-recoverable deformations of the tissue during a stenting procedure (which cannot be captured by the applied hyperelastic constitutive behavior for the vessel and plaque) will influence the tissue stress levels, as already mentioned by Holzapfel et al. [37], and probably also influence the interaction between the stent and the vessel wall.

3.3. Folded balloon

Recently, De Beule and colleagues [32] accomplished to model the free expansion of the *Cypher* (Cordis) stent by applying a uniform pressure up to 1.5 N/mm^2 on the inner surface of a tri-folded balloon. Accounting for the presence of the balloon in its actual tri-folded shape reveals a particular deployment pattern as depicted in Figure 9. The numerical results correspond very well to pressure/diameter data provided by the manufacturer as the maximum percent difference in diameter is an overestimation of only 5.7% occuring at a pressure of 0.6 N/mm^2. Discarding the balloon from the analysis and applying the same pressure directly on the stent's inner surface leads to an underestimation of the diameter of 62.6% at the same pressure (Figure 10). The validated model is then used to study the effect of the crimping process (on the delivery catheter) on the stent expansion behavior. This allowed the authors to conclude that the influence of the residual stresses induced by the crimping phase on the overall stent expansion behavior is minor, but nevertheless influences the maximum von Mises stress and nominal strain. Depending on the context in which to use the developed mathematical models, the crimping phase can be discarded from the analyses in order to speed them up. In addition, the effect of the size dependency of the material properties and the impact of the magnitude of the yield stress on the stent expansion behavior are investigated by De Beule and co-workers [33]. Murphy et al. [18] examined small 316L stainless steel strut specimens (60–500 μm) under uniaxial tensile loading conditions in an experimental study. The results showed a size dependency of the failure strain

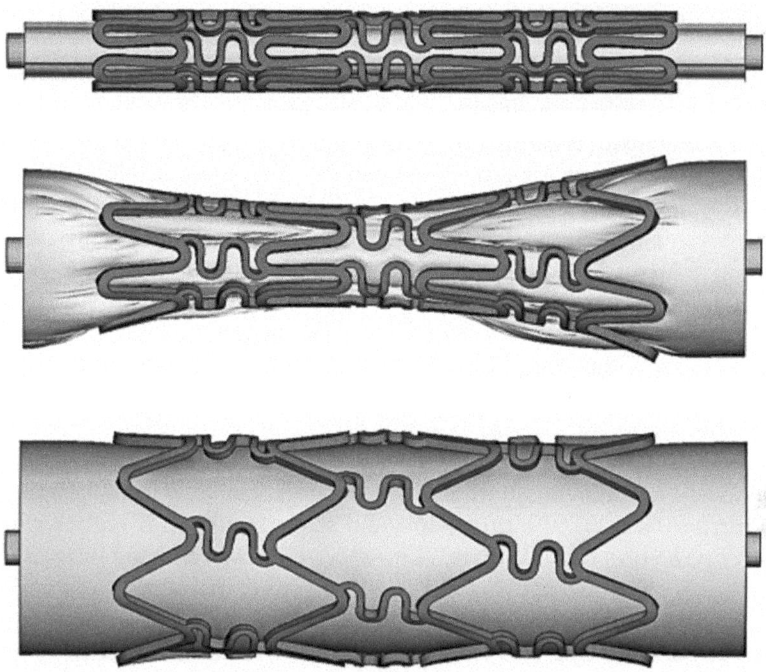

Figure 9 Pressure-driven *Cypher* stent expansion (folded balloon). Top: no pressure; Middle: low pressure range; and Bottom: high pressure range.

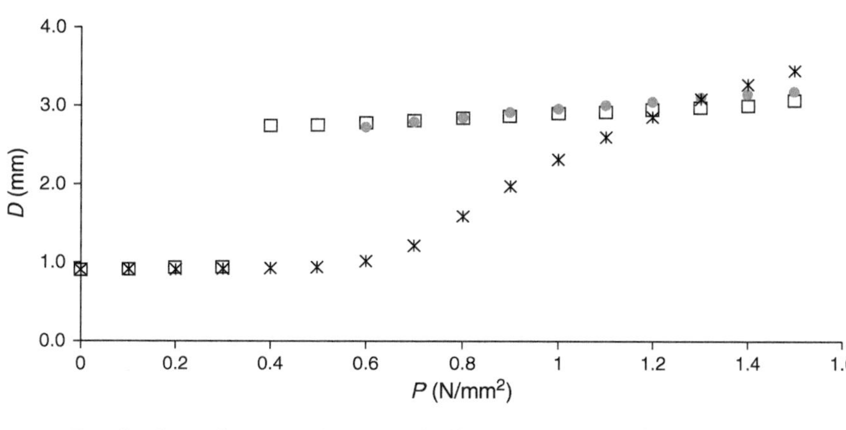

Figure 10 Results from the FE analyses (No balloon: crosses; and tri-folded balloon: empty squares) and data provided by the manufacturer (solid circles). Simulation results accounting for the folded balloon compare favorably with manufacturer data.

(i.e., smaller struts fail earlier than larger struts). In the numerical model, the maximum value of the nominal strain in the expanded stent does not exceed the critical values defined by Murphy et al; moreover, the critical values are nowhere exceeded in the whole stent during the expansion. Finally, there seems to be no consensus in literature regarding the mechanical properties to describe the (in) elastic material behavior of 316L stainless steel (SS 316L) (see Table 1). Values for the modulus of Young vary little from 190,000 [24] to 209,000 N/mm^2 [26], though the reported values for the yield stress range from 205 N/mm^2 [23] – even further reduced to 105 N/mm^2 [30] – to 375 N/mm^2 [18]. Presumably, this wide range in reported yield stress values is caused by the pre-treatment (e.g., hot rolling, cold finishing, annealing, electropolishing) of the tested experimental specimens. Consequently, the yield stress should be obtained from tensile tests on specimens extracted from batches that are eventually used to manufacture the stent structure. Because of this discrepancy in the description of the constitutive material behavior, the authors examined the impact of the magnitude of the yield stress on the stent expansion behavior. As the stent expansion is driven by plastic deformation, an increase (from 205 to 375 N/mm^2) in yield stress results in a higher resistance to expansion. To the best of the author's knowledge, these validated numerical models are the first that take realistic (folded) balloon–stent interaction into account. As these numerical results – accounting for the presence of the balloon in its actual folded shape – correspond very well with pressure/diameter data provided by the

Table 1 Material properties SS316L

Group	Young Modulus (N/mm^2)	Yield stress (N/mm^2)	Material reference
Auricchio [23]	196,000	205	Metals handbook (American Society for Metals)
Etave [24]	190,000	?	Standard stress/strain curve for annealed material
Migliavacca [25]	196,000	205	Auricchio et al. [23]
McGarry [26]	209,000	264	Uniaxial tensile test
Gu [27]	196,000	205	Auricchio et al. [23]
Lally [29]	200,000	–	–
Migliavacca [30]	196,000	105	Kinematic hardening yield stress reduction (crimping)
Liang [38]	201,000	330	Tensile tests on wires
Hall [41]	196,000	290	–
Wang [42]	201,000	?	Tensile test on annealed wire
Bedoya [34,35]	200,000	–	–
De Beule [31–33]	196,000 [196,000]	205 [375]	Auricchio et al. [23] [Tensile test on (stent) struts with varying width (Murphy et al. [18])]
Wu [44]	201,000	280	Standard true stress/strain curve for annealed material
Takashima [43]	200,000	315	–

manufacturer, this approach could be the basis for new realistic computational models of angioplasty procedures. These numerical models will offer interesting insights into the mechanical aspects of such procedures and play a key role in the design of new devices (e.g., dilatation catheters, stents).

3.4. Summary

For each of the above-mentioned studies; the aim, the balloon inflation methodology, and the validation strategy are briefly summarized in Table 2.

Table 2 Chronological overview of balloon-expandable stent modeling

Group	Research	Balloon	Validation
Rogers [5]	2D analysis of balloon/artery interaction. Contact pressure and surface are dependent on balloon compliance, stent geometry, and pressure load	Linear elastic membrane	In vivo and in vitro experiments (qualitatively)
Dumoulin [22]	Free stent expansion: evaluation and characterization of some mechanical stent properties	No balloon	Examination of in vitro experiment
Holzapfel [36]	Mechanisms of angioplasty	Elastic cylinder	Literature (principle of luminal gain)
Auricchio [23]	Vascular reaction to *Palmaz-Schatz* stent expansion: influence of modified stent geometry	No balloon	Literature (stent mechanical properties)
Etave [24]	Seven mechanical characteristics of *Palmaz-Schatz* and *Freedom* stent	Theoretical analysis	Not reported
Migliavacca [25]	Mechanical behavior of *Palmaz-Schatz* stent: modification of stent design. Comparison Carbostent and Multi-Link Tetra stent	No balloon	None
Holzapfel [37]	3D stress state in patient-specific stenosis with anisotropic material properties during balloon expansion and stent deployment: influence of constitutive material model simplifications	Rigid cylinder	None (proposes appropriate in vitro benchmark test)

(Continued)

Table 2 (*Continued*)

Group	Research	Balloon	Validation
McGarry [26]	Deployment characteristics (recoil, foreshortening, and fatigue safety factor) of *NIR* stent using a 2D model of a unit cell (recoil, foreshortening) based on micromechanical modeling. Stent material behavior described by classical and crystal plasticity theory	No balloon	Literature (non-uniform grain level deformation and residual stresses)
Gu [27]	Mechanical performance of the covered *Palmaz-Schatz* microstent (e.g., longitudinal shortening, elastic recoil, deployment pressure vs. covering thickness)	No balloon	In vivo and in vitro experiments (qualitatively)
Holzapfel [28]	3D interaction between different stent models (*Multi-Link Tetra, Niroyal-Elite* and *Inflow-Gold-Flex*) and a patient-specific wall model with anisotropic material properties. Quantification of the change in the mechanical vascular environment during stent placement and comparison of the effects of different parameters to identify optimal stent designs. Linked to restenosis	No balloon	Literature (degree of restenosis)
Lally [29]	Stress levels within a symmetric atherosclerotic artery provoked by *S7* and *NIR* stent. Linked to restenosis	No balloon	Literature (degree of restenosis)
Migliavacca [30]	Mechanical behavior of *BX Velocity* coronary stent	No balloon	Data manufacturer and experimental test.
Liang [38]	Simulation of balloon-driven stent expansion in stenosed artery	Hyperelastic cylinder	None

Table 2 (*Continued*)

Group	Research	Balloon	Validation
Ballyk [39]	Evaluation of the impact of stent oversizing on resultant arterial wall stress concentrations and examination of the concept of a "stress threshold" for neointimal hyperplasia development	Rigid cylinder	Literature (locations of intimal proliferation after stent deployment)
De Beule [31]	Introduction of asymmetry in the *Palmaz-Schatz* stent design to avoid dogboning	No balloon	None
Marrey [40]	New methodology for quantitative prediction of the fatigue life of cobalt-chromium stents	Semi-rigid cylinder	Experimental tests on stent (wire) specimens
Hall [41]	Comparison of element technologies for modeling stent expansion	Semi-rigid cylindrical surface	None
Wang [42]	Transient expansion behavior and design optimization of coronary stents: reduction of dogboning by adapting stent geometry and balloon length	Linear elastic cylinder	In vitro experiments (qualitatively)
Bedoya [34]	Monitoring of the radial displacements achieved by the stent during systole and diastole and the stress levels within a symmetric artery provoked by different stent designs. Linked to restenosis	No balloon	None
Wu [44]	Stent expansion in both straight and curved vessel	Rigid cylinder	None
De Beule [32]	Effect of crimping the *Cypher* stent on a trifolded balloon on its expansion behavior. Quantification of the impact of balloon neglect on the stent expansion behavior	Elastic trifold	Compliance chart manufacturer

(*Continued*)

Table 2 (*Continued*)

Group	Research	Balloon	Validation
De Beule [33]	Effect of the size dependency of the material and the impact of the magnitude of the yield stress on the *Cypher* stent expansion behavior	Elastic trifold	Compliance chart manufacturer
Timmins [35]	Development of an algorithm to optimize stent design, allowing for consideration of competing wall stress, lumen gain, and cyclic deflection	No balloon	None
Takashima [43]	Simulation and experimental observation of contact conditions between stents and artery models.	Elastic cylinder	Qualitative and quantitative (contact area ratio)

4. Alternative Methods for Biomechanical Modeling of Stents

As stated in the introduction, this review focuses on Finite Element Method (FEM)-based studies dealing with the mechanical behavior of (balloon expanding/ expanded) stents. The study of the mechanical properties of the *Palmaz-Schatz* and *Freedom* stents by Tan and his colleagues [45] was not included in the survey as the expansion methodology is not described. In addition, the FEM can be applied to study other important mechanical stent characteristics, such as tissue prolapse, flexibility, strut micromechanics, and also self-expandable stents. Furthermore, the impact of the stent design (e.g., strut shape, interstrut distance) on the blood flow patterns can be examined using Computational Fluid Dynamics (CFD). Finally, the kinetics of the drug release – in the case of DES – can also be examined numerically. Some studies dealing with these issues are summarized below.

4.1. FEM – Prolapse, flexibility, and strut micromechanics

To analyze tissue prolapse in cardiovascular stents, Prendergast et al. [46] performed uni- and biaxial experiments on human femoral artery and porcine aortic vascular tissue to develop a hyperelastic constitutive model of vascular tissue suitable for implementation in a finite element model. To study the deflection of tissue between the struts of the stent (i.e., prolapse), four stent designs (*BeStent 2*, Medtronic AVE; *Niroyal*, Boston Scientific; *Velocity*, Cordis; *Tetra*, Guidant) were expanded in vitro to determine their repeating-unit dimensions. This geometrical information was used to generate a finite element model of the vascular tissue supported within a repeating unit of the stent. This methodology is proposed as a way to compare stents relative to their potential for restenosis and as a basis for a biomechanical design of a stent repeating-unit that would minimize restenosis. The stent measurements were obtained by taking pictures of the fully expanded stent under a microscope and the images were then projected onto a 3.5 mm cylinder to define the geometry of one repeating unit of each stent design. Consequently, the blood vessel is modeled as a symmetric concentric cylindrical tube. The most important simplification made by the authors is that the prolapse is not investigated by taking the stent expansion process into account. Furthermore, the effect on the prolapse of taking a realistic (patient-specific) artery with a stenosis into account is not (yet) examined.

Both Petrini and her colleagues [47] and Mori and Saito [48] explored the advantages of the finite element method in order to investigate stent performance in terms of flexibility. Petrini et al. considered two 3D models resembling two new-generation coronary stents (*BX Velocity*, Cordis and *Carbostent*, Sorin Biomedica), both in the expanded and in the unexpanded configuration. The main limitations of the work are the absence of an experimental validation and the absence of the delivery system during the flexibility analysis in the unexpanded configuration. Mori and Saito applied the four-points bending test to evaluate stent flexibility, and employed the finite element method to assess the effect of stent structure on flexibility in stents with differing link structures under compressive loading conditions. The flexibility predicted from their simplified 2D finite element analysis correlated with the bending stiffness of the stents. The main limitation of this study is that the flexibility is examined only in the stent's expanded configuration, where the stent also should possess flexibility for maneuverability through tortuous vasculature to the site of occlusion in the unexpanded configuration.

The use of finite element analysis to predict the mechanical failure of stent struts was investigated by Savage et al. [49]. A 2D computational micromechanics approach was considered involving an explicit representation of the grain structure in the steel struts to accurately represent the constitutive behavior of the material, as question arises as to the suitability of using bulk material properties at the size scale of stent struts. The predictions of the failure strains in virtual uniaxial tensile tests on strut specimens are in good qualitative agreement with the experimental results of Murphy and his colleagues [18], i.e., the observation of a similar trend in the variation of the failure strain with strut thickness. In order to produce better quantitative results, full 3D models and models with an explicit damage mechanism seem necessary before even thinking of studying the (micro)mechanical behavior of entire stents in their natural environment. Furthermore, these extended models might require very high computational times as already stated by the authors.

4.2. FEM – Self-expandable stents

The first one to report a numerical study on the behavior of self-expandable stent structures was Whitcher [50], who estimated the in vivo structural behavior and fatigue properties of nitinol stents. Migliavacca and colleagues [51] analyzed the stress state induced on the vascular wall, by the expansion of both balloon- and self-expandable stents using the finite element method, as the modified mechanical stress state is in part responsible for the restenosis process. Comparing the stainless steel balloon expandable stent with the shape memory alloy (Nitinol) self-expandable one, the latter induces lower stresses, but on the other hand, its lower stiffness induces a lower capability to restore artery lumen and to counteract arterial elastic recoil. However, because of the strong dependency of the material properties on the shape memory alloy composition and thermomechanical treatment, this study does not make any attempt to give quantitative indications. In a more general computational study of the shape memory alloy (SMA) behavior in biomedical applications, Petrini and her colleagues [52] performed an experimental and virtual stent crush test. Again, the exact knowledge of the real characteristic material parameters is missing, but the observed agreement between the experimental and the numerical results is encouraging and shows the capability of the implemented material model to describe the fully 3D behavior of SMA devices. A novel and interesting concept of progressively expanding shape memory stents was proposed and studied by Theriault et al. [53]. To prevent restenosis caused by intimal hyperplasia, related to arterial wall injury due to the violent penetration of the stent structure, this study proposes a nitinol self-expanding stent with a progressive expansion system activated by the creep effect of polyethylene rings. This device will impose a smooth and gradual contact between the stent and the endothelium, giving the artery the opportunity to adapt to the presence of the device. A modeling strategy was proposed to obtain the final geometry of the stent laser cut from a small tube and to examine the behavior of the prosthesis during surgery and over the 4 weeks following the operation. The numerical results demonstrate that a compromise can be reached between a limited expansion prior to the inflation of an expandable balloon and a significant expansion by creep of the polymer rings. However, experimental validation of the mechanical behavior of the prosthesis is missing and the interaction of the device with a (patient specific) stenosis has not yet been investigated. In a virtual study by Wu and colleagues [54], two superelastic stents were delivered into a stenotic carotid bifurcation through a sheath and self-expanded in the internal and common carotid artery. The stent with shorter struts had less malapposition areas, a higher luminal gain, caused more tortuosity changes and induced higher maximum vessel stresses. The actual stent release from the catheter is simplified by assuming a uniform diameter increase of the sheath and contact between the stent struts during compression of the stent in the sheath is neglected. Furthermore, only one kind of nitinol material properties was modeled, whereas the actual (and highly adaptable) nitinol properties can influence the stent mechanical properties greatly. The above-mentioned assumptions can (only) be justified through a missing targeted experimental validation.

As summarized above, the studies in engineering literature dedicated to the analysis of the mechanical properties of nitinol self-expanding stents and to their interaction with the vessel wall focus on tubular stents. In contrast, the number of virtual studies related to wire stents is rather limited. This is surprising because of the wide application field of these devices [55–58]. Jedwab and Clerc [59] developed an experimentally validated theoretical model of a self-expanding metallic wire stent with the aim of predicting the various mechanical properties of the stent. Wang and Ravi-Chandar [60,61] developed a mathematically rigorous model, based on the theory of slender rods, that describes the response to internal and external loading conditions of a metallic braided wire stent, and the obtained results were validated with experimental data. In addition, Canic et al. [62] performed an analytical study in order to analyze the mechanical properties of two bare-metal braided stent grafts and to compare their response to hemodynamic forces. Furthermore, Brand and his colleagues [63] developed an analytical approach for the mechanical interaction of the self-expanding Cardiocoil (Medtronic InStent) stent with the stenosed artery. All these studies show good results, but at the same time, they demonstrate the necessity to build complex analytical models to examine the mechanical behavior of wire stents. Furthermore, these analytical models seem challenging to analyze the expansion of wire stents in patient-specific anatomical geometries and seem unable to capture the mechanical behavior of innovative bioabsorbable braided stents [64]. In an attempt to partially justify the lack of finite element-based wire stent studies, one should consider the complexity of building the geometrical model of such a wire stent using classical CAD methodologies and subsequent (numerical) analysis of this family of stents.

4.3. CFD–drug elution and immersed FEM

Frank and his colleagues [65] have given a brief overview of Computational Fluid Dynamics (CFD) research studies dealing with stent design. CFD approaches are well suited for obtaining detailed information on stent flow patterns, and the strong dependence of flow stagnation (possibly related to restenosis) on stent-strut spacing has been clearly demonstrated. He and co-workers [66] examined detailed flow characteristics by estimating the Wall Shear Stress (WSS) in the near-strut region of realistic stent designs using 3D CFD. Their results also indicated that stent design is crucial in determining the fluid mechanical environment in an artery. The results of the 3D CFD study of LaDisa et al. [67] supported the hypothesis that circumferential vascular deformation after stent implantation imparts distinctive alterations in WSS that are not detected by using a standard circular model of the vascular cross section. In addition, Benard and his colleagues [68] investigated the effects of blood (rheological) properties on changes in intra-stent flow with a 3D CFD model. Balakrishnan et al. [69] coupled computational fluid dynamics and a mass transfer model to predict drug deposition for single and overlapping drug-eluting stents. The simulations correlated with in vivo effects and revealed that drug deposition occurs not only beneath regions of arterial contact with the stent strut, but surprisingly also beneath standing drug pools created by strut disruption of flow. To conclude, the mechanical behavior of stents during and after implantation were studied by Gay et al. [70] using a fluid–structure interaction computational technique. This complex fluid and deformable structure interaction problem was tackled using an immersed finite element method (IFEM) and allows to study the flow patterns during the deployment of a stent. However, the proposed modeling strategy is not (yet) validated, and the effect of realistic stent expansion by balloon unfolding on the flow patterns remains unknown.

5. Future Prospects

Future developments in stent modeling will most likely include further integration of innovative (braided and tubular) stent designs and materials in realistic patient-specific stenosis models. Such integrated models may even further raise its share in the stent design phase and eventually enter the clinical practice to optimize the coronary revascularization procedure for a specific patient (e.g., as a presurgical planning tool).

In the "hunt" for the ideal stent, new stent materials [e.g., cobalt-chromium, superelastic shape-memory alloys, (biodegradable) polymers, magnesium.] are emerging. Each of these materials has a specific constitutive behavior and consequently requires a specific innovative stent design. These inventive stent designs and materials can and will further be tested and evaluated "in numero" [40,52,71].

The current (r)evolution in medical imaging, providing scientists with even more detailed and accurate geometrical information for the vessel wall and plaque architecture, in combination with new insights into the constitutive material behavior of soft (biological) tissues and calcifications will allow further optimization of the strategy proposed by Holzapfel and his colleagues [36]. In addition, the recent advances in both image reconstruction, allowing to transfer these detailed medical images and material data into an accurate mathematical model, and simulation tools (e.g., complex contact definitions [32,33]) will only speed-up this optimization. Furthermore, validation of these numerical results with innovative (targeted) experiments [43] will only encourage the use of the "in numero" strategy. Therefore, the idea of a surgeon implanting a patient-specific stent, selected by means of a numerical presurgical planning tool, no longer seems science fiction.

6. Conclusion

From this review, it is obvious that the finite element method offers numerous possibilities in the optimization of (coronary) revascularization procedures. The application of such a numerical approach in this specific biomechanical research domain is quite recent (1997) and has known an enormous evolution the last few years. Undoubtedly, all previously mentioned papers have contributed to the current level of understanding the mechanics of both stent designs and angioplasty procedures, and the quality of these studies should be evaluated taking into account the available prescience and computational facilities (at the moment of publication). In the early years, numerical simulations were based on Palmaz–Schatz (Johnson & Johnson)-like stents and simplified symmetrical stenosed arteries, mainly due to their straightforward geometry (e.g., the pioneering study published by Auricchio [23]). Furthermore, the Palmaz–Schatz stent was the only stent used in a multitude of large clinical studies at that time. Since then this numerical (r)evolution has been characterized by some important milestones, such as (i) the introduction of patient-specific stenosis models by Holzapfel and coworkers in 2000 [36], (ii) the introduction of next generation stent geometries by Migliavacca et al. in 2002 [25] and the introduction of stent design-optimization techniques by Timmins and his colleagues [35].

An astonishing observed fact in numerous studies is the lack of experimental evidence for the obtained numerical results, creating a missing-link with reality and provoking an (understandable) skepticism with respect to numerical models and to the conclusions drawn from them. The little validation that is performed is often merely qualitative, and thus not always applicable to interpret and verify the numerical results. The only consistent quantitative validation regarding the stent-free expansion is from Migliavacca et al. [30], showing the considerable discrepancy between numerical results and reality when discarding the presence of the balloon. Regarding the vascular reaction to stent deployment, it should be nuanced that measuring the — by the stent (expansion) induced — stresses in the vessel wall in an experimental setup is a huge challenge. For such specific numerical studies, it makes sense to compare the results with clinical findings in literature (e.g., restenosis rates). Nevertheless, caution should be paid to the adopted methodology of the cited clinical trials (i.e., investigated stent designs, patient recruitment), and specifically designed new clinical follow-ups could be required for thorough validation of the developed hypotheses from the numerical results. In addition, it should be feasible to examine the deployment characteristics of a stent using, for example, X-ray (Nano) Microtomography [72]. Furthermore, the impact of the assumptions that were made (e.g., balloon neglect or simplification, applied constitutive material models), on the obtained results should be investigated thoroughly, as already indicated by Holzapfel et al. [37].

The reported different methodologies to model the free expansion of a balloon-expandable stent should be evaluated and compared to assess the optimal free expansion methodology for finite element-based stent design. Such a comparison can help define the context in which neglecting the folded shape of the balloon is acceptable.

ACKNOWLEDGMENTS

The author gratefully acknowledges the valuable support and assistance from Rudy Van Impe (Ph.D.), Pascal Verdonck (Ph.D.), Benedict Verhegghe (Ph.D.), Patrick Segers (Ph.D.), Stéphane Carlier (MD, Ph.D.), Peter Mortier, Denis Van Loo, and Bert Masschaele (Ph.D.).

REFERENCES

[1] A. H. Association, Heart Disease and Stroke Statistics – 2007 Update, Dallas, TX: American Heart Association (2007) 1–43.

[2] D. E. Kandzari, J. E. Tcheng, J. P. Zidar, Coronary artery stents: evaluating new designs for contemporary percutaneous intervention, Catherization and Cardiovascular Interventions 56 (4) (2002) 562–576.

[3] A. Kastrati, J. Mehilli, J. Dirschinger, J. Pache, K. Ulm, H. Schuhlen, M. Seyfarth, C. Schmitt, R. Blasini, F. J. Neumann, A. Schomig, Restenosis after coronary placement of various stent types, Am J Cardiol 87 (1) (2001) 34–39.

[4] A. C. Morton, D. Crossman, J. Gunn, The influence of physical stent parameters upon restenosis, Pathol Biol (Paris) 52 (4) (2004) 196–205.
[5] C. Rogers, D. Y. Tseng, J. C. Squire, E. R. Edelman, Balloon-artery interactions during stent placement: a finite element analysis approach to pressure, compliance, and stent design as contributors to vascular injury, Circ Res 84 (4) (1999) 378–383.
[6] R. Mongrain, J. Rodes-Cabau, Role of shear stress in atherosclerosis and restenosis after coronary stent implantation, Rev Esp Cardiol 59 (1) (2006) 1–4.
[7] M.-C. Morice, P. W. Serruys, J. E. Sousa, J. Fajadet, E. Ban Hayashi, M. Perin, A. Colombo, G. Schuler, P. Barragan, G. Guagliumi, F. Molnar, R. Falotico, A randomized comparison of a sirolimus-eluting stent with a standard stent for coronary revascularization, N Engl J Med 346 (23) (2002) 1773–1780.
[8] J. E. Sousa, M. A. Costa, A. Abizaid, F. Feres, A. C. Seixas, L. F. Tanajura, L. A. Mattos, R. Falotico, J. Jaeger, J. J. Popma, P. W. Serruys, A. G. M. R. Sousa, Four-year angiographic and intravascular ultrasound follow-up of patients treated with sirolimus-eluting stents, Circulation 111 (18) (2005) 2326–2329.
[9] P. W. Serruys, M. J. B. Kutryk, A. T. L. Ong, Coronary-artery stents, N Engl J Med 354 (5) (2006) 483–495.
[10] C. Kaiser, H. P. Brunner-La Rocca, P. T. Buser, P. O. Bonetti, S. Osswald, A. Linka, A. Bernheim, A. Zutter, M. Zellweger, L. Grize, M. E. Pfisterer, Incremental cost-effectiveness of drug-eluting stents compared with a third-generation bare-metal stent in a real-world setting: randomised Basel Stent Kosten Effektivitats Trial (BASKET), Lancet 366 (9489) (2005) 921–929.
[11] R. Virmani, G. Guagliumi, A. Farb, G. Musumeci, N. Grieco, T. Motta, L. Mihalcsik, M. Tespili, O. Valsecchi, F. D. Kolodgie, Localized hypersensitivity and late coronary thrombosis secondary to a sirolimus-eluting stent: should we be cautious? Circulation 109 (6) (2004) 701–705.
[12] M. Perry, S. Oktay, J. C. Muskivitch, Finite element analysis and fatigue of stents, Minim Invasive Ther Allied Technol 11 (4) (2002) 165–171.
[13] J. Mackerle, Finite element modelling and simulations in cardiovascular mechanics and cardiology: a bibliography 1993–2004, Comput Methods Biomech Biomed Engin 8 (2) (2005) 59–81.
[14] FDA Center for Devices and Radiological Health, Non Clinical Tests and Recommended Labeling for Intravascular Stents and Associated Delivery Systems – Guidance for Industry and FDA Staff (2005) 1–48.
[15] UGCT, http://www.UGCT.UGent.be.
[16] M. Saab, Applications of high-pressure balloons in the medical device industry, Medical Device & Diagnostic Industry Magazine, NH, USA (1999).
[17] S. Timoshenko, Strength of Materials (Part I: Elementary Theory and Problems), Van Nostrand Reinhold, New York, p. 45, Chap. 2 (1955).
[18] B. P. Murphy, P. Savage, P. E. McHugh, D. F. Quinn, The stress-strain behavior of coronary stent struts is size dependent, Ann Biomed Eng 31 (6) (2003) 686–691.
[19] C. Lally, P. Prendergast, D. Kelly, Stents, Wiley Encyclopaedia of Biomedical Engineering, John Wiley & Sons 2007.
[20] J. Squire, Dynamics of endovascular stent expansion, Ph.D. thesis, Massachusetts Institute of Technology, US (2000).
[21] H. Takebayashi, G. S. Mintz, S. G. Carlier, Y. Kobayashi, K. Fujii, T. Yasuda, R. A. Costa, I. Moussa, G. D. Dangas, R. Mehran, A. J. Lansky, E. Kreps, M. B. Collins, A. Colombo, G. W. Stone, M. B. Leon, J. W. Moses, Nonuniform strut distribution correlates with more neointimal hyperplasia after sirolimus-eluting stent implantation, Circulation 110 (22) (2004) 3430–3434.
[22] C. Dumoulin, B. Cochelin, Mechanical behaviour modelling of balloon-expandable stents, J Biomech 33 (11) (2000) 1461–1470.
[23] F. Auricchio, M. Di Loreto, E. Sacco, Finite element analysis of a stenotic artery revascularization through a stent insertion, Comput Methods Biomech Biomed Engin 4 (2001) 249–263.
[24] F. Etave, G. Finet, M. Boivin, J. C. Boyer, G. Rioufol, G. Thollet, Mechanical properties of coronary stents determined by using finite element analysis, J Biomech 34 (8) (2001) 1065–1075.

[25] F. Migliavacca, L. Petrini, M. Colombo, F. Auricchio, R. Pietrabissa, Mechanical behavior of coronary stents investigated through the finite element method, J Biomech 35 (6) (2002) 803–811.
[26] J. McGarry, B. O'Donnell, P. McHugh, J. G. McGarry, Analysis of the mechanical performance of a cardiovascular stent design based on micromechanical modelling, Comp Mater Sci 31 (2004) 421–438.
[27] L. Gu, S. Santra, R. A. Mericle, A. V. Kumar, Finite element analysis of covered microstents, J Biomech 38 (6) (2005) 1221–1227.
[28] G. A. Holzapfel, M. Stadler, T. C. Gasser, Changes in the mechanical environment of stenotic arteries during interaction with stents: computational assessment of parametric stent designs, J Biomech Eng 127 (1) (2005) 166–180.
[29] C. Lally, F. Dolan, P. J. Prendergast, Cardiovascular stent design and vessel stresses: a finite element analysis, J Biomech 38 (8) (2005) 1574–1581.
[30] F. Migliavacca, L. Petrini, V. Montanari, I. Quagliana, F. Auricchio, G. Dubini, A predictive study of the mechanical behaviour of coronary stents by computer modelling, Med Eng Phys 27 (1) (2005) 13–18.
[31] M. De Beule, R. Van Impe, B. Verhegghe, P. Segers, P. Verdonck, Finite element analysis and stent design: reduction of dogboning, Technol Health Care 14 (4–5) (2006) 233–241.
[32] M. De Beule, P. Mortier, R. Van Impe, B. Verhegghe, P. Segers, P. Verdonck, Plasticity in the mechanical behaviour of cardiovascular stents during stent preparation (crimping) and placement (expansion), Key Eng Mater 340–341 (2007) 847–852.
[33] M. De Beule, P. Mortier, J. Belis, R. Van Impe, B. Verhegghe, P. Verdonck, Plasticity as a lifesaver in the design of cardiovascular stents, Key Eng Mater 340–341 (2007) 841–846.
[34] J. Bedoya, C. A. Meyer, L. H. Timmins, M. R. Moreno, J. E. Moore, Effects of stent design parameters on normal artery wall mechanics, J Biomech Eng 128 (5) (2006) 757–765.
[35] L. H. Timmins, M. R. Moreno, C. A. Meyer, J. C. Criscione, A. Rachev, J. E. J. Moore, Stented artery biomechanics and device design optimization, Med Biol Eng Comput 45 (5) (2007) 505–513.
[36] G. A. Holzapfel, C. A. J. Schulze-Bauer, M. Stadler, Mechanics of angioplasty: wall, balloon and stent, Mechanics in Biology – AMD-Vol. 242 and BED-Vol. 46 (2000) 141–156.
[37] G. A. Holzapfel, M. Stadler, C. A. J. Schulze-Bauer, A layer-specific three-dimensional model for the simulation of balloon angioplasty using magnetic resonance imaging and mechanical testing, Ann Biomed Eng 30 (6) (2002) 753–767.
[38] D. K. Liang, D. Z. Yang, M. Qi, W. Q. Wang, Finite element analysis of the implantation of a balloon-expandable stent in a stenosed artery, Int J Cardiol 104 (3) (2005) 314–318.
[39] P. D. Ballyk, Intramural stress increases exponentially with stent diameter: a stress threshold for neointimal hyperplasia, J Vasc Interv Radiol 17 (7) (2006) 1139–1145.
[40] R. V. Marrey, R. Burgermeister, R. B. Grishaber, R. O. Ritchie, Fatigue and life prediction for cobalt-chromium stents: a fracture mechanics analysis, Biomaterials 27 (9) (2006) 1988–2000.
[41] G. J. Hall, E. P. Kasper, Comparison of element technologies for modeling stent expansion, J Biomech Eng 128 (5) (2006) 751–756.
[42] W.-Q. Wang, D.-K. Liang, D.-Z. Yang, M. Qi, Analysis of the transient expansion behavior and design optimization of coronary stents by finite element method, J Biomech 39 (1) (2006) 21–32.
[43] K. Takashima, T. Kitou, K. Mori, K. Ikeuchi, Simulation and experimental observation of contact conditions between stents and artery models, Med Eng Phys 29 (3) (2007) 326–335.
[44] W. Wu, W.-Q. Wang, D.-Z. Yang, M. Qi, Stent expansion in curved vessel and their interactions: a finite element analysis, J Biomech, doi:10.1016/j.jbiomech.2006.11.009 (2007).
[45] L. B. Tan, D. C. Webb, K. Kormi, S. T. Al-Hassani, A method for investigating the mechanical properties of intracoronary stents using finite element numerical simulation, Int J Cardiol 78 (1) (2001) 51–67.
[46] P. J. Prendergast, C. Lally, S. Daly, A. J. Reid, T. C. Lee, D. Quinn, F. Dolan, Analysis of prolapse in cardiovascular stents: a constitutive equation for vascular tissue and finite-element modelling, J Biomech Eng 125 (5) (2003) 692–699.

[47] L. Petrini, F. Migliavacca, F. Auricchio, G. Dubini, Numerical investigation of the intravascular coronary stent flexibility, J Biomech 37 (4) (2004) 495–501.
[48] K. Mori, T. Saito, Effects of stent structure on stent flexibility measurements, Ann Biomed Eng 33 (6) (2005) 733–742.
[49] P. Savage, B. P. O'Donnell, P. E. McHugh, B. P. Murphy, D. F. Quinn, Coronary stent strut size dependent stress-strain response investigated using micromechanical finite element models, Ann Biomed Eng 32 (2) (2004) 202–211.
[50] F. Whitcher, Simulation of in vivo loading conditions of nitinol vascular stent structures, Comput Struct 64 (5/6) (1997) 1005–1011.
[51] F. Migliavacca, L. Petrini, P. Massarotti, S. Schievano, F. Auricchio, G. Dubini, Stainless and shape memory alloy coronary stents: a computational study on the interaction with the vascular wall, Biomech Model Mechanobiol 2 (4) (2004) 205–217.
[52] L. Petrini, F. Migliavacca, P. Massarotti, S. Schievano, G. Dubini, F. Auricchio, Computational studies of shape memory alloy behavior in biomedical applications, J Biomech Eng 127 (4) (2005) 716–725.
[53] P. Theriault, P. Terriault, V. Brailovski, R. Gallo, Finite element modeling of a progressively expanding shape memory stent, J Biomech 39 (15) (2006) 2837–2844.
[54] W. Wu, M. Qi, X. Liu, D. Yang, W. Wang, Delivery and release of nitinol stent in carotid artery and their interactions: a finite element analysis, J Biomech, doi:10.1016/j.jbiomech.2007.02.024 (2007).
[55] R. Morgan, A. Adam, Use of metallic stents and balloons in the esophagus and gastrointestinal tract, J Vasc Interv Radiol 12 (3) (2001) 283–297.
[56] A. R. Assali, S. Sdringola, A. Moustapha, M. Rihner, A. E. Denktas, M. A. Lefkowitz, M. Campbell, R. W. Smalling, Endovascular repair of traumatic pseudoaneurysm by uncovered self-expandable stenting with or without transstent coiling of the aneurysm cavity, Catheter Cardiovasc Interv 53 (2) (2001) 253–258.
[57] E. M. Walser, B. Robinson, S. A. Raza, O. S. Ozkan, E. Ustuner, J. Zwischenberger, Clinical outcomes with airway stents for proximal versus distal malignant tracheobronchial obstructions, J Vasc Interv Radiol 15 (5) (2004) 471–477.
[58] S. Resnick, V. Rome, R. Vogelzang, Use of a partially deployed wallstent to act as an inferior vena cava filtration device during coil embolization of a high-flow arteriovenous fistula, J Vasc Interv Radiol 17 (2 Pt 1) (2006) 369–372.
[59] M. R. Jedwab, C. O. Clerc, A study of the geometrical and mechanical properties of a self-expanding metallic stent–theory and experiment, J Appl Biomater 4 (1) (1993) 77–85.
[60] R. Wang, K. Ravi-Chandar, Mechanical response of a metallic aortic stent – Part I: Pressure-diameter relationship, J Appl Mech 71 (2004) 697–705.
[61] R. Wang, K. Ravi-Chandar, Mechanical response of a metallic aortic stent – Part II: A beam on elastic foundation model, J Appl Mech 71 (2004) 706–712.
[62] S. Canic, K. Ravi-Chandar, Z. Krajcer, D. Mirkovic, S. Lapin, Mathematical model analysis of Wallstent and AneuRx – dynamic responses of bare-metal endoprosthesis compared with those of stent-graft, Tex Heart I J 32 (4) (2005) 502–506.
[63] M. Brand, M. Ryvkin, S. Einav, L. Slepyan, The cardiocoil stent-artery interaction, J Biomech Eng 127 (2) (2005) 337–344.
[64] J.-P. Nuutinen, C. Clerc, P. Tormala, Theoretical and experimental evaluation of the radial force of self-expanding braided bioabsorbable stents, J Biomater Sci Polym Ed 14 (7) (2003) 677–687.
[65] A. O. Frank, P. W. Walsh, J. E. J. Moore, Computational fluid dynamics and stent design, Artif Organs 26 (7) (2002) 614–621.
[66] Y. He, N. Duraiswamy, A. O. Frank, J. E. J. Moore, Blood flow in stented arteries: a parametric comparison of strut design patterns in three dimensions, J Biomech Eng 127 (4) (2005) 637–647.
[67] J. F. J. LaDisa, L. E. Olson, I. Guler, D. A. Hettrick, J. R. Kersten, D. C. Warltier, P. S. Pagel, Circumferential vascular deformation after stent implantation alters wall shear stress evaluated with time-dependent 3D computational fluid dynamics models, J Appl Physiol 98 (3) (2005) 947–957.

[68] N. Benard, R. Perrault, D. Coisne, Computational approach to estimating the effects of blood properties on changes in intra-stent flow, Ann Biomed Eng 34 (8) (2006) 1259–1271.
[69] B. Balakrishnan, A. R. Tzafriri, P. Seifert, A. Groothuis, C. Rogers, E. R. Edelman, Strut position, blood flow, and drug deposition: implications for single and overlapping drug-eluting stents, Circulation 111 (22) (2005) 2958–2965.
[70] M. Gay, L. Zhang, W. K. Liu, Stent modeling using immersed finite element method, Comput Methods Appl Mech Eng 195 (2006) 4358–4370.
[71] H. Müller, Development of metallic bioabsorbable intravascular implants, ESVB 2005, New Technologies in vascular biomaterials – Fundamentals about stents; eds. Chafke, N. and Durand, B. and Kretz, J. G. (2005).
[72] Stent-Research-Unit, http://www.stent-IBiTech.UGent.be.

CHAPTER 3

Signal Extraction in Multisensor Biomedical Recordings

Vicente Zarzoso, Ronald Phlypo, Olivier Meste, *and* Pierre Comon

Contents

1. Introduction 96
   1.1. Aim and scope of the chapter 96
   1.2. Mathematical notations 97
2. Genesis of Biomedical Signals 98
   2.1. A biomedical source model 98
   2.2. Cardiac signals 101
   2.3. Brain signals 105
3. Multi-Reference Optimal Wiener Filtering 109
   3.1. Non-invasive fetal ECG extraction 109
   3.2. Optimal Wiener filtering 110
   3.3. Adaptive noise cancellation 112
   3.4. Results 113
4. Spatio-Temporal Cancellation 115
   4.1. Atrial activity extraction in atrial fibrillation 115
   4.2. Spatio-temporal cancellation of the QRST complex in AF episodes 117
5. Blind Source Separation (BSS) 123
   5.1. The isolation of interictal epileptic discharges in the EEG 123
   5.2. Modeling and assumptions 125
   5.3. Inherent indeterminacies 127
   5.4. Statistical independence, higher-order statistics and non-Gaussianity 127
   5.5. Independent component analysis 129
   5.6. Algorithms 131
   5.7. Results 133
   5.8. Incorporating prior information into the separation model 136
   5.9. Independent subspaces 138
   5.10. Softening the stationarity constraint 138
   5.11. Revealing more sources than sensor signals 138
6. Summary, Conclusions and Outlook 139
Acknowledgments 140
References 141

Advances in Biomedical Engineering
ISBN 978-0-444-53075-2

Abstract

A wide variety of biomedical problems require the extraction of signals of interest from recordings corrupted by other physiological activity signals, noise and interference. The task is facilitated when multiple sensors on different locations record simultaneously the biomedical phenomenon under examination. This chapter provides an overview of signal processing techniques to extract the signal(s) of interest and cancel the interference by exploiting the spatial diversity available in the multisensor measurements. The focus is on three different approaches, namely, multi-reference optimal Wiener filtering, spatio-temporal cancellation and blind source separation. The working assumptions and main algorithmic implementations of these techniques are discussed, and their usefulness is demonstrated on real recordings issued from various biomedical signal processing problems.

Keywords: array signal processing, blind source separation, electroencephalogram, electrocardiogram, higher-order statistics, independent component analysis, multisensor measurements, optimal Wiener filtering, principal component analysis, second-order statistics, spatial diversity, synchronized averaging

1. INTRODUCTION

1.1. Aim and scope of the chapter

Physiological phenomena are typically associated with measurable variations of physical magnitudes such as electric potential, giving rise to biomedical signals. Nevertheless, the measurement of signals of physiological or clinical interest is often hindered by other biomedical activities, noise and interference. In recordings taken from the mother's skin during pregnancy, the electrical activity from the fetal heartbeat can be masked by the stronger maternal cardiac activity. In electrocardiogram (ECG) recordings from atrial fibrillation sufferers, the electrical activity from the atria appears mixed with that from the ventricles. In electroencephalogram (EEG) recordings from epileptic patients, epileptic discharges and the brain's background activity contribute simultaneously to the signals measured by scalp electrodes, and can be further corrupted by artifacts such as eye blinks or body movements.

These and many other similar scenarios call for signal processing techniques to extract the desired signal – the fetal heartbeat, the atrial activity, the epileptic-related brain source, etc. – from the artifact-corrupted measurements. Traditional techniques such as frequency filtering or synchronized averaging and subtraction can be used to process the output of a single sensor. However, in many applications multiple sensors at different locations are available to record simultaneously the physiological phenomenon under analysis. The spatial diversity of the multisensor recordings can then be exploited to significantly improve the signal extraction and interference cancellation performance. In particular, the proper use of spatial diversity allows the separation of signals with overlapping frequency spectra. The main idea is to combine judiciously the multisensor observations, an operation known as spatial filtering, in order to estimate the signal of interest directly, or indirectly by first estimating and then cancelling the interference. Specific approaches differ in the manner in which spatial filters are designed.

Designs can be based on elaborate signal models trying to explain accurately the physiological phenomena under study, or simpler models sacrificing precision for mathematical tractability.

Three spatial filtering approaches are surveyed in this chapter. The first approach (Section 3) is arguably the most classical and consists of the application of optimal Wiener's filter to the scenario of multiple reference signals [1]. The technique assumes the availability of some observations containing information correlated with the artifact, but uncorrelated with the desired signal. The interference present in the corrupted sensor output is estimated by filtering these artifact-correlated recordings. The desired signal is finally obtained by subtracting the estimated interference. This technique exploits the often plausible assumption of uncorrelation between the desired signal and the artifacts and relies on the second-order statistics of the data.

The second approach, referred to as spatio-temporal cancellation (Section 4), can be considered as an extension of the conventional synchronized averaging and subtraction technique to the multisensor case [2]. This approach constructs a spatial template of the interfering waveforms by first detecting their occurrence in the recording of each lead. By relying on the repetitive character of the interference, synchronized averaging helps remove other activity, such as the desired signal, uncorrelated with the interfering waveform. Before subtraction, the estimated template is optimally scaled in amplitude and aligned in time to every occurrence in the recordings.

Blind Source Separation (BSS), a rather general array processing approach not limited to biomedical applications, is addressed at the end of this chapter (Section 5). In this recent technique [3–6], the observed signals are considered as unknown linear mixtures of some unknown sources representing the underlying biological activity. The source waveforms can be estimated at the separator output by spatial filters recovering a known property of the sources such as mutual independence. Measuring these properties typically involves statistical information at orders higher than two (higher-order statistics). By isolating the estimated sources of interest and their corresponding mixing coefficients, the contributions of the desired biomedical activity to the recordings can be easily reconstructed free from noise and interference. BSS can be seen as a generalization of Wiener's noise cancelling approach without pure references.

For didactic purposes, each approach is first motivated by one of the biomedical problems briefly presented at the beginning of this chapter. The mathematical details of the signal processing techniques are then developed and illustrated by application results on real recordings. A summary and some concluding remarks on the perspectives and further applications of the presented techniques (Section 6) bring the chapter to an end.

First, Section 2 addresses the modeling of the biomedical signals considered in this chapter. The signal processing techniques developed in Sections 3–5 are based on these models.

1.2. Mathematical notations

Throughout the rest of the chapter, scalars, vectors, and matrices are represented by lightface (a, A), boldface lowercase ($\mathbf{a}$) and boldface uppercase ($\mathbf{A}$) letters, respectively. $\mathbf{I}_n$ is the $n \times n$ identity matrix. Notation $\|\mathbf{a}\|$ denotes the Euclidean norm of vector **a**. Superscript $(\cdot)^{T}$ is the transpose operator; hence, $\mathbf{a}^{T}\mathbf{b}$ is the Euclidean scalar product between vectors **a** and **b**, whereas $\mathbf{a}^{T}\mathbf{a} = \|\mathbf{a}\|^{2}$. Symbol $E\{\cdot\}$ stands for the mathematical expectation.

2. Genesis of Biomedical Signals

The modeling of the generation of bioelectrical signals and their propagation across the human body plays an important role in biomedical signal processing. The assumed model inspires, and to a large extent determines, the signal processing methods that may be proposed to solve a given biomedical problem. The application of the proposed methods on biomedical signals will generate results whose analysis and validation will provide enhanced clinical information and physiological understanding about the biomedical problem under study. In turn, this improved knowledge will help refine the models and methods, and so on, thus closing the methodology cycle illustrated in Figure 1. This section summarizes the signal models on which the extraction techniques presented later in this chapter implicitly rely.

2.1. A biomedical source model

At the cellular level, communications involving cerebral or muscular neurons occur through the release and uptake of ions called neurotransmitters. This form of communication, taking place at the extracellular space where two or more neurons connect (the so-called synaptic clefts), introduces an oriented current and an associated electromagnetic field. The model used hereafter to represent this bioelectrical activity is a dipole and its corresponding potential field. The sum of dipoles can be considered as another equivalent dipole, so that a totally analogous model holds, but at different scales, from the cellular to the organ level. This section addresses the dipole model and shows its importance in bioelectrical signal modeling. The concepts of stationarity and instantaneous linear propagation are also briefly discussed.

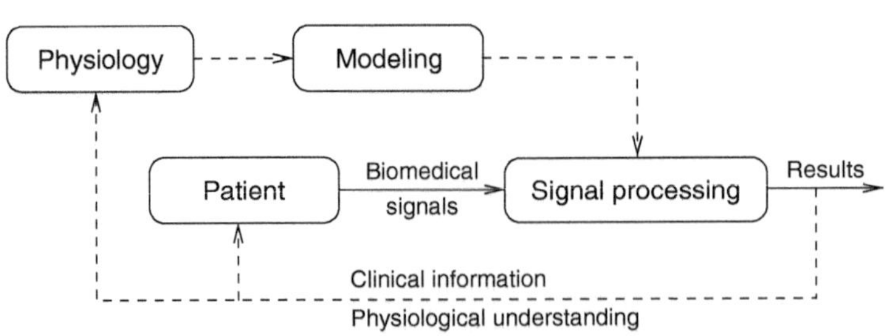

Figure 1 Biomedical signal processing methodology.

2.1.1. The dipole model in a homogeneous infinite medium

The relevance of the dipole model lies in that it defines realistic linear generative models for the signals issued from bioelectrical phenomena. Under such models, these signals can be processed by simple linear operations, such as the linear spatial filters discussed in this chapter. The dipole model is often very accurate in practice, since the potential created by any electrical charge distribution accepts the so-called multipole expansion where the dipole term is dominant when the total charge is null.

As is well known in electromagnetism, a dipole consists of two opposite point charges q and $-q$ situated at a small distance d of one another (see Figure 2). The corresponding dipole field can then be expressed as the superposition of the field created by both point charges. Let us assume that the dipole center is located at position $\mathbf{r}_0$, and let $\mathbf{d}$ be the vector oriented from the negative to the positive charge, with $\|\mathbf{d}\| = d$. Vector $\mathbf{p}(\mathbf{r}_0) = q(\mathbf{r}_0)\mathbf{d}$ represents the corresponding dipole moment. In a homogeneous medium, the electric potential (in Volts) observed at a point $\mathbf{r}_1$ sufficiently distant from the dipole can be written as

$$\Phi(\mathbf{r}_1) = \frac{k}{\|\mathbf{r}\|^3}\mathbf{r}^{\mathrm{T}}\mathbf{p}(\mathbf{r}_0) \qquad (1)$$

where $\mathbf{r} = (\mathbf{r}_1 - \mathbf{r}_0)$ is the observation vector relative to the dipole's position. In the above equation, $k = (4\pi\varepsilon_0)^{-1}$ and ε_0 stands for the permittivity constant of the medium. This expression relates $\Phi(\mathbf{r}_1)$ and $\mathbf{p}(\mathbf{r}_0)$ through a simple function of the dipole moment's orientation relative to the observed position $\mathbf{r}$, up to a multiplicative constant. This relationship is thus linear in the dipole moment $\mathbf{p}(\mathbf{r}_0)$.

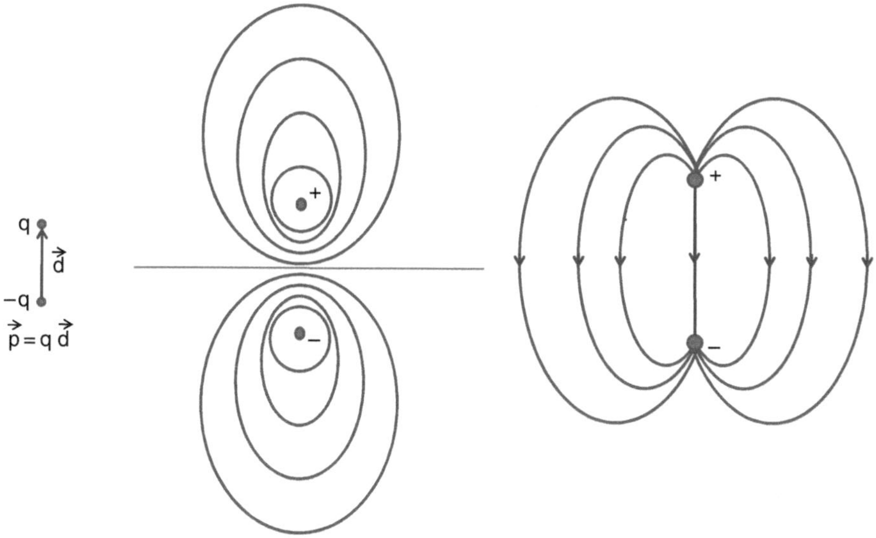

Figure 2 A pictorial representation of a dipole moment **p** (left), its equipotential lines (middle), and its current flow (right).

Equation (1) can be rewritten to obtain $\Phi(\mathbf{r}_1)$ as a function of the electrical charge $q(\mathbf{r}_0)$ under the constraint that $\mathbf{d}$ and $\mathbf{r}$ are fixed:

$$\Phi(\mathbf{r}_1) = c(||\mathbf{r}||, k)\mathbf{r}^{\mathrm{T}}\mathbf{p}(\mathbf{r}_0) = c(||\mathbf{r}||, k)(\mathbf{r}^{\mathrm{T}}\mathbf{d})q(\mathbf{r}_0) = h(\mathbf{r}, \mathbf{d}, k)q(\mathbf{r}_0). \qquad (2)$$

The potential at point $\mathbf{r}_1$ is then expressed as a linear function of the electrical charge $q(\mathbf{r}_0)$ if the dipole's orientation $\mathbf{d}$ and its position $\mathbf{r}$ relative to the measurement point are held fixed. These stationary conditions thus result in an instantaneous linear relationship between the charge distribution and the generated potential. Likewise, under these conditions, the electric potential generated by a dipole distribution $\mathbf{p}(\mathbf{r}_0^i) = q(\mathbf{r}_0^i)\mathbf{d}_i$, $1 \leq i \leq M$, is given by $\Phi(\mathbf{r}_1) = \Sigma_{i=1}^{M}\ h(\mathbf{r}_i,\ \mathbf{d}_i,\ k)\ q(\mathbf{r}_0^i)$, with $\mathbf{r}_i = (\mathbf{r}_1 - \mathbf{r}_0^i)$.

2.1.2. The dipole in a heterogeneous medium

Most bioelectrical media are not homogeneous, and thus the simple relation between Φ and q does not hold for all $\mathbf{r}_1$ and $\mathbf{r}_0$. However, it is possible to decompose the total medium into piecewise homogeneous volumes (PHVs). Each PHV can be considered as a closed volume of a bioelectrical medium with quasi-homogeneous electrical properties (such as a single ε_0 for the whole volume). For the PHV which encloses the dipole, the potentials can be calculated at its volume surface. This charged surface can then be seen as an ensemble of dipoles which, in turn, can be enclosed by another volume. The simple analytical relation between the potential and the charge distribution of Eq. (1) no longer exists but for a fixed dipole and a fixed observation vector, Eq. (2) still holds, even if we cross the boundaries of homogeneity. The proof of this linear relation, which is omitted here due to lack of space, can be found in Gulrajani [7]. Since the analytical model that would result from this nested structure of PHVs does not have a straightforward solution, an appropriate h for each $\mathbf{r}$, $\mathbf{d}$, and k can be obtained using numerical models based on sampled 3D spaces [8].

2.1.3. Not everything is a dipole: more realistic models

However, not all bioelectrical activity can be modeled accurately by considering a single dipole and its potential field. In some cases, a dipole layer or a multi-dipole configuration would be more suitable approximations. More elaborate models proposed in the literature range from coupled dipoles [9] to distributed source models [10], including dipole lines and dipole surfaces/volumes.

On the other hand, the currents generated in the synaptic clefts do not directly reflect as such at the cutaneous electrodes used for measuring biopotentials. A sufficiently high magnitude with respect to the noise (e.g., the quantization noise introduced by any digital system, the background noise, etc.) demands the simultaneous firing of groups of neurons. If the group of neurons are assumed to fire synchronously at fixed positions and with fixed orientations, the activity of the ensemble can be modeled by a single dipole in the far-field approximation (see Section 2.1.1).

Hence, whether represented as a single dipole or otherwise, the following properties of electrical signal propagation hold in most cases:

1. Far-field observations ($\|\mathbf{r}\| \gg \|\mathbf{d}\|$) simplify the field equations to simple projections of the dipole moment onto the observation vector, i.e., the linear model as described above.
2. Most bioelectrical phenomena of interest take place in the spectral band below 100 Hz. Capacitive effects and thus time delays and phase shifts are negligible in this frequency range. As a result, the propagation of biomedical signals across the body can be assumed to be instantaneous.

In conclusion, the bioelectrical activity measured at a fixed location in space can be accurately described by the instantaneous linear model.

2.2. Cardiac signals

2.2.1. The heart

The heart is an organ composed of four cavities, two atria and two ventricles (Figure 3), surrounded by muscle tissue (the myocardium) whose coordinated activity ensures the circulation of blood to all parts of the body. The heart consists of two separate blood pumps maintaining a double circulatory system. In the pulmonary circulation, the right ventricle pumps blood depleted from oxygen and other nutrients to the lungs through the pulmonary arteries for oxygenation. The oxygenated blood is then carried back to the left atrium through the pulmonary veins. In the systemic circulation, the left ventricle pumps the oxygenated blood through the aorta to the rest of the body. The deoxygenated blood returns to the right atrium through the venae cavae. Hence, the right-hand side of the heart acts as

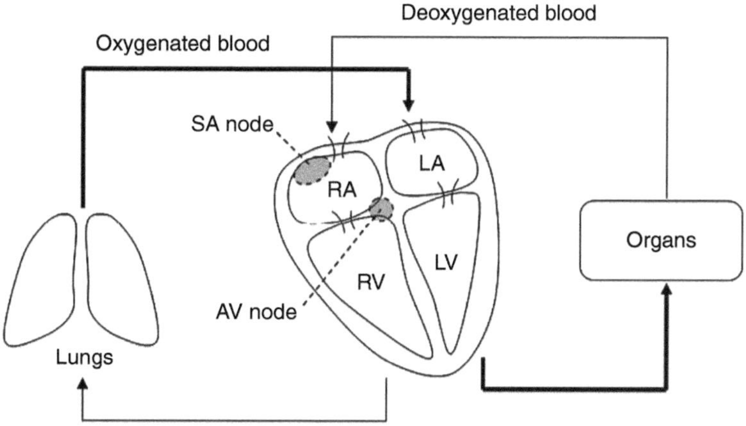

Figure 3 Schematic depiction of the human heart. The two smaller cavities at the top (RA, LA) represent the right and left atria, respectively; the larger cavities at the bottom (RV, LV) are the ventricles.

a pump for deoxygenated blood, carrying it from the body to the lungs, whereas the left-hand side pumps oxygenated blood from the lungs to the body.

The mechanical function of the heart is triggered by the electrical activation of the myocardium cells. The cutaneous potentials generated by the electrical activity of the myocardium were first recorded at the end of the nineteenth century. Their measurement has nowadays become a well-established common clinical practice for the diagnosis of many heart conditions, as these are closely linked to the shape and/or timing of the observed potential waveforms.

2.2.2. Normal heart activity: sinus rhythm

A healthy heart is characterized by a regular, well-organized electro-mechanical activity known as normal sinus rhythm (NSR), which results in an efficient cardiac muscle contraction (heartbeat), assuring an optimal blood circulation across the body [11]. The heartbeat cycle is initiated by specialized cells forming the so-called sino–atrial (SA) node, located at the right atrium (Figure 3). The SA node is modulated by the autonomous nervous system and fires at a regular rate of about 60–80 times per minute. The electrical impulse propagates in an orderly manner across the atrial muscle tissue, causing the depolarization and subsequent contraction of the atria. To allow completion of blood transfer from atria to ventricles, the electrical depolarization front gets delayed at the atrio–ventricular (AV) node, situated at the top of the ventricles. Propagation then resumes through the specialized fibres of the His bundle, the bundle branches, and Purkinje's system, which rapidly depolarize the ventricles and make them strongly contract from the bottom upwards, thus allowing an efficient pumping of blood to the lungs and the rest of the body.

On the skin surface, this coordinated electro-mechanical activity is reflected by potential variations or signals with characteristic shape and timing, known as waves and complexes (Figure 4). Ventricular depolarization causes the impulsive QRS complex, whereas ventricular repolarization and the subsequent relaxation of the ventricular muscles are associated with the T wave. Depolarization of the atria generates the P wave; however, their repolarization is thought to take place during ventricular contraction, so that the corresponding wave is masked by the strong QRS complex.

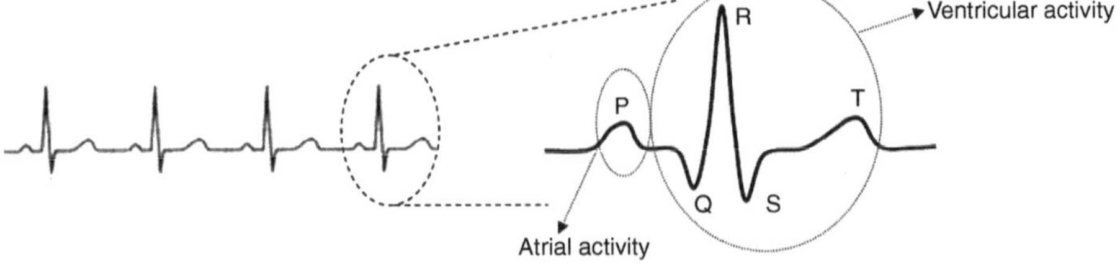

Figure 4 Typical waveform of a surface potential recording of cardiac activity. The detail shows the PQRST complex.

Table 1 Heart activity during normal sinus rhythm

Electrophysiology	Mechanical function	Surface ECG
SA node fires causing atrial depolarization	Atrial contraction	P wave
AV node slows down propagation	Completion of blood transfer from atria	Isoelectric line
Propagation through His bundle, bundle branches, Purkinje system	Ventricular contraction	QRS complex
Ventricular repolarization	Ventricular relaxation	T wave

Table 1 summarizes the different stages of the NSR cycle, including the electrophysiological activity, the associated mechanical functions, and the related electrical waves observed on the skin-surface potentials.

2.2.3. The standard electrocardiogram (ECG)

In measuring the electrical activity of the heart, conventional medical practice considers the output of a number of electrodes located on predefined skin positions, as depicted in Figure 5. The simultaneous recording of electric potentials between such electrodes is known as the standard 12-lead ECG. Bipolar limb leads I, II, and III measure the voltage difference between electrodes located in the right arm, the left arm, and the left foot. The augmented leads, aVR, aVL, and aVF, measure the voltage difference between the same electrodes and a reference point known as Wilson's central terminal (WCT). Finally, the precordial leads V1–V6 output the voltage difference between the electrodes situated across the chest and WCT.

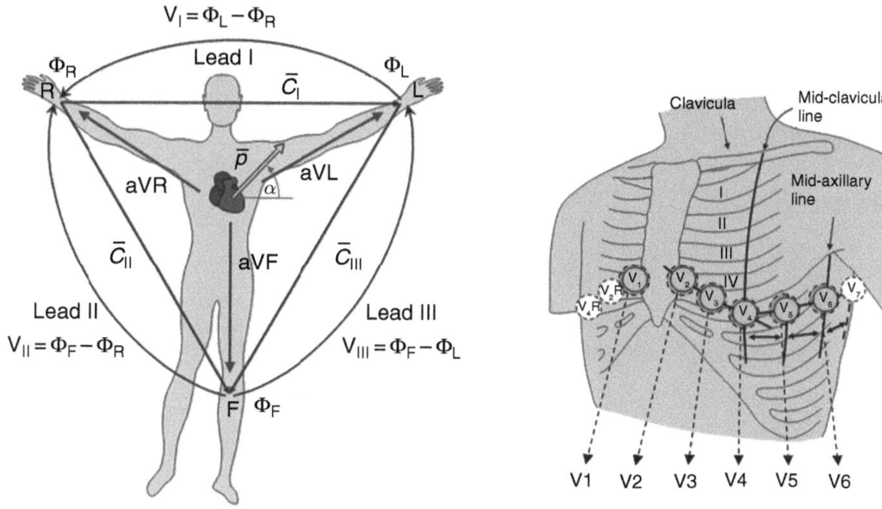

Figure 5 Electrode positions in the standard 12-lead ECG [12].

2.2.4. Electrical model of the heart

Modeling the heart electrical activity is a complex task. Nevertheless, a first-order approximation along the lines of Section 2.1 turns out to provide satisfactory results in many cases, as will be seen throughout the rest of the chapter. This approximation is obtained by assuming that the cardiac activity can be represented by a global current dipole with fixed location but varying amplitude and orientation [12–14]. Mathematically, this is described by a time-varying dipole moment vector $\mathbf{p}(t)$ with coordinates $p_x(t)$, $p_y(t)$, and $p_z(t)$ in a 3D Euclidean space, or $\mathbf{p}(t) = [p_x(t), p_y(t), p_y(t)]^T$.

This approximation further assumes that the body behaves as a homogeneous conductor. The body tissues can be considered to be purely resistive, and any capacitive or inductive effects are neglected. Propagation delays are also negligible, given the bandwidth of the cardiac signals (typically under 100 Hz) and the propagation speed of electrical signals across the body. As a result, the potential $\mathbf{\Phi_r}$ (Volts) measured at a fixed point $\mathbf{r} = [r_x, r_y, r_z]^T$ relative to the dipole's center can be approximated, up to a constant factor, by the scalar product [cf. Eq. (1)]:

$$\Phi_\mathbf{r}(t) \propto \mathbf{r}^T\mathbf{p}(t) = r_x p_x(t) + r_y p_y(t) + r_z p_z(t).$$

Vector $\mathbf{r}$ is called lead vector, and the equivalent dipole $\mathbf{p}$ is referred to as the electric heart vector [12]. The electric potential at any point in space can thus be expressed as a linear combination of the electrical heart vector components. The coefficients of the linear combinations are given by the lead-vector components (Figure 6). The above scalar product can also be regarded as a projection of the cardiac vector onto the lead vector. Consequently, the ECG lead outputs correspond to specific projections of the electric heart vector along directions defined by the electrode locations. Theoretically, the resulting electrical field is completely characterized by three orthogonal leads, X, Y, and Z, forming the so-called Frank lead system. The Frank system constitutes the basic tool for the measurement of the electric heart vector, or vectorcardiography [12].

According to Section 2.1.2, a similar model holds under the more realistic assumption of the body as a heterogeneous medium. Note that the electric heart vector models the behaviour of the heart as a whole, since the equivalent global dipole reflects the sum of individual dipoles with different charges and orientations. More elaborate models of the ECG are reviewed in van Oosterom [15].

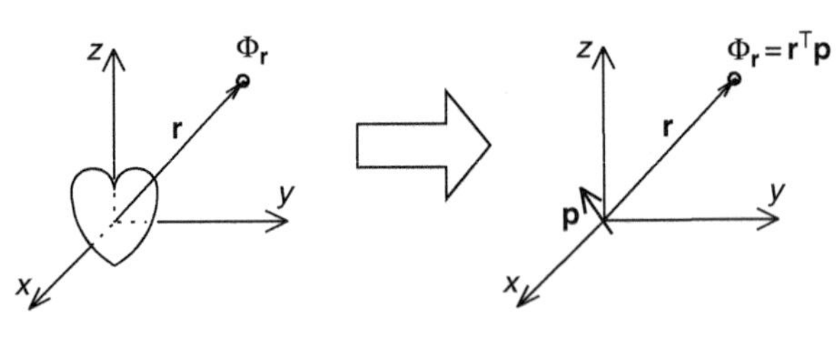

Figure 6 The cardiac current dipole or lead-vector model.

2.3. Brain signals

2.3.1. The brain

The brain is the central processing unit of the human body. Its high processing power makes it capable of performing real-time data monitoring and processing as well as body-function control tasks. This power lies in the high parallelism provided by the huge amount of specialized interconnections at the cellular level. The functions of the brain can roughly be classified in three main parts:

1. Processing of the sensory information received through afferent fibres.
2. Transmitting information to the body via efferent fibres, both consciously and unconsciously (voluntary, respectively autonomous body functions).
3. Stocking and processing information: memory, cognition, emotion, and attention.

Most information to be processed from the sensory organs enters the brain through the spinal cord which is reached via the spinal (afferent) nerves. Subsequently, the sensory signals are passed to an intermediate neuronal processing layer from where they are distributed to specialized processing areas at the cortical layer. Zones of the major cortical lobes are thus directly related to specific sensory information as shown in Figure 7 and Table 2. The steering of body parts, either consciously or

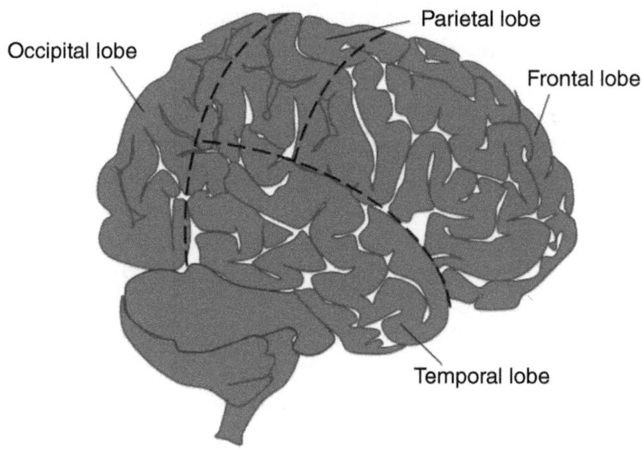

Figure 7 The cortical lobes of the brain.

Table 2 The four main brain lobes and their principal functions

Lobe	Main function
Frontal	Conscious thought
Parietal	Processing and integration of sensory information; visuospatial processing
Occipital	Sight
Temporal	Face and scene processing as well as olfactory and audio signals

unconsciously, is mainly governed by the brain. Exceptions are the well-known reflexes which only use the spinal cord and nerves and the autonomous body muscles. The remainder of the brain functions, concerning cognition, memory, emotion, and attention, are probably the least understood. Thanks to its high degree of parallelism and autodidactic capabilities, the human brain is able to perform highly complex tasks far beyond those of currently available computer power.

2.3.2. The standard electroencephalogram (EEG)

The EEG denotes the electrical signals recorded from electrodes placed on a patient's scalp. The recorded potential differences are generated by the cerebral electrical activity and are often contaminated by external interference. The EEG, whose first registration dates back to 1928 by Hans Berger, is still used in daily clinical practice and experimental psychology. Whereas alternative modern techniques such as positron emission tomography (PET), single photon emission tomography (SPECT), and magnetic resonance imaging (MRI) offer a much higher spatial resolution, the EEG – together with its magnetic counterpart, the magnetoencephalogram (MEG) – is mainly praised for its very high temporal resolution, which can easily reach below 1 ms, and its cost-effectiveness.

Originally employed in clinical environments to detect tumorous tissue in the brain, nowadays its use is essentially limited to the recognition of brain states associated with epilepsy, the different sleep stages, and cerebral mortality. In experimental environments, the EEG is used to record cerebral responses to external stimuli, or event-related potentials (ERPs). However, the electrodes' sensitivity to weak electrical fields is limited to the volume surrounding the electrode, which hampers the registration of weaker signals from deeper brain areas. The volume analyzed by the EEG thus mainly reduces to the cortical layer, the cerebral centre of higher level processing. To maximize the amount of cortical activity captured, the American Clinical Neurophysiology Society recommend the so-called 10–20 International System of scalp electrode placement [16] (see Figure 8).

Traditionally, the clinically relevant activity is mainly obtained through a decomposition in frequency bands (e.g., by means of a Fourier Transform). The most important bands are briefly summarized in Table 3. In an experimental setting (e.g., involving ERPs), the information of interest is contained in the temporal delay or latency, Δt, between the presented stimulus and the cerebral response. Examples are summarized at the bottom of Table 3. Next to timing and frequency information, the topography or spatial distribution of the signals plays an important role. Despite the poor spatial resolution offered by the 10–20 International System, this information helps localize roughly the processing areas responsible for the cerebral activity measured by the scalp electrodes.

2.3.3. Electrical model of the brain

The brain is a complex structure composed of a highly interconnected network of specialized cells: information processing and transmission units (neurons) and cells

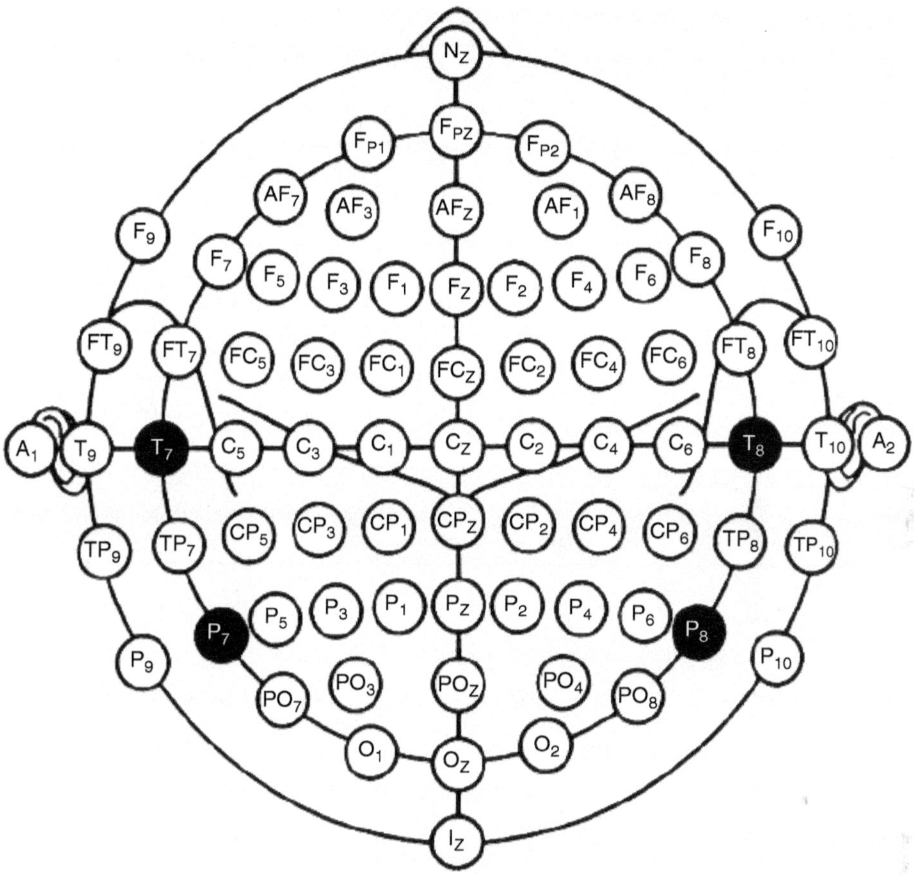

Figure 8 Scalp electrode positioning in the 10–20 International System [16].

providing support, nutrition, and protection (glial cells). Communication between neurons occurs through release and uptake of ions, the so-called neurotransmitters. Since the brain is constructed of about 10^{10}–10^{11} neuronal cells and each cell is believed to receive information from 10^3 to 10^5 directly connected neurons, it is easy to understand why its information processing capacity outperforms today's most powerful computers. Modeling of the brain is thus a highly complex task, and research into how the brain works is still mainly limited to the analysis of recorded signals (see the feedback cycle in Figure 1), although some efforts have been made to model the brain as a huge neural network [18,19]. Analyzing the activity of a single neuron and its contribution to the whole may prove useful in understanding the global operation of the brain. However, the single-neuron activity will not be addressed in the sequel because of its limited clinical significance. Physicians and psychologists are rather interested in areas of the brain that function as a group. Data obtained by EEG or (functional) MRI recordings provide information about this activity taking place in specific areas.

Table 3 Typical waveforms found in EEG and ERP studies

Name	Characterization	Brain state
δ-waves	f=0–4 Hz	Stage III and IV sleep, young persons Underlying lesions
θ-waves	f=4–8 Hz	Upon waking and just before falling asleep Drowsiness
α-waves	f=8–15 Hz	Relaxed, alert state of consciousness Originates from occipital and parietal regions
μ-waves	in α-band	Over the motor areas of the central cortex Attenuated during (imaginary) movement Sensorimotor rhythm
β-waves	f=12–30 Hz	Varying frequencies: busy or anxious thinking, active concentration Dominant frequencies: drug effects, pathology
Low γ-waves	f=30–60 Hz	Higher mental tasks
High γ-waves	f=60–200 Hz	E.g., over Broca's area: imaginary singing [17]
Spikes + waves	Transients	Often correlated with epileptic activity
P300	$\Delta t \approx 300$ ms	Categorization and simple decision making
N400	$\Delta t \approx 300$–500 ms	Semantics

2.3.4. Artifacts in the EEG

Due to the nature of the recordings, EEG signals are very prone to electric and magnetic interference from outside the brain. Depending on their origin, these artifacts can be roughly classified as physiological or environmental noises. Table 4 gives an overview of the most common artifacts found in the EEG.

Artifact removal is a real challenge since the spectral overlap between the cerebral signals (Table 3) and the artifacts is significant and, as a result, ordinary frequency filters cannot resolve the separation task in most cases. More elaborate signal-processing techniques such as those described in this chapter are thus necessary.

Table 4 Common artifacts in the EEG and their origin

Origin	Characteristics
Physiological artifacts	
Ocular movements: blinking, turning, saccade	Slow evolving wave ($f<2$ Hz) Fixed topography (electrical dipole of the eye)
Muscular activity	Higher frequency part ($f>15$ Hz) Not deterministic Not a single stationary source
Cardiac activity	Prominent QRS complex
Environmental artifacts	
Power line	50 or 60 Hz component, quasi-sinusoidal
Impedance mismatch, bad electrode connection	Discontinuity, floating potential Low frequencies
Electromagnetic incompatibility: e.g., mobile phone, current switching	Strongly dependent on nuisance source

3. Multi-Reference Optimal Wiener Filtering

3.1. Non-invasive fetal ECG extraction

During pregnancy, it is important to monitor the fetus' heart in order to verify its well being, predict possible conditions and, if applicable, propose a prompt treatment. Currently, the monitoring of the fetal heart is mainly based on the measurement of the fetal heart-rate using ultrasounds. Clearly, more detailed information can be gathered, as for an adult, by measuring the surface potentials generated by the fetal myocardium. Invasive techniques such as scalp electrodes present potential dangers to both mother and fetus (e.g., they can provoke a premature birth), and are thus restricted to the period just before delivery (*antepartum*) or to delivery itself. The risk of complications notably decreases with the use of non-invasive techniques, which can also be employed over several gestational stages.

A natural non-invasive approach is based on maternal cutaneous electrodes. Figure 9 shows the signals simultaneously recorded by eight leads on different parts of a pregnant woman's skin. The first five leads have been measured from the

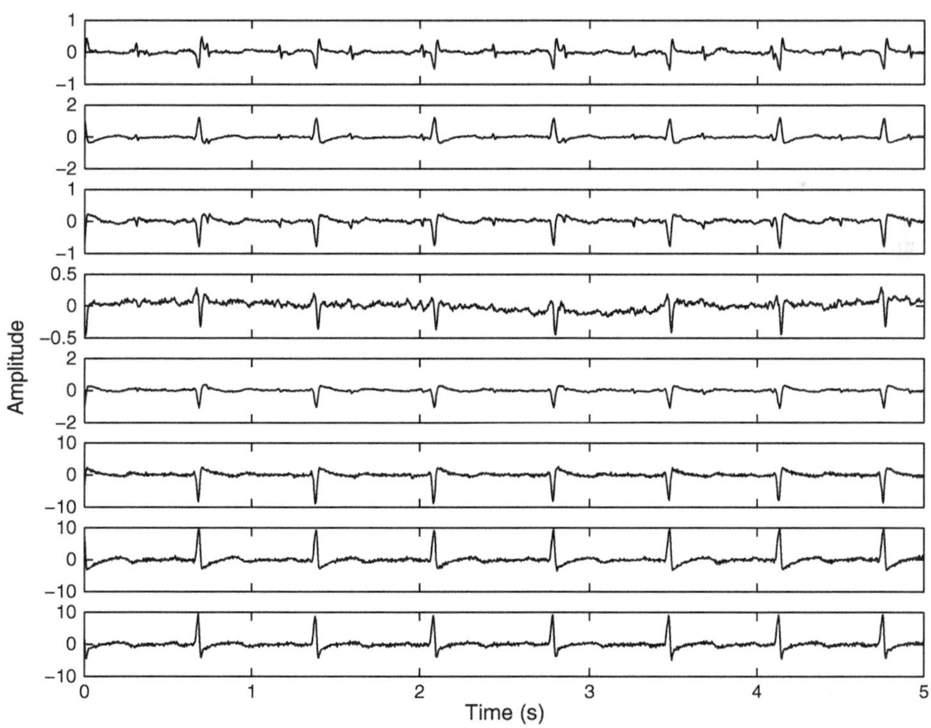

Figure 9 An eight-lead surface potential recording taken from a pregnant woman. The first five leads have been recorded from the mother's abdomen, while the last three have been measured from her chest. Signals have been sampled at 500 Hz. Only the relative amplitudes are important in the vertical axes.

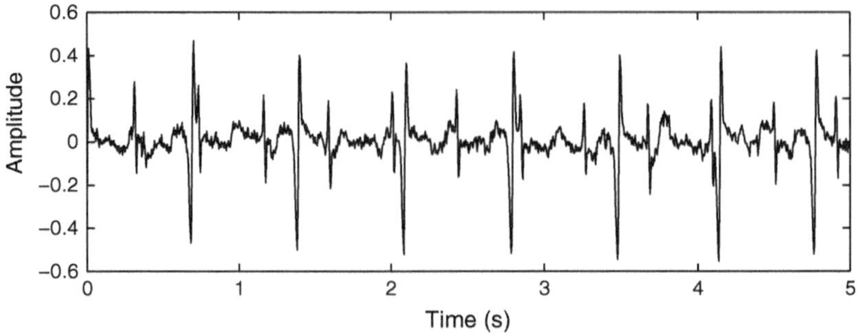

Figure 10 The first lead of the recording of Figure 9. Since mother's and fetus' heart activities are not synchronized, the former may sometimes mask the latter, as occurs around time instants 1, 3 and 4 s.

abdominal region. The fetal cardiac activity (low-amplitude peaks) appears mixed to that of the mother, which has much stronger amplitude. Since mother and fetus present different heart-rates, the fetal heartbeat is sometimes masked by the maternal one. This overlap can be clearly perceived around time instants 1, 3, and 4 s in the enlarged version of the first lead shown in Figure 10. In addition, other sources of noise and interference such as maternal respiration (causing quite probably the baseline wander in the fourth lead), muscle activity, mains coupling, etc., also hinder the observation of the fetal cardiac signal. Signal processing techniques are thus necessary to extract the fetal ECG (FECG) from the maternal cutaneous recordings.

The extraction technique presented in this section relies on the measurement of another set of signals – so-called reference inputs – correlated to the noise and interference, but uncorrelated with the desired signal. The last three recordings in Figure 9 are an example of such a set. They have been obtained from the mother's thoracic region, far from the fetal heart, so that they mainly contain the maternal heartbeat components without contribution from the fetal heartbeat. The idea consists of trying to estimate the interfering maternal components present in the abdominal leads by filtering the chest leads. The estimated interference is then cancelled (subtracted) from the abdominal leads, leaving only the desired fetal heartbeat signal.

3.2. Optimal Wiener filtering

Let us denote by $x(t)$ the signal observed at an abdominal electrode output. This signal, known as primary input to the canceller, can be considered as the mixture of the desired fetal heartbeat signal $d(t)$ and the interference (mainly due to the maternal heartbeat) represented by $b(t)$:

$$x(t) = d(t) + b(t).$$

The aim is to suppress the interference present in the primary signal in order to obtain the desired signal. The desired signal and the interference are assumed to be uncorrelated, which can be expressed mathematically as $\mathrm{E}\{d(t)b(t)\} = 0$. It is further assumed that multiple reference signals $\{z_k(t)\}_{k=1}^{R}$ are available. In the example above, these are the thoracic leads, so $R = 3$. The reference signals are also assumed to be uncorrelated with the desired signal $d(t)$ but correlated with the interference $b(t)$.

According to the lead-vector model recalled in Section 2.2.4, the surface potential at any point in space can be expressed, in a first-order approximation, as the linear combination of the cardiac current dipole components. Consequently, the maternal heart contributions present in the recorded leads can be considered as different linear combinations of the maternal dipole components. As a result of this approximation, the maternal contribution to each of the primary leads may be estimated as a particular linear combination of the reference leads. The coefficients of the linear combination can be stacked in a column vector $\mathbf{w} = [w_1\ w_2 \ldots w_R]^{\mathrm{T}}$. The application of this vector, via the scalar product, on the reference leads constitutes a spatial filtering operation whose output is given by

$$y(t) = \sum_{k=1}^{R} w_k z_k(t) = \mathbf{w}^{\mathrm{T}}\mathbf{z}(t)$$

where $\mathbf{z}(t) = [z_1(t)\ z_2(t) \ldots z_R(t)]^{\mathrm{T}}$. The estimated maternal contribution is then subtracted from the primary lead. The remaining signal at the canceller output can be expressed as

$$s(t) = x(t) - y(t) = x(t) - \mathbf{w}^{\mathrm{T}}\mathbf{z}(t). \tag{3}$$

Figure 11 depicts the multi-reference Wiener canceller set-up for $R = 3$ reference signals.

The optimal Wiener canceller finds the coefficient vector $\mathbf{w}$ by minimizing the output power

$$\psi(s) \triangleq \mathrm{E}\{s^2(t)\}. \tag{4}$$

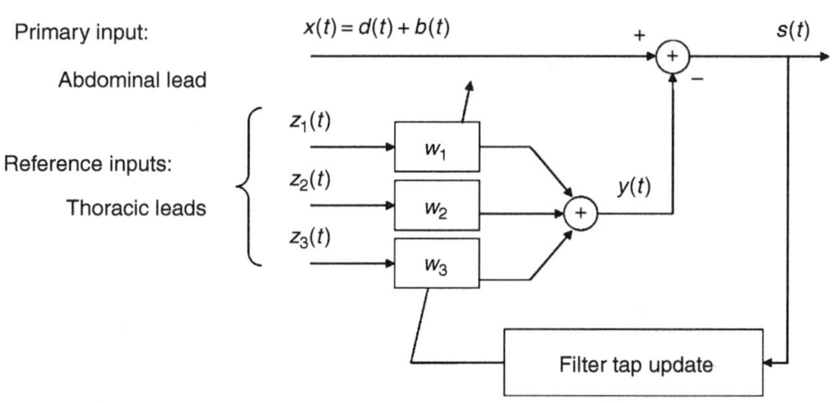

Figure 11 Schematic diagram of the multi-reference noise canceller.

To understand why this criterion provides the desired results under the working assumptions, one may expand the canceller output power as

$$E\{s^2(t)\} = E\{[d(t)+b(t) - y(t)]^2\} = E\{[b(t) - y(t)]^2\} + E\{d^2(t)\}.$$

The uncorrelation of $d(t)$ with both $b(t)$ and $\mathbf{z}(t)$ [and hence with $y(t)$] has been invoked to arrive at the last equality. Consequently, minimizing $\psi(s)$ is equivalent to making the spatial filter output $y(t)$ as close as possible, in the mean square error sense, to the interference $b(t)$. In turn, since $[s(t)-d(t)] = [b(t)-y(t)]$, making $y(t)$ close to $b(t)$ is tantamount to minimizing the difference between the canceller output $s(t)$ and the desired signal $d(t)$.

The minimization of $\psi(s)$ is carried out with respect to the spatial filter coefficients stored in vector $\mathbf{w}$. The gradient of $\psi(\mathbf{w})$ is given by

$$\nabla\psi(\mathbf{w}) = -2E\{s(t)\mathbf{z}(t)\} \tag{5}$$

and is null at the optimal solution. Hence, the second-order statistical information contained in the reference signals is "exhausted" by the optimal spatial filter, as it decorrelates them with the canceller output.

Combining Eqs (3) and (5), the so-called optimal Wiener–Hopf solution is readily obtained as

$$\mathbf{w}_{\text{opt}} = \mathbf{R}_{zz}^{-1}\mathbf{r}_{xz} \tag{6}$$

where $\mathbf{R}_{zz} = E\{\mathbf{z}(t)\mathbf{z}(t)^{\text{T}}\}$ is the reference-signal covariance matrix, and $\mathbf{r}_{xz} = E\{x(t)\mathbf{z}(t)\}$ represents the cross-correlation vector between the primary lead and the reference inputs.

In practice, mathematical expectations are replaced by their sample estimates. Assuming that signal blocks of T samples are available, we can approximate:

$$\mathbf{R}_{zz} \approx \frac{1}{T}\sum_{n=0}^{T-1} \mathbf{z}(nT_s)\mathbf{z}(nT_s)^{\text{T}} \quad \mathbf{r}_{xz} \approx \frac{1}{T}\sum_{n=0}^{T-1} x(nT_s)\mathbf{z}(nT_s) \tag{7}$$

where T_s denotes the sampling period.

3.3. Adaptive noise cancellation

Obtaining the optimal Wiener–Hopf filter requires the solution of the linear system (6), which can be found off-line by standard linear algebra routines. Off-line algorithms converge after a finite number of iterations executed over the entire available data (the block of samples in the recording). Alternatively, the solution to Eq. (6) can be computed on-line. On-line algorithms involve an explicit recursion over time, either sample by sample (sample-wise) or over consecutive signal blocks (block-wise). As opposed to off-line algorithms, the number of iterations necessary for convergence may not be bounded, particularly in the presence of non-stationarities. The simplest adaptive algorithm consists of descending the surface of criterion (4) by moving the filter vector in small steps along the direction opposite to the gradient. In this gradient-descent implementation, the spatial filter weights are updated as

$$\mathbf{w}(k+1) = \mathbf{w}(k) - \mu \nabla \psi(\mathbf{w}(k)). \tag{8}$$

Once the steady-state is reached, the update term in Eq. (8) causes oscillations of the spatial filter coefficients around the optimal solution $\mathbf{w}_{\text{opt}}$ (6), a phenomenon known as misadjustment. The choice of parameter μ, referred to as adaption coefficient or step size, establishes a well-known trade-off between convergence speed and accuracy: μ should be chosen small enough to prevent divergence and to obtain an accurate final solution (low misadjustment), but large enough to avoid slow convergence.

Alternative implementations differ in how gradient (5) is estimated in update rule (8). Batch, block, or windowed methods estimate the gradient by averaging over the observation window samples, as in Eq. (7). Stochastic methods drop the expectation operation altogether and use a one-sample approximation of the gradient, as in the conventional least mean squares (LMS) algorithm [20]. This yields Widrow's multi-reference adaptive noise cancelling method [1]:

$$\mathbf{w}(k+1) = \mathbf{w}(k) - \mu'[x(kT_s) - \mathbf{w}(k)^{\mathrm{T}}\mathbf{z}(kT_s)]\mathbf{z}(kT_s) \tag{9}$$

with $\mu' = 2\mu$. This rough approximation of the gradient generally leads to slower convergence and increased misadjustment relative to the block-gradient implementation. Common belief states that the one-sample algorithm presents improved tracking capabilities. As a matter of fact, better performance can be expected from the block algorithm if the signals remain stationary over the observation window.

The above development can easily be generalized to the use of spatio-temporal filters, as in Widrow's pioneering contribution [1]. Each reference input is convolved with the impulse response of a Pth-order finite impulse response filter, with $P > 0$, before adding the outputs to form interference estimate $y(t)$. However, the heart-dipole model of Section 2.2.4 suggests that $P = 0$ suffices for the application in hand, thus reducing the required processing of the reference inputs to a purely spatial filtering operation.

3.4. Results

The multi-reference optimal Wiener spatial filtering method reviewed in Section 3.2 is applied in turn to each of the five primary (abdominal) inputs of the recording shown in Figure 9. The last three signals, recorded from the thoracic electrodes, are employed as reference inputs to the canceller ($R = 3$). The resulting output signals are represented by the dark lines of Figure 12; for comparison, the outputs are superimposed to the original primary inputs plotted in a lighter line. The canceller is successful in that it considerably reduces the maternal ECG interference from all abdominal leads. Nevertheless, residual maternal peaks remain in the time waveforms. FECG contributions are clearly visible in the first three outputs, although the second and third signals are rather noisy. The fetal heartbeat peaks can hardly be perceived over the noise components in the output signal from the fourth and fifth abdominal leads.

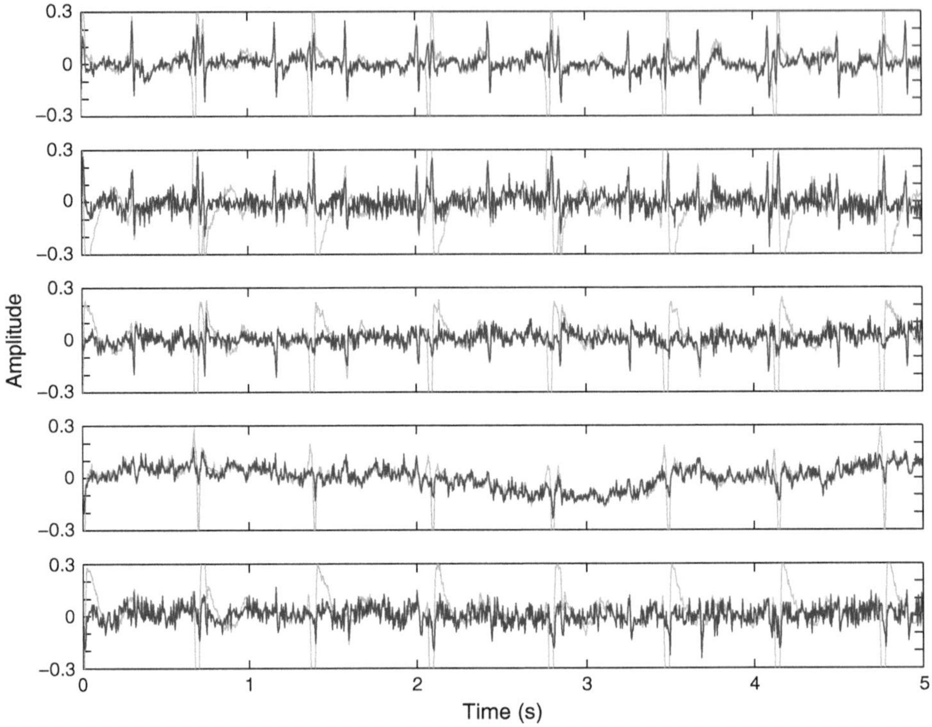

Figure 12 FECG contributions (dark lines) estimated by multi-reference optimal Wiener filtering of the primary leads (light-grey lines) in the recording of Figure 9.

The fourth output shows an important baseline wandering, arguably due to mother's respiration, which could have been eliminated by high-pass filtering the abdominal leads before passing them as primary inputs to the canceller. As the reference leads contain no component correlated with the respiration, this artifact cannot be estimated from the reference signals and cancelled from the primary input. Here lies precisely the major limitation of the Wiener filtering technique: an artifact cannot be cancelled from the primary input unless it contributes to the reference inputs as well. Conversely, the desired signal may be partially suppressed from the primary input if it leaks into the reference inputs [1]. Consequently, an appropriate electrode placement is of paramount importance for this approach to provide satisfactory results. As will be seen later in this chapter, more elaborated approaches such as BSS are able to surmount these shortcomings. Otherwise, the multi-reference Wiener canceller is conceptually simple and computationally affordable.

The performance limits of the Wiener noise canceller are analyzed in Comon and Pham [21]. Further results obtained by this technique are shown and discussed in Refs [1,22] and references therein. The BSS approach of Section 5 is applied to the FECG extraction problem in De Lathauwer et al. [6]. A comparison between multi-reference optimal Wiener filtering and BSS in the context of this biomedical application is carried out in Zarzoso and Nandi [22].

4. Spatio-Temporal Cancellation

4.1. Atrial activity extraction in atrial fibrillation

A cardiac arrhythmia is a disorder in the regular well-coordinated activity of the heart. Atrial fibrillation (AF) is an abnormal rhythm involving the atria, the entry chambers of the heart (see Figure 3). AF is the most prevalent cardiac arrhythmia; its risk increases with age, affecting up to 8% of people over 80 years old. In this frequent condition, the electrical impulses originated by the SA node in NSR (Section 2.2.2) are replaced by the disorganized generation and irregular conduction of impulses in the atria. Impulses passing through the AV node into the ventricles cause a heartbeat. Extra heartbeats that would not appear in NSR are referred to as ectopic beats. The result is an irregular cardiac rhythm, less efficient than in a healthy heart since the atria are not completely filled when atrial contraction occurs. Due to the blocking property of the AV node, the high rate of these impulses does not correspond to the contraction rate of the ventricles.

AF may be continuous (persistent or permanent AF) or alternating with periods of normal heart rhythm (paroxysmal AF). The natural tendency of AF is to become a chronic condition, leading to an increased risk of death. Figure 13 plots two beat occurrences recorded on the V_2-lead from a healthy patient. The waves of interest P, Q, R, S, and T are clearly visible, and so is the silent period of isopotential electric activity. Figure 14 shows leads $V_1 - V_3$ from an AF sufferer. The P wave

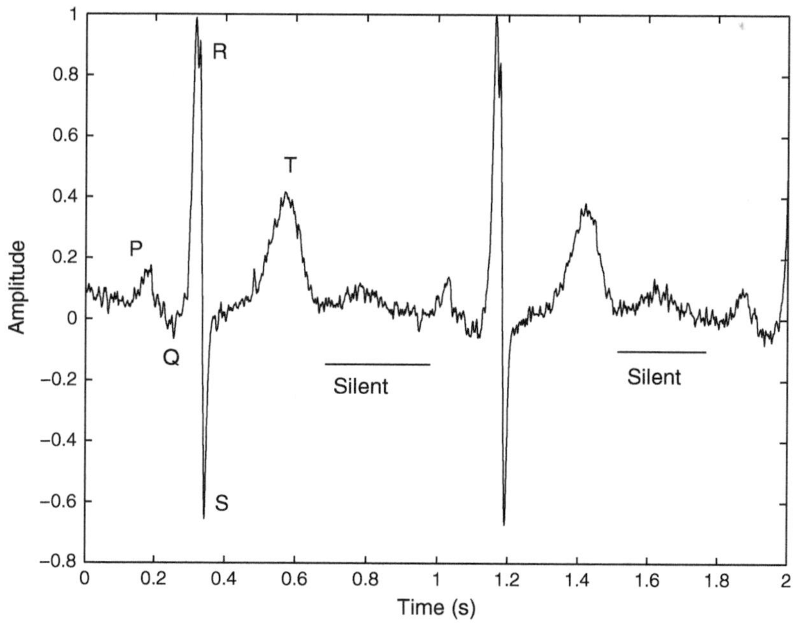

Figure 13 ECG and corresponding waves on lead V_2 from a healthy patient (sampling frequency: 1000 Hz).

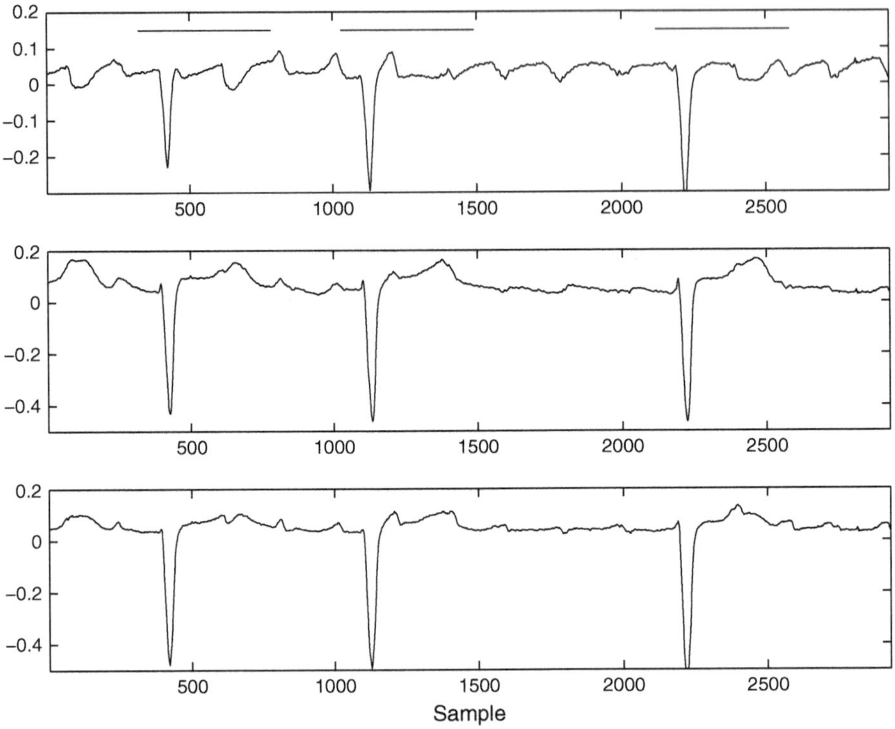

Figure 14 AF patient's ECG recording. Only leads $V_1 - V_3$ (top to bottom) among the recorded leads $V_1 - V_6$, I, II, and III are plotted. The synchronous segments containing the Q, R, S, and T waves are represented by horizontal lines (sampling frequency: 1000 Hz).

has been replaced by a continuously present activity that corresponds to the fibrillation signal – the so-called F wave – originated in the atria.

The characterization of the atrial activity (AA) signal plays an important role not only in understanding the genesis and self-perpetuation of AF, but also in quantifying the effect of evolving therapies such as new antiarrhythmic drugs and catheter ablation techniques. Clinically, the accurate estimation of the AF signal has a twofold purpose. It has been shown that both frequency [23] and time-frequency analysis [24] can establish a link between the main frequency of the AA signal and the probability of spontaneous cardioversion (return to NSR) of this arrhythmic disorder. Temporal analysis of AA can lead to a better understanding of the cardiac electrical pathway [25,26], which can prove helpful in designing new pacemakers and implantable cardioverter devices [25]. Unfortunately, the stronger QRST complex masks the AA precisely over the time intervals where the latter could provide more significant information about key physiological phenomena associated with AF such as the ectopic activation of the AV node. Hence, this problem calls for signal processing techniques to extract the AA in the surface ECG.

4.2. Spatio-temporal cancellation of the QRST complex in AF episodes

Among the possible extraction techniques, spatio-temporal cancellation is a generic methodology based on a model for the propagation of electrical signal from the heart to the leads on the body surface. Temporal information is also taken into account by this model. The method is usually expressed in the sampled-time domain and requires judiciously placed electrodes on the torso in order to comprise all the information contained in the cardiac electrical field. Theoretically, orthogonal leads X, Y, and Z should be sufficient to completely characterize this field (Section 2.2.4). The method relies on the estimation of the time occurrence of ventricular activity (VA) allowing the segmentation and synchronization of the wave of interest, i.e., the identification of Q, R, S, and T wave segments. Time synchronization is not critical since a realignment step can be added to the global process. Such a step will not be dealt with in the sequel.

In the presence of AF, each lead exhibits a mixture of consecutive transient signals (the Q, R, S, and T waves of VA) added to the AA spread over time. The latter is no longer a transient signal, as opposed to the P wave in NSR (see Figure 14). The AA in AF is assumed to be spectrally stationary at least over one beat and can be approximated by a sawtooth-shaped periodic signal. In the sampled-time domain, the *i*th beat segmented in a given lead is represented by a column vector made up of N samples:

$$\mathbf{x}_i = [x_i(0)\ x_i(T_s)\ \ldots\ x_i((N-1)T_s)]^{\mathrm{T}}. \tag{10}$$

Note the column-wise arrangement of samples in the framework of spatiotemporal or temporal cancellation, in contrast with the row-wise sample placement in BSS approaches (see Section 5). For the sake of consistency with the literature on the topic, this section will adhere to the column-wise arrangement of Eq. (10).

The essential idea behind spatio-temporal cancellation is to obtain an estimate of the average VA, and then subtract it, after appropriate scaling and time shift, from the actual beats observed in the lead outputs. Specific methods are detailed next. Their differences lie in the manner in which the average VA is modeled and estimated, as well as in how it is fitted to the observed signals before subtraction. A common implicit assumption is the uncorrelation between AA and VA.

4.2.1. Single-lead processing

Recording the ECG with a multilead system allows the best characterization of the overall cardiac electrical activity. However, only few leads may be available in some particular cases. The paroxysmal stage of AF recorded by a Holter system is a typical example. An attempt to deal with the single-lead case is made in Castells et al. [27]. The proposed method assumes that, in the aligned (e.g., on the R peak) observed beats, the vectors of sampled VA lie on a subspace orthogonal to that associated with the AA vectors. This orthogonality is a direct consequence of the assumed uncorrelation between AA and VA. The corresponding subspaces are spanned by some sampled-time basis functions (or source vectors) estimated by the Principal

Component Analysis (PCA) of the observed data. The QRST cancellation and AA extraction is achieved by retaining the eigenstructure associated with the AA.

Mathematically, this method accepts the following development, alternative to Castells et al. [27]. An observation matrix $\mathbf{X}$ is built of B consecutive beats:

$$\mathbf{X} = [\mathbf{x}_1 \;\; \mathbf{x}_2 \ldots \mathbf{x}_B] \in \mathbb{R}^{N \times B}. \tag{11}$$

Its PCA can be performed from the sample correlation matrix, $\mathbf{R}_x = \mathbf{X}\mathbf{X}^{\mathrm{T}}/B$, but deeper insights can be gained by using the Singular Value Decomposition (SVD) [28] of the data matrix, $\mathbf{X} = \mathbf{U}\mathbf{\Sigma}\mathbf{V}^{\mathrm{T}}$, with theoretically equivalent yet numerically more reliable results. Matrix $\mathbf{U} \in \mathbb{R}^{N \times N}$ contains the left singular vectors, matrix $\mathbf{V} \in \mathbb{R}^{B \times B}$ the right singular vectors, and $\mathbf{\Sigma} \in \mathbb{R}^{N \times B}$ is a diagonal matrix with the singular values along its main diagonal. An important property of the SVD is that the singular vectors are orthonormal:

$$\mathbf{U}^{\mathrm{T}}\mathbf{U} = \mathbf{I}_N \quad \mathbf{V}^{\mathrm{T}}\mathbf{V} = \mathbf{I}_B. \tag{12}$$

Under the assumptions of Castells et al. [27], this decomposition can be split into three terms associated with the VA, the AA, and the noise, respectively:

$$\mathbf{X} = \mathbf{U}_{\mathrm{VA}}\mathbf{\Sigma}_{\mathrm{VA}}\mathbf{V}_{\mathrm{VA}}^{\mathrm{T}} + \mathbf{U}_{\mathrm{AA}}\mathbf{\Sigma}_{\mathrm{AA}}\mathbf{V}_{\mathrm{AA}}^{\mathrm{T}} + \mathbf{U}_{\mathrm{N}}\mathbf{\Sigma}_{\mathrm{N}}\mathbf{V}_{\mathrm{N}}^{\mathrm{T}}. \tag{13}$$

In the above expression, we have $\mathbf{U} = [\mathbf{U}_{\mathrm{VA}}, \mathbf{U}_{\mathrm{AA}}, \mathbf{U}_{\mathrm{N}}]$, $\mathbf{V} = [\mathbf{V}_{\mathrm{VA}}, \mathbf{V}_{\mathrm{AA}}, \mathbf{V}_{\mathrm{N}}]$ and $\mathbf{\Sigma} = \mathrm{diag}(\mathbf{\Sigma}_{\mathrm{VA}}, \mathbf{\Sigma}_{\mathrm{AA}}, \mathbf{\Sigma}_{\mathrm{N}})$. The basis functions accounting for each activity appear in the columns of matrices $\mathbf{U}_{\mathrm{VA}}$, $\mathbf{U}_{\mathrm{AA}}$, and $\mathbf{U}_{\mathrm{N}}$. Uncorrelation between VA and AA within one beat is expressed as the orthogonality between the corresponding basis functions stored in $\mathbf{U}_{\mathrm{VA}}$ and $\mathbf{U}_{\mathrm{AA}}$, or $\mathbf{U}_{\mathrm{VA}}^{\mathrm{T}}\,\mathbf{U}_{\mathrm{AA}} = \mathbf{0}$, which follows immediately from property (12) and decomposition (13). Matrices $\mathbf{V}_{\mathrm{VA}}$, $\mathbf{V}_{\mathrm{AA}}$, and $\mathbf{V}_{\mathrm{N}}$ represent the linear weights whereby the basis functions contribute to each beat. For instance, the ith element of the jth column vector of $\mathbf{V}_{\mathrm{VA}}$ defines the contribution of the jth ventricular basis function into the ith beat. It is expected that the VA can be explained by the singular vectors associated with the largest singular values $\mathbf{X}$ (dominant eigenvalues of $\mathbf{R}_x$), as this activity contributes to the observed data with the highest variance. Accordingly, the noise is typically associated with the least significant eigenvalues. The AA subspace is identified from the remaining eigenstructure, and the AA contribution to the recording is then reconstructed as

$$\hat{\mathbf{X}}_{\mathrm{AA}} = \mathbf{U}_{\mathrm{AA}}\mathbf{\Sigma}_{\mathrm{AA}}\mathbf{V}_{\mathrm{AA}}^{\mathrm{T}}.$$

Due to the orthonormality of the singular vectors, the above AA estimate can be computed from the observed data matrix in two fashions:

$$\hat{\mathbf{X}}_{\mathrm{AA}} = \mathbf{X}\mathbf{V}_{\mathrm{AA}}\mathbf{V}_{\mathrm{AA}}^{\mathrm{T}} \tag{14}$$

$$\hat{\mathbf{X}}_{\mathrm{AA}} = \mathbf{U}_{\mathrm{AA}}\mathbf{U}_{\mathrm{AA}}^{\mathrm{T}}\mathbf{X}. \tag{15}$$

These equivalent forms lead to alternative interpretations. On the one hand, Eq. (14) defines the orthogonal projection (regression) of each beat vector sample (row of $\mathbf{X}$) on the column space of $\mathbf{V}_{AA}$. Right-multiplication of $\mathbf{X}$ by a column of $\mathbf{V}_{AA}$ yields a weighted average of the available beat time segments. This averaging is analogous to spatial filtering but computed across beats instead of across leads. On the other hand, Eq. (15) represents the orthogonal projection of each beat time segment (column of $\mathbf{X}$) on the column space of $\mathbf{U}_{AA}$. Left-multiplication of $\mathbf{X}$ by a column transpose of $\mathbf{U}_{AA}$ defines the scalar product of an observed beat with an AA basis function. From this perspective, the AA reconstruction can be interpreted as the linear regression of the beat segments on the AA basis function subspace estimated via PCA.

The ability of the SVD to capture the VA among the most dominant singular vectors of the observed data matrix is guaranteed if this activity contributes with the highest variance to the recording. As a consequence, this approach will fail when applied on leads exhibiting a low contribution of VA with respect to other signal components. In addition, telling the AA from the noise by considering only their relative power is generally a difficult task, so that the AA basis vectors need to be selected by hand.

Further examples of the application of PCA techniques in the field of AF analysis are given in Castells et al. [29].

4.2.2. Multilead processing

Unlike the previous case, a recording system composed of L leads ($L > 1$) is often available in practice (e.g., the standard ECG), so that AA extraction techniques can exploit both time and space features of the recorded dataset. Mainly two approaches taking advantage of the spatio-temporal properties of the ECG have been proposed in the literature [2,30]. Their key assumption is to consider the AA signal as uncorrelated with the VA. This assumption allows us to condense the VA of all leads and all beats into a single set of source vectors.

The observation matrix for the ith cardiac beat, $\mathbf{X}_i \in \mathbb{R}^{N\times L}$, is made up of samples $\mathbf{x}_{ij}$ from each lead j for the ith beat, $\mathbf{X}_i = [\mathbf{x}_{i1}\ \mathbf{x}_{i2}\ \ldots\ \mathbf{x}_{iL}]$. This matrix is modeled as

$$\mathbf{X}_i = \mathbf{V}\mathbf{M}_i + \mathbf{A}_i + \mathbf{N}_i. \tag{16}$$

The columns of $\mathbf{V}$ contain the VA source vectors that span the space of the observed ventricular activity in all leads. Each column of mixing matrix $\mathbf{M}_i$ determines how the signal recorded in the corresponding lead is spanned by the VA sources. Matrix $\mathbf{A}_i$ describes the AA present in each lead during the beat. The remaining activity is considered as noise and is represented by matrix $\mathbf{N}_i$. Remark that the AA is not assumed to follow a generative mixture model like that of VA, but appears as a separate matrix term in Eq. (16). This is one of the fundamental differences with BSS approaches to AA extraction [31,32].

Care should be taken when forming the observation matrix in order to include all the waves of interest (Q, R, S, and T) synchronized on the most energetic one, usually the R wave. To account for recording bias due to sensor imbalance and baseline wander, the model (16) should also incorporate a term $\mathbf{B}\mathbf{c}_i$, where $\mathbf{B}_i = [\mathbf{1}\ \mathbf{n}]$, with $\mathbf{1} = [1\ 1\ \ldots\ 1]^{\mathrm{T}}$ and $\mathbf{n} = [1\ 2\ \ldots\ N]^{\mathrm{T}}$ [30]. The baseline wander is approximated by a straight line in this manner. The AA extraction problem in the context of spatio-temporal cancellation aims at estimating matrix $\mathbf{A}_i$ from the observation matrix $\mathbf{X}_i$ in model (16).

In the first place, the estimation of the ventricular contribution over the leads can be addressed in two different manners [2,30], although in both cases an averaging over all beats is performed. Due to the hypothesis of uncorrelation between the AA and the VA signals, and according to model (16), the average beat

$$\overline{\mathbf{X}} = \frac{1}{B}\sum_{i=1}^{B} \mathbf{X}_i \tag{17}$$

will approximately produce a matrix $\mathbf{V}\overline{\mathbf{M}}$, with $\overline{\mathbf{M}} = \frac{1}{B}\sum_{i=1}^{B}\mathbf{M}_i$. An example of average beat calculation over $B = 50$ segments is given in Figure 15 (plot on the

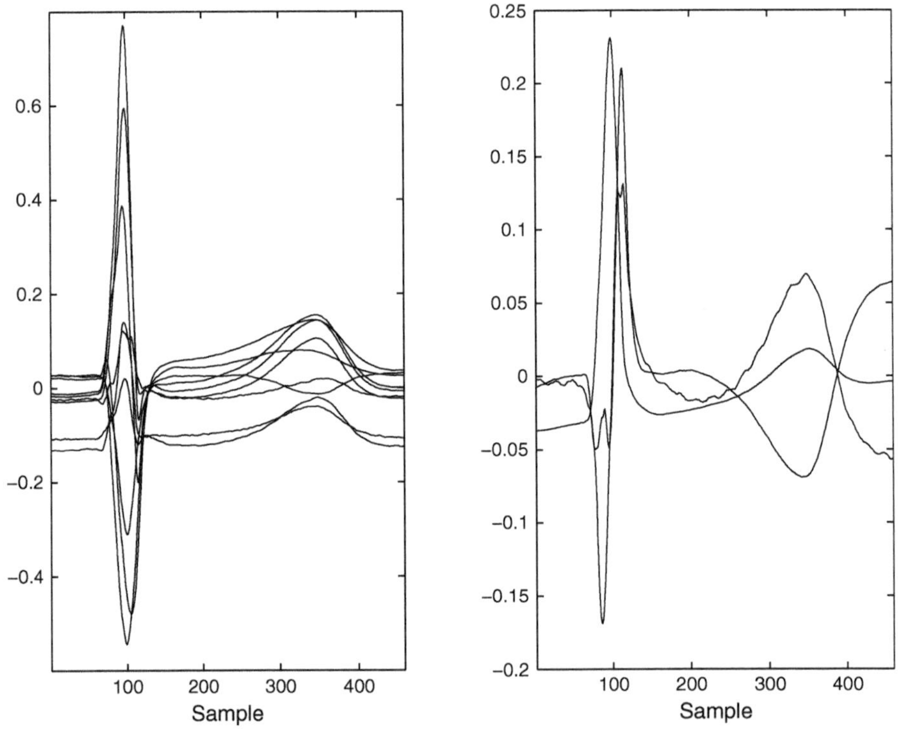

Figure 15 Average beats in AF episodes. (Left) Average of all segments of the AF patient's ECG of Figure 14 for the nine recorded leads. These average beats are stored in the columns of matrix $\overline{\mathbf{X}}$. (Right) The three main principal vectors obtained by the PCA-based approach of Meste and Serfaty [30] from the same recording. These vectors form matrix $\mathbf{V}$ containing the average VA.

left-hand side). This approximation is valid as far as the AF contributions are zero mean and independent for different beats. The same assumptions apply to the observation noise. The methods presented in Stridh and Sörnmo [2] and Meste and Serfaty [30] differ from that point on. In the former, the VA in $\mathbf{V}$ is constituted by a selection of vectors from the average beat $\overline{\mathbf{X}}$. In the latter, an SVD-based approach is employed to summarize the information of all the available leads into a reduced set of basis vectors (see the plot on the right-hand side of Figure 15). Note that, unlike the second method, the first exploits a priori information about the most suitable leads. When the model accounts for the recording bias, only the unit vector appears in the average [30], since the baseline wander averaged over the different segments is assumed to be null.

Recall that the quantity of interest in Eq. (16) is the AA activity modeled by $\mathbf{A}_i$. From the model definition, it is clear that its estimation relies on the calculation of the mixing matrix $\mathbf{M}_i$. The estimation of $\mathbf{M}_i$ is not straightforward since it will itself be biased by the presence of the AA term. Two methods to overcome this difficulty are presented next.

The calculation of $\mathbf{M}_i$ serves to illustrate the inherent differences between physical-based and blind models. Several models taking into account the observation of the electrical field from the heart through potential differences recorded on the body surface are considered in Stridh and Sörnmo [2]. The physiologically most plausible model assumes that the electrical potential recorded by the leads at a given time instant is a rotation of the row vectors of $\mathbf{V}$ in the $\mathbb{R}^L$ space. Accordingly, the following structure is assumed for the mixing matrix:

$$\mathbf{M}_i = \mathbf{D}_i\mathbf{Q}_i \quad \text{with} \quad \mathbf{D} = \text{diag}(d_1, \ldots, d_L) \quad \text{and} \quad \mathbf{Q}\mathbf{Q}^{\mathrm{T}} = \mathbf{I}. \tag{18}$$

Clearly, the orthonormality constraint on $\mathbf{Q}$ will reduce the variance in the estimation of $\mathbf{M}_i$, as the former is defined by $L(L-1)/2$ rotation angles only (one planar rotation per coordinate pair). But this model, even if physically justified, presents a lack of flexibility since the linear combination coefficients are computed from a rigid model for the observation variability. The solution providing the highest flexibility, at the expense of an increased variance, is to assume that matrix $\mathbf{M}_i$ has no specific structure. As an additional advantage, this solution allows the individual processing of the leads, yielding the following AA estimate for the ith beat observed at the jth lead:

$$\hat{\mathbf{a}}_{ij} = (\mathbf{I} - \mathbf{V}(\mathbf{V}^{\mathrm{T}}\mathbf{V})^{-1}\mathbf{V}^{\mathrm{T}})\mathbf{x}_{ij}. \tag{19}$$

This result corresponds to the least squares (LS) solution to

$$\hat{\mathbf{m}}_{ij} = \arg \min_{\mathbf{m}_{ij}} ||\mathbf{x}_{ij} - \mathbf{V}\mathbf{m}_{ij}||^2 \tag{20}$$

and can also be regarded as a regression onto the lower dimensional column subspace of $\mathbf{V}$ (deflation) once its redundancy has been reduced through the SVD-based procedure of Meste and Serfaty [30]. This estimate is obviously suboptimal since the covariance matrix of the term $\mathbf{A}_i + \mathbf{N}_i$ is not diagonal, which will produce spurious oscillations in the reconstructed AA signal at the locations of the R wave. To reduce this transient error, we can assume that it weakly affects the cross-correlation function of the initial AA estimate (19). Under this hypothesis, a "stationary" covariance matrix $\hat{\mathbf{C}}_\mathbf{a}$ can be computed and plugged into a best linear unbiased estimator [30] as

$$\hat{\mathbf{a}}_{ij} = (\mathbf{I} - \mathbf{V}(\mathbf{V}^{\mathrm{T}}\hat{\mathbf{C}}_{\mathrm{a}}^{-1}\mathbf{V})^{-1}\mathbf{V}^{\mathrm{T}}\hat{\mathbf{C}}_{\mathrm{a}}^{-1})\mathbf{x}_{ij} \tag{21}$$

which stems from the weighted LS minimization problem:

$$\hat{\mathbf{m}}_{ij} = \arg\min_{\mathbf{m}_{ij}} \, (\mathbf{x}_{ij} - \mathbf{V}\mathbf{m}_{ij})^{\mathrm{T}}\hat{\mathbf{C}}_{\mathrm{a}}^{-1}(\mathbf{x}_{ij} - \mathbf{V}\mathbf{m}_{ij}). \tag{22}$$

This solution is optimal in the maximum likelihood sense if the distribution of the AA (plus the noise) is Gaussian, which is a reasonable hypothesis in most AF recordings. The result obtained by this method for the AF patient's dataset of Figure 14 is plotted in Figure 16, where the AA is extracted with negligible artifacts

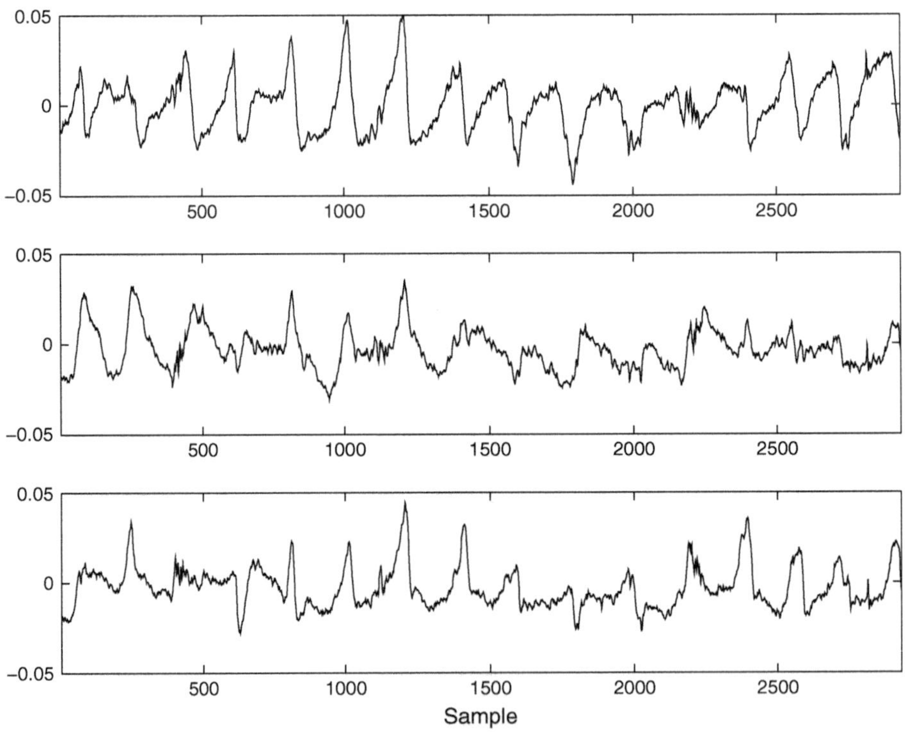

Figure 16 The AA reconstructed by estimator (21) on leads $V_1 - V_3$ (top to bottom) from the ECG of Figure 14.

and in agreement with the AA visible in the original recording. An alternative to AA covariance matrix estimation is proposed in Stridh and Sörnmo [2], where the AA in the TQ interval, i.e., the ventricular silent segment (see Figure 13), is linearly approximated in the QT interval assuming a repetitive constant pattern.

4.2.3. A simple alternative: average beat subtraction (ABS)

The traditional ABS method neglects the spatial properties of multisensor cardiac signal recordings. The synchronous averaged beat $\overline{\mathbf{X}}$ in Eq. (17) is employed as a reference [33] and is optimally shifted in time before being subtracted from each observed beat. The original method can be extended to the multilead case by solving the minimization problem [2]:

$$\hat{\delta}_i = \arg\min_{\delta_i} ||\mathbf{X}_i - \overline{\mathbf{X}}_{\delta i}||_F^2 \tag{23}$$

where $(.)_\delta$ stands for the δ-sample delay operator. This formulation based on the Frobenius norm can be seen as the minimization of the sum of individual Euclidean norms over all leads, that is,

$$\hat{\delta}_i = \arg\min_{\delta_i} \sum_{j=1}^{L} ||\mathbf{x}_{ij} - \overline{\mathbf{x}}_{j\delta i}||^2 = \arg\max_{\delta_i} \sum_{j=1}^{L} \mathbf{x}_{ij}^{\mathrm{T}} \overline{\mathbf{x}}_{j\delta i} \tag{24}$$

where $\overline{\mathbf{x}}_j$ denotes the jth column of the average beat $\overline{\mathbf{X}}$, i.e., the average beat of the jth lead. The scalar product in Eq. (24) can be considered as an approximation of the cross-correlation function of the observed signal and the average beat in the corresponding leads at lag δ_i. Once the optimum delay is estimated, the AA in each lead is obtained by the subtraction:

$$\hat{\mathbf{a}}_{ij} = \mathbf{x}_{ij} - \overline{\mathbf{x}}_{j\hat{\delta}_i} \tag{25}$$

Note that, at each beat, this method does not seek a delay for each lead, but finds the optimal delay over all leads. This option yields more accurate results under the sensible assumption of negligible propagation delays between the leads (Section 2.2.4). The technique can be simply improved by scaling the average beat as $a_i\overline{\mathbf{x}}_{j\hat{\delta}_i}$. In such a case, the amplitude factor a_i needs to be incorporated into Eq. (24) and its optimal value estimated accordingly.

5. Blind Source Separation (BSS)

5.1. The isolation of interictal epileptic discharges in the EEG

About 0.5–1% of the population suffers from epilepsy, a neurological disorder associated with recurrent unprovoked seizures. These seizures are often linked with cognitive absences or state alterations, tonic or clonic movements or convulsions.

These symptoms are a direct reaction to an abnormal temporal synchronization of electrical activity in the brain. Treatment in the form of drug administration helps control the seizures but does not tackle the epilepsy origin and, as a consequence, cannot be seen as a cure. However, only 75% of the treated patients react to anticonvulsive or antiepileptic drugs with a significant reduction of the seizure frequency or even full seizure control. For the remaining 25%, surgical interventions are proposed such as implantation of vagus nerve stimulators, deep brain stimulators, or even an anterior temporal lobotomy. In the latter two cases, gathering presurgical information about the origin of the epileptic seizures is of uttermost importance.

The origin of the epileptic spike can be deduced from the EEG or from MRI studies at the onset of the ictal activity (the activity that is visible in the EEG during a seizure, also known as the clinical seizure). It can also be detected between ictal activities if Interictal Epileptic Discharges (IEDs) are present (see Table 5). Theoretically, this activity may be clearly visible in MRI, since the onset region presents an important electrical peak and thus demands a high provision of oxygen, resulting in an augmented blood oxygen level. In practice, however, these activities are of short duration (IEDs) or are paired with body movements (seizure). Unfortunately, analyzing the former is incompatible with the low temporal resolution of the MRI, while measuring the latter is in addition ethically questionable. Several studies have been conducted to trigger MRI scans on the spikes [34], to correlate simultaneously registered EEG and functional MRI (fMRI) [35], or even EEG, MRI, and SPECT datasets [36].

Consequently, the EEG arises as an indispensable tool for the localization of epileptic activity. However, seizure recordings are often contaminated by artifacts caused by muscular activity or body movements. Moreover, during ictal onset, there exists a quick transition from localized activity (few electrodes) to a generalized disruption of the potential field, and thus an obvious choice for localization is to rely on the IEDs, which are believed to share the same spatial origin as the seizure onset – in contrast to the global seizure activity – and are (quasi-)stationary over time [38]. IEDs in the EEG cover spikes, sharp waves, spike-and-wave complexes and related phenomena, which are all characterized by sharp transitions, electrical fields that extend beyond a single electrode, disruption of the background activity, and a negative cerebral surface potential. The third characteristic points out the need to separate the IED activity from the background activity to increase the localization accuracy.

Figure 17 shows an EEG segment recorded from an epileptic patient. This bipolar recording is obtained as the potential differences between scalp electrodes (cf. Fig. 8). IEDs can be perceived around time instants 1, 6, 7, 9, 10, and 11 s, and appear most clearly on electrodes situated around the temporal area on the right-

Table 5 Statistics of epilepsy [37]. Figures represent the probability of the given event. Screened patients have at least once been referred to hospital for an examination.

Case	Unscreened (%)	Screened (%)
Epilepsy	0.5–1	50
Epilepsy given IED	50	93
IED given epilepsy	90	

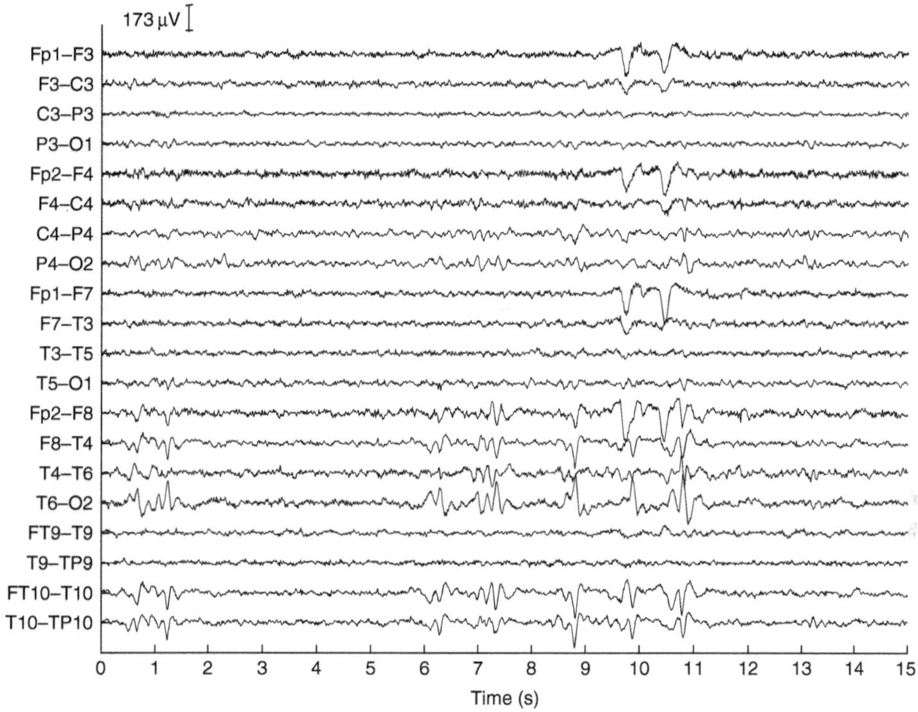

Figure 17 An EEG fragment from an epileptic patient. Signals have been sampled at 200 Hz. IEDs appear around time instants 1, 6, 7, 9, 10, and 11 s. Ocular artifacts are visible over the last two IEDs.

hand side of the brain. An ocular artifact contributes strongly to leads located on the frontal area, near the eyes. The figure illustrates that the EEG comprises the contribution of various physiological activities with possible temporal and spatial overlap. The ocular artifact corrupts the IED activity over the 10- to 11-s interval, thus hampering its subsequent analysis (e.g., localization, counting) by a physician.

As epileptic spikes are spontaneous untriggered events, we may reasonably suppose that the IEDs are independent of the ocular artifact and the background activity. Moreover, their localization is spread over different electrodes and they correlate highly with the seizure onset zone. Due to their transient character, the associated distribution is clearly non-Gaussian, with significant probability of large amplitude values leading to heavy or long tails in the probability density function. These features make IED activity an excellent candidate for extraction by BSS techniques based on independent component analysis.

5.2. Modeling and assumptions

As in the extraction of IED activity from the EEG, the assumption that some sensors receive contributions from a single source is not always realistic in practice. In the general case, all measurements may possibly contain contributions of several sources. The technique of Section 3, which strongly relies on the existence of reference signals free from the source of interest, may yield poor results in such scenarios. Similarly, the spatio-temporal cancellation approach of Section 4 becomes inadequate in situations where a quasi-periodic source such as the VA is not encountered.

These remarks motivate the simplest instance of the BSS problem, where the sensor signals follow the linear instantaneous statistical model:

$$\mathbf{x}(t) = \mathbf{A}\mathbf{s}(t). \tag{26}$$

The components of vector $\mathbf{x}(t) = [x_1(t)\ x_2(t)\ \ldots\ x_L(t)]^{\mathrm{T}}$ denote the L signals recorded, whereas the components of vector $\mathbf{s}(t) = [s_1(t)\ s_2(t) \ldots s_M(t)]^{\mathrm{T}}$ are the M unknown sources. $\mathbf{A}$ is an unknown $L \times M$ mixing matrix reflecting how the sources contribute to the sensors. Because it is not known in advance whether some sensors receive contributions from a reduced number of sources or not, matrix $\mathbf{A}$ has no known structure or sparsity pattern. The above model can also be decomposed as

$$\mathbf{x}(t) = \sum_{m=1}^{M} \mathbf{a}_m s_m(t) \tag{27}$$

where $\mathbf{a}_m$ is the mth column vector of $\mathbf{A}$. This equation signifies that each source contributes to the sensors through the projection of its amplitude on the associated column of the mixing matrix.

The goal of BSS is to estimate the realizations of source vector $\mathbf{s}(t)$ from the corresponding realizations of observed vector $\mathbf{x}(t)$. To this end, an $M \times L$ separating matrix filter $\mathbf{B}$ is built such that its outputs

$$\mathbf{y}(t) = \mathbf{B}\mathbf{x}(t) \tag{28}$$

are estimates of the sources. Note that the ith row of $\mathbf{B}$ defines a spatial filter acting on the sensor output $\mathbf{x}(t)$ to produce the separator output $y_i(t)$, the ith component of vector $\mathbf{y}(t)$.

In order to solve this problem, additional assumptions are necessary. Depending on the assumptions, the BSS problem is of variable difficulty and can be solved more or less easily. We shall limit our scope to the following:

A1. Matrix $\mathbf{A}$ is of full rank M, and there are at least as many sensors as sources ($L \geq M$).
A2. Sources $s_m(t)$ are mutually statistically independent.
A3. At least $(M-1)$ sources are non-Gaussian, in the sense that they admit a non-zero finite cumulant of order four.

Under assumption A1, Eq. (26) models a so-called overdetermined mixture, where the available spatial diversity is sufficient to recover the sources by applying a linear transformation on the observed data. Indeed, the sources $\mathbf{s}$ can be recovered by a linear oblique projection of the observation vector $\mathbf{x}$ onto the columns of $\mathbf{A}$. This projection is carried out by separating matrix $\mathbf{B}$ in Eq. (28). Assumptions A2–A3 allow the design of BSS techniques exploiting the property of statistical independence between the sources. This is an important property, as it is plausible in numerous applications and leads to computationally efficient practical BSS algorithms such as that presented in Section 5.6. Independence can be seen as a generalization of conventional second-order uncorrelation, the property exploited by the extraction techniques of Sections 3–4. Section 5.4 will explain in more detail the concept of independence (assumption A2), together with its connection with cumulants and non-Gaussianity (assumption A3).

5.3. Inherent indeterminacies

It is clear that if the components of $\mathbf{s}(t)$ undergo an arbitrary permutation (re-ordering), assumptions A1–A3 are still fulfilled and the observation $\mathbf{x}(t)$ in Eq. (26) is not altered as long as the columns of $\mathbf{A}$ are re-arranged accordingly. The same result holds if a source is scaled by a constant factor α and the corresponding mixing matrix column by $1/\alpha$. These simple remarks show that there are inherent indeterminacies in the BSS problem if addressed under assumptions A1–A3: the source signals and the mixing matrix columns can be estimated up to permutation and scale ambiguities. Conversely, it can be shown that these three assumptions are sufficient to establish identifiability of matrix $\mathbf{A}$ up to the above ambiguities and to estimate matrix $\mathbf{B}$ [3]. Note that they are not necessary, for other assumptions such as non-stationarity or colorness (time coherence) of the sources also allow identifiability under certain circumstances; see [39] and references therein.

Our goal is thus to estimate one representative $(\mathbf{A}, \mathbf{s}(t))$ of a whole equivalence class of solutions. Note that the number of solutions may be reduced to a finite number by imposing each estimated source to have unit variance. The remaining indeterminacy eventually reduces to sign and permutation.

As time coherence will be ignored in the sequel, all pertinent information is contained in the probability distribution of the processed signals, whose samples are considered as realizations of random variables. Consequently, the time dependency denoted by index t becomes irrelevant and will be dropped for convenience.

5.4. Statistical independence, higher-order statistics and non-Gaussianity

The components of random vector $\mathbf{s} \in \mathbb{R}^M$ are said to be independent if and only if their joint probability density function can be decomposed as the product of their marginal densities:

$$p_s(\mathbf{s}) = \prod_{m=1}^{M} p_{s_m}(s_m). \tag{29}$$

Dealing with probability distributions is usually hard. Instead, independence can easily be measured using statistical parameters known as cumulants [40,41]. The cumulants of random vector **s**, denoted by $C_{s,ijk\ldots} = \mathrm{cum}(s_i, s_j, s_k, \ldots)$, where s_i is the ith component of **s**, are defined as the coefficients of the Taylor expansion of its second characteristic function $\Psi(\mathbf{w}) = \log \mathrm{E}\{\exp(j\mathbf{w}^T\mathbf{s})\}$, where j is the imaginary unit. Omitting mathematical details beyond our scope, it will suffice to mention here that the cumulants are linked to the moments through relationships that are often non-linear. For *zero-mean* variables, we have the following expressions for the cumulants up to fourth order:

$$
\begin{aligned}
&- \text{order 1:} \quad && C_{s,i} = \mathrm{E}\{s_i\} = 0 \\
&- \text{order 2:} \quad && C_{s,ij} = \mathrm{E}\{s_i s_j\} \\
&- \text{order 3:} \quad && C_{s,ijk} = \mathrm{E}\{s_i s_j s_k\} \\
&- \text{order 4:} \quad && C_{s,ijk\ell} = \mathrm{E}\{s_i s_j s_k s_\ell\} - \mathrm{E}\{s_i s_j\}\{s_k s_\ell\} - \mathrm{E}\{s_i s_k\}\mathrm{E}\{s_j s_\ell\} \\
& && \qquad - \mathrm{E}\{s_i s_\ell\}\mathrm{E}\{s_j s_k\}.
\end{aligned}
\tag{30}
$$

Cumulants are said to be marginal if they involve a single component of **s**, while cross-cumulants involve more than one component. The second-order cross-cumulant is the conventional covariance. Cumulants of order higher than two are referred to as higher-order cumulants and, along with the higher-order moments, constitute the higher-order statistics (HOS). The normalized fourth-order marginal cumulant, $C_{s,iiii}/C_{s,ii}^2$, is also called kurtosis.

The cross-cumulants of a random vector with independent components are null at all orders. Full independence is too stringent an assumption and is indeed not necessary in practice to accomplish the separation. The concept can be relaxed by defining independence at order r, which only requires null cross-cumulants up to that order. In particular, independence at order 2 reduces to the classical second-order uncorrelation. Note that higher-order independence is a stricter property as it implies, in particular, second-order uncorrelation, whereas the converse is not true: uncorrelated signals may not necessarily be independent. The term 'higher-order' will be omitted in the sequel when talking about independence, as will be 'second-order' when referring to uncorrelation. As will be seen later (Section 5.5.1), decorrelating the observations (i.e., making them uncorrelated at the separator output) is generally insufficient to achieve the separation if the temporal coherence of the sources is not (or cannot be) exploited.

All the higher-order cumulants of a Gaussian random variable are zero, so that these statistical parameters arise as natural measures of non-Gaussianity. From the connection between independence and cumulants just recalled, it follows that imposing uncorrelation on Gaussian variables renders them independent without enforcing any constraints involving HOS. As a result, recovering the property of independence is insufficient to separate these type of signals [3]. Indeed, independence-exploiting techniques are unable to perform a complete separation if more than one source is Gaussian; hence, the need for assumption A3. HOS-based methods can separate the non-Gaussian sources, while the Gaussian sources will remain mixed at the separator output. This limitation will be illustrated in Section 5.5.2 with the particular separation criterion of [3] (see also Section 5.9).

5.5. Independent component analysis

It is possible to estimate the source signals by first identifying the mixing matrix (blind identification) before inverting the mixture. More common in biomedical applications are inverse approaches. These are formulated as the maximization of a so-called contrast criterion and are thus expected to be more robust to additive noise than blind identification approaches. Inverse approaches look for a separating matrix $\mathbf{B}$ such as to maximize a contrast $\Upsilon(\mathbf{B})$. A contrast is a function measuring a known property of the sources. From this perspective, BSS is approached as a property recovery problem: recover the source waveforms by recovering the source property. The maximization of a contrast function guarantees the separation, so that the contrast is said to discriminate among the sources having that property.

If source and noise distributions are known, one can rely on the Maximum Likelihood (ML) principle. Otherwise, one can use a contrast measuring the statistical independence between the components of the separator output (28). The maximization of statistical independence gives rise to the *Independent Component Analysis* (ICA) of the observed data [3]. The most natural criterion for ICA is the Mutual Information (MI):

$$\Upsilon_{\mathrm{MI}}(\mathrm{B}) = \int \mathrm{p}_{\mathrm{y}}(\mathbf{u}) \log \frac{p_{\mathrm{y}}(\mathbf{u})}{\prod_m p_{y_m}(u_m)} d\mathbf{u}. \tag{31}$$

According to the definition of independence expressed by Eq. (29), the above criterion is null if and only if the separator output is made up of independent components. The MI involves probability densities that are unknown; this is difficult to handle even if they can be estimated, especially in large dimensions. Moreover, maximizing Eq. (31) with respect to $\mathbf{B}$ can only be done iteratively [42], which may raise problems due to the multimodality of the likelihood function, in addition to the huge computational complexity. These shortcomings explain why several other more practical criteria have been proposed in the literature. Many of these alternative criteria capitalize on the ability of cumulants to measure independence (Section 5.4). The kurtosis-based contrast [3], which is probably the most widely used, will be presented in Section 5.5.2. Let us first point out the inability of second-order techniques to accomplish the separation.

5.5.1. Principal component analysis (PCA)

Since it is desired to obtain independent components y_m, $1 \leq m \leq M$, in particular they must also be uncorrelated (Section 5.4). An efficient technique to obtain uncorrelated variables via a linear transformation is PCA.

For simplicity and without restricting the generality, assume that vector $\mathbf{x}$ is zero-mean, and denote $\mathbf{R}_x = \mathrm{E}\{\mathbf{x}\mathbf{x}^{\mathrm{T}}\}$ its covariance matrix. Then there exists an $M \times L$ full rank matrix $\mathbf{W}$ such that

$$\mathbf{W}\mathbf{R}_x\mathbf{W}^{\mathrm{T}} = \mathbf{I}_M. \tag{32}$$

Recall that the diagonal elements of the covariance matrix contain the variance, whereas its off-diagonal entries contain the covariance of the random vector components. Hence, an identity covariance matrix means that the corresponding random vector has uncorrelated unit-variance components, as desired. Matrix $\mathbf{W}$ defines a *whitening filter* and is not uniquely defined. For instance, denote $\mathbf{V}$ the set of eigenvectors of $\mathbf{R}_x$, and $\mathbf{\Lambda}$ the diagonal matrix containing its eigenvalues. We have, by definition, $\mathbf{R}_x = \mathbf{V}\mathbf{\Lambda}\mathbf{V}^{\mathrm{T}}$; this is the Eigenvalue Decomposition (EVD) of $\mathbf{R}_x$. One possibility is to choose $\mathbf{W} = \mathbf{\Lambda}^{-1/2}\mathbf{V}^{\mathrm{T}}$. Actually, any matrix of the form $\mathbf{W} = \mathbf{Q}\mathbf{\Lambda}^{-1/2}\mathbf{V}^{\mathrm{T}}$, where $\mathbf{Q}$ is orthogonal, i.e., $\mathbf{Q}\mathbf{Q}^{\mathrm{T}} = \mathbf{I}_M$, will obviously also satisfy equality (32).

Once $\mathbf{W}$ is chosen, define the *standardized variable* as $\tilde{\mathbf{x}}(t) = \mathbf{W}\mathbf{x}(t)$. By construction, this variable has a unit covariance matrix. Even if the components of $\tilde{\mathbf{x}}(t)$ have unit variance and are uncorrelated, they are not necessarily independent, as explained in Section 5.4. It indeed remains to find another transformation improving statistical independence, but keeping the covariance matrix to identity and thus the uncorrelation between the output components. Such a transformation must be an orthogonal matrix. Hence, the separating filter may be thought as the product between $\mathbf{W}$ and an orthogonal matrix $\mathbf{Q}$: $\mathbf{B} = \mathbf{Q}\mathbf{W}$. It is now clear that PCA does not allow the identification of the mixing matrix, since an indeterminacy up to a multiplicative orthogonal matrix is much larger than the inherent permutation and scaling indeterminacy described in Section 5.3. The identification of matrix $\mathbf{Q}$ requires tools beyond second-order statistics.

5.5.2. Kurtosis-based independent component analysis

Matrix $\mathbf{Q}$ can be estimated by performing the ICA at order 4 of the whitened observations, that is, by imposing fourth-order independence between the components of the separator output $\mathbf{y}$. According to Section 5.4, fourth-order independence can be achieved by minimizing the sum of squared fourth-order cross-cumulants of $\mathbf{y}$. Under the whitening constraint, i.e., under orthonormal transformations, this is equivalent to maximizing the sum of squared fourth-order marginal cumulants (kurtosis):

$$\Upsilon_{\mathrm{CoM2}}(\mathbf{Q}) = \sum_{i=1}^{M} C_{y,iiii}^{2}. \tag{33}$$

By virtue of Section 5.4, this criterion can also be interpreted as the maximization of non-Gaussianity of the separator output components. Under assumptions A1–A3, Eq. (33) is proven to be a contrast function [3] and is referred to as CoM2 contrast.

As opposed to Eq. (31), the above criterion can be explicitly written as a function of $\mathbf{Q}$ (and hence $\mathbf{B}$) by replacing the output cumulants by

$$C_{y,ijk\ell} = \sum_{pqrs=1}^{M} Q_{ip} Q_{jq} Q_{kr} Q_{\ell s} C_{\tilde{x},pqrs} \tag{34}$$

where Q_{ij} denotes the element (i, j) of matrix $\mathbf{Q}$. This equality expresses the multilinearity property of cumulants [40]: a linear transformation $\mathbf{Q}$ applied on the data $\tilde{\mathbf{x}}$ induces a multilinear transformation on its cumulants. This transformation is given by Eq. (34) at order 4 but is easily extended at any order. Taking into account this relationship, we see that Eq. (33) is actually a polynomial of degree 8 in the entries of $\mathbf{Q}$. Its particular form allows us to find its absolute maximum with the help of specific algorithms [3], involving the rooting of polynomials of degree at most 4, as will be seen subsequently (Section 5.6.3).

Section 5.4 recalled that the higher-order marginal cumulants, and in particular the kurtosis, of Gaussian signals are null. For sources of this kind, relation (34) shows that the separator-output kurtoses are also null for any $\mathbf{Q}$, and thus contrast function (33) becomes ineffective to carry out the separation.

Criterion (33) may be seen as an approximation of MI, when expanding (in the Edgeworth sense) the densities about the closest Gaussian [3]. This is a way to establish a connection between Eqs (31) and (33). It is also possible to make a connection between MI (31) and ML [4].

5.6. Algorithms

Let us now turn to numerical methods able to implement ICA in practice. Instead of surveying superficially several methods, we shall concentrate on the CoM2 algorithm [3]. Various other techniques may be used in the present context, and relevant references may be found in Kachenoura et al. [43], including iterative algorithms [5].

Let us assume that we are given T realizations of $\mathbf{x}$, denoted by $\mathbf{x}(nT_s)$, $0 \leq n \leq (T-1)$. The CoM2 algorithm, devised in 1991, proceeds in two stages. First, realizations of the standardized random variable $\tilde{\mathbf{x}}$ are computed and, in a second stage, an orthogonal separating matrix $\mathbf{Q}$ is estimated.

5.6.1. Sample spatial whitening

Since $\mathbf{x}$ is zero-mean, a consistent estimate of the covariance matrix is given by the sample covariance

$$\hat{\mathbf{R}}_x = \frac{1}{T} \sum_{n=0}^{T-1} \mathbf{x}(nT_s)\mathbf{x}(nT_s)^{\mathrm{T}}.$$

As seen in Section 5.5.1, a whitening filter is given by $\hat{\mathbf{W}} = \hat{\mathbf{\Lambda}}^{-1/2} \hat{\mathbf{V}}^{\mathrm{T}}$ where $\hat{\mathbf{R}}_x = \hat{\mathbf{V}}\hat{\mathbf{\Lambda}}\hat{\mathbf{V}}^{\mathrm{T}}$ denotes the EVD of $\hat{\mathbf{R}}_x$. Standardized realizations are then eventually obtained as $\tilde{\mathbf{x}}(t) = \hat{\mathbf{W}}\mathbf{x}(t)$.

However, several improvements are suitable. Without going into the details, one can mention the following:

1. The computational complexity can be reduced by avoiding the explicit computation of the sample covariance. In fact, the $M \times L$ whitening matrix $\mathbf{W}$ is such that $\tilde{\mathbf{X}}\tilde{\mathbf{X}}^{\mathrm{T}} = T\mathbf{I}_M$, where $\tilde{\mathbf{X}} = \mathbf{W}\mathbf{X}$. For the purpose of computing $\tilde{\mathbf{X}}$, the "economical" SVD of the data matrix can be used. By economical, it is meant that the singular vectors associated with null singular values are not computed. It takes the form $\mathbf{X} = \mathbf{U}\boldsymbol{\Sigma}\mathbf{V}^{\mathrm{T}}$, where the diagonal matrix $\boldsymbol{\Sigma}$ is of dimension $M \times M$. The M dominant right singular vectors then yield the standardized data matrix up to a scale factor, as $\tilde{\mathbf{X}} = \sqrt{T}\mathbf{V}^{\mathrm{T}}$. The $M \times L$ whitening filter is then given by $\mathbf{W} = \sqrt{T}\boldsymbol{\Sigma}^{-1}\mathbf{U}^{\mathrm{T}}$. A Gram–Schmidt orthogonalization is sometimes employed inside the economical SVD calculation.
2. Note that if the data matrix is not full rank, i.e., if $L > M$, only the range space is kept, which reduces the dimension of the observation space. In the EVD approach, this would correspond to computing the pseudo-inverse of matrix $\boldsymbol{\Lambda}$ when it is singular.
3. In the presence of additive noise with known covariance matrix, the effect of the noise can be reduced by computing an unbiased estimate of the whitening filter, which is aimed at whitening the source part. In other words, it suffices to compute the whitening filter so that it reduces the noiseless covariance matrix to identity, instead of the observation covariance matrix.

5.6.2. Decomposition of the orthogonal separating matrix

The CoM2 contrast function is a rational function in many variables. Since finding the absolute maximum of such a function is not easy, it is convenient to decompose the orthogonal matrix $\mathbf{Q}$ into the product of plane rotations. A plane rotation is a rotation acting in a plane defined by two coordinates, say $\tilde{x}_i$ and $\tilde{x}_j$. In an M-dimensional space, such a rotation can be represented by an $M \times M$ matrix which differs from the identity only by four entries, namely its (i, i), (i, j), (j, i), and (j, j) entries. These four entries can be compactly stored in a 2×2 matrix $\mathbf{G}[i, j]$, with $i < j$, which forms a Givens rotation, defined by a single parameter: its angle, or the tangent of its angle.

The idea is then to proceed as in the Jacobi sweeping algorithm devised for the diagonalization of symmetric matrices [28]: maximize sequentially a series of contrast functions in one variable, instead of maximizing a single contrast function in many variables. The length of the series must be at least $M(M-1)/2$, that is, the number of free parameters in the $M \times M$ rotation matrix, which constitutes one *sweep* of the algorithm. In the CoM2 algorithm, convergence is typically reached after at most $\lceil\sqrt{M}\rceil$ sweeps.

5.6.3. Calculation of an optimal plane rotation

It now remains to show how the 2×2 problem is solved by maximizing the CoM2 contrast. Let t denote the tangent of the angle of the Givens rotation $\mathbf{G}[i, j]$ acting in the $(\tilde{x}_i, \tilde{x}_j)$ plane. We have

$$\mathbf{G}[i,j] = \frac{1}{\sqrt{1+t^2}}\begin{bmatrix} 1 & t \\ -t & 1 \end{bmatrix}.$$

The contrast criterion Υ_{CoM2} defined in Eq. (33) turns out to be a rational function in t of degree 8, as shown by the multilinearity property (34). It has been proven in Comon [3] that, because of its particular form, *all* stationary values of this contrast criterion can be found by rooting a polynomial of degree 4, which can be done in an entirely algebraic (non-iterative) manner. By plugging back these values in the contrast expression, we are able to select the *absolute* maximum of Υ_{CoM2}, within a *finite* number of operations. Since 1992, the source code of this algorithm can be downloaded from `www.i3s.unice.fr/~pcomon`.

The coefficients of the above rational function involve sample cumulants of the pair $(\tilde{x}_i, \tilde{x}_j)$, or of the pair (y_i, y_j) previously computed. Sample cumulants can be estimated from sample moments either in a batch manner, by averaging over the T realizations in the available signal block, or recursively (on-line) if a real-time implementation is preferred.

5.7. Results

We apply the CoM2 cumulant-based BSS algorithm of Comon [3] (summarized in Section 5.6) to the interictal EEG fragment shown in Figure 17. The algorithm separates the measurements into the maximally independent sources of Figure 18.

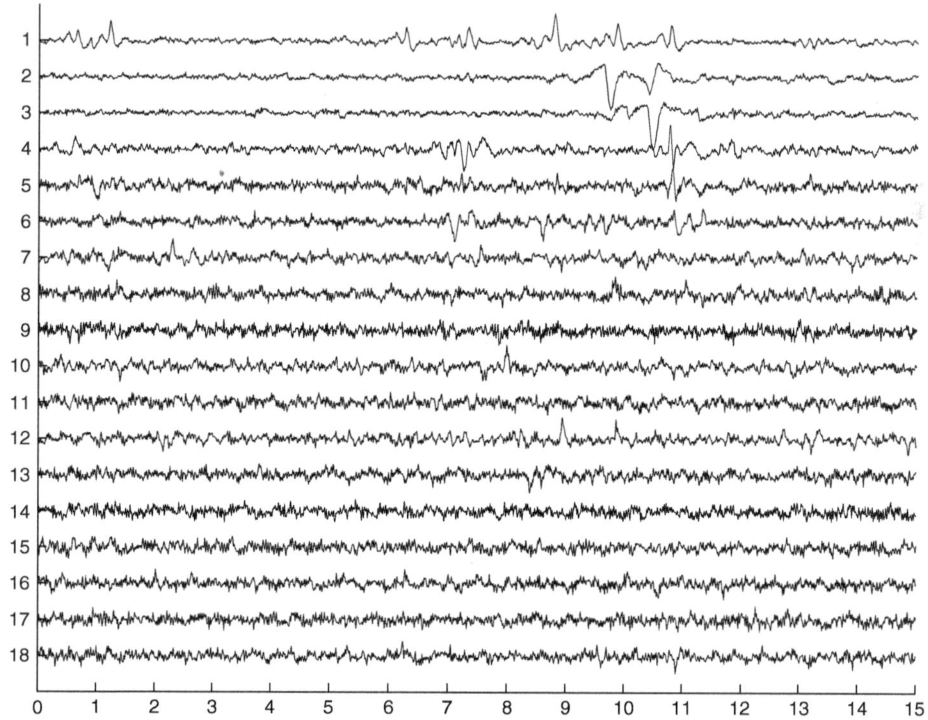

Figure 18 Brain signal extraction in the EEG: decomposition of the EEG fragment of Figure 17 into its source signals by the CoM2 algorithm of [3] (Section 5.6). Sources 2 and 3 represent the ocular artifact. Amplitudes are normalized to yield unit power sources.

Due to Kirchhoff's Voltage Law, the bipolar configuration introduces linear dependencies in the recordings, so that the dimension of the observed signal subspace is actually smaller than the number of leads in Figure 17. The number of sources that can be found by conventional (overdetermined) BSS techniques is bounded by this dimension, which explains why fewer sources than observations are recovered in this example. In Figure 18, sources 2 and 3 seem to condense the ocular activity. To suppress their contribution to the observed EEG segment, these sources can be neglected in Eq. (27) when projecting the source amplitudes back on the mixing matrix columns to reconstruct the recordings. The signals so obtained are free from ocular artifact, as shown in Figure 19. Background activity and noise may be further reduced by selecting only the IED sources in Figure 18 and reconstructing their contribution to the original recordings.

By way of comparison, the source signals estimated by PCA (Section 5.5.1) are shown in Figure 20. PCA-based separation is achieved by maximizing the variance of the extractor output subject to uncorrelation constraints on the previously extracted sources. Since the IEDs and the ocular artifact contribute to the recording with similar amplitudes, PCA is unable to separate them correctly. Indeed, both types of activity appear mixed among the first five sources of Figure 20, and the ocular artifact suppression is negatively affected as a consequence. These results

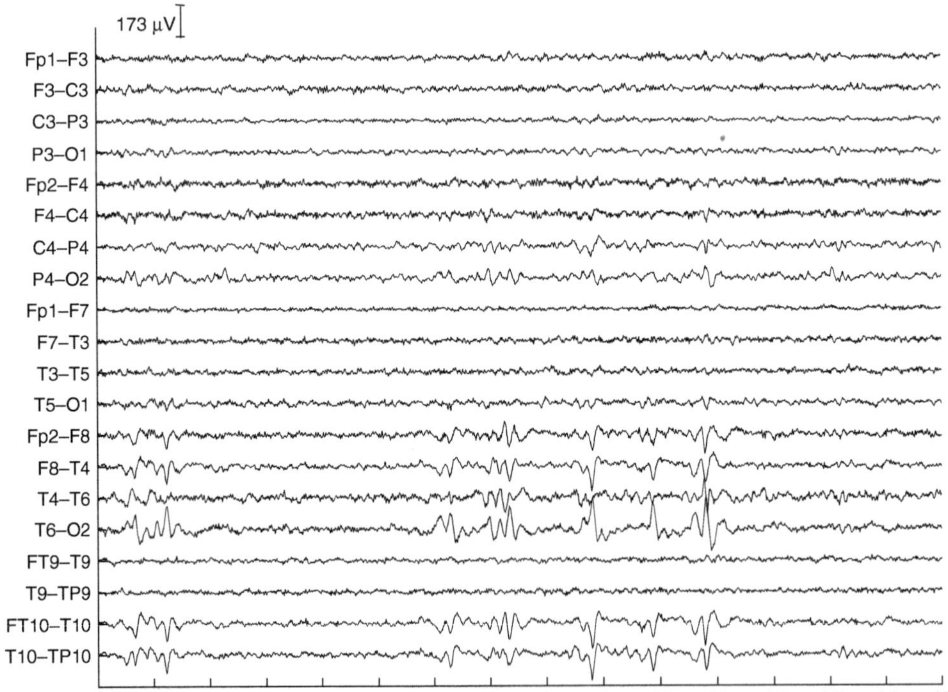

Figure 19 Reconstruction of the EEG segment of Figure 17 from the estimated sources of Figure 18 without the ocular artifact sources.

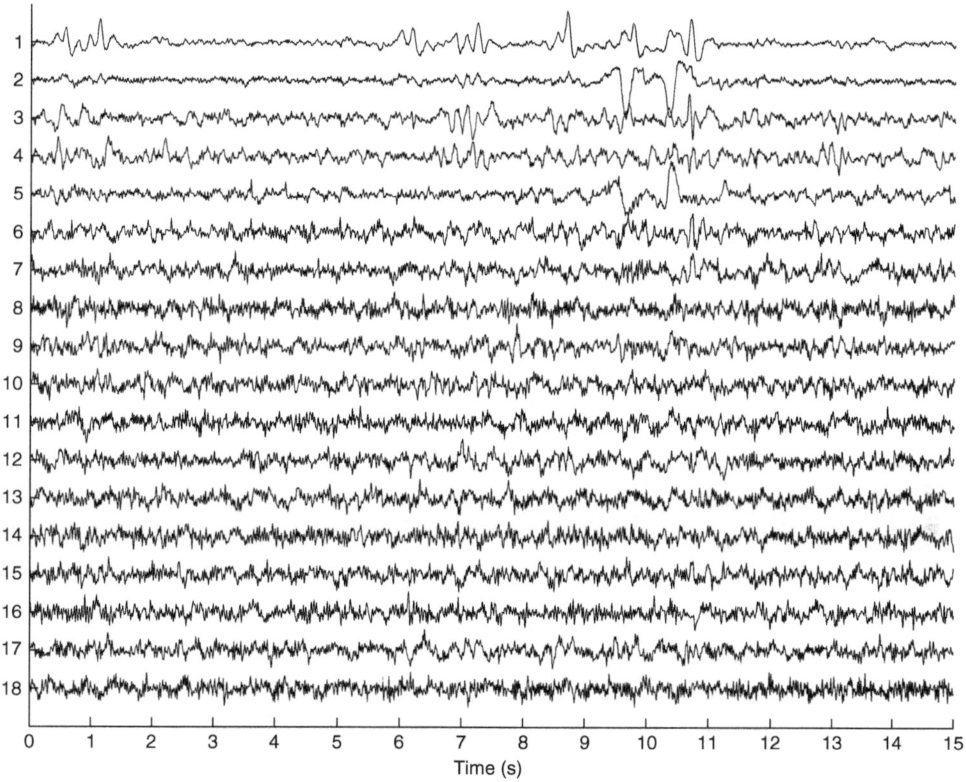

Figure 20 Brain signal extraction in the EEG: decomposition of the EEG fragment of Figure 17 into its source signals by PCA (Section 5.5.1). Ocular artifact and IED activity appear mixed among the first five sources. Amplitudes are normalized to yield unit power sources.

demonstrate that relying only on second-order statistical measures like variance and uncorrelation as separation parameters, though useful in specific scenarios, fails to yield physiologically or clinically significant information in the general case.

As illustrated by the above example, a decomposition into independent components using standard ICA techniques can shed new light onto EEG processing [44,45]. Indeed, ICA is a very powerful technique since it requires little a priori information to obtain interpretable results. The reduced amount of prior information renders the approach very robust to modeling errors. The BSS approach also proves useful in the processing of other types of biomedical data, such as in the ECG problems treated earlier in the chapter; see, e.g., [6,22,31,32].

Care should be taken when implementing the theory of ICA in practical settings. The instantaneous linear overdetermined mixture model resolves mixtures of sources in the case where the mixture is constant over the observed window length. Also, the number of sources that can be extracted is limited by the number of sensors used to record the activity (or, rather, by the dimension of the observed signal subspace, as seen in the previous example). Although the standard algorithms usually perform well and provide useful clinical and physiological information, more advanced methods circumventing the above limitations can be developed by replacing, altering or adding assumptions to the separation criteria. In what follows, we briefly summarize some promising developments adding on the capabilities of classical ICA in multisensor biomedical data analysis. This overview does not pretend to be exhaustive.

5.8. Incorporating prior information into the separation model

The above-mentioned techniques for BSS make little to no use of prior knowledge of the source signals except for assumptions A1–A3 concerning their statistical independence and non-Gaussian character. By inserting some prior knowledge into the objective of ICA, the standard ambiguities mentioned in Section 5.3 can be alleviated while improving the performance of the resulting techniques. Recent methods have been proposed by inserting priors into the ICA contrast and/or into the ICA-updating algorithm. The following sections shortly highlight some interesting contributions to this topic.

5.8.1. Temporal reference

The use of a reference signal in conjunction with independence has been proposed in Adib et al. [46]. The method in Lu and Rajapakse [47] is based on the so-called FastICA algorithm [5,48], but includes a Lagrangian term measuring the distance to the reference. The method is reminiscent of the optimal Wiener filtering technique of Section 3 in the exploitation of a reference signal. Its major drawback is precisely the generation of that signal, ideally the source of interest and thus not available a priori. The major advantage of ICA with reference signal is the reduction in computational cost, as a full decomposition with a posteriori selection is reduced into a single signal extraction. Since only the source of interest will be found at the separator output, the permutation ambiguity inherent to classical implementations of ICA is alleviated or even avoided.

5.8.2. Spatial reference

In a given source separation problem, the contribution of a source signal to the sensors is determined by the mixing matrix column associated with that source, as signified mathematically by Eq. (27). A spatial map quantifies the typical configuration of this contribution. In biomedical signal processing, the idea of exploiting a source spatial map seems more practical than using a reference signal, as much more a priori information is available for spatial maps than for time variations [49,50]. Since many processes captured by a multisensor recording system display a rather fixed and known topography, this information can be readily incorporated into ICA contrasts. This yields source separation techniques with spatial maps adapted to epileptic spikes [49] and artifacts [51], among others, thus simplifying the identification of these activities.

In the context of ocular artifact suppression in the EEG, the benefits of exploiting a spatial reference are illustrated in Figure 21. The top plot shows an EEG recording clearly corrupted by eye blinks. The application of the signal extraction method of Phlypo et al. [51], based on a library of spatial maps adapted

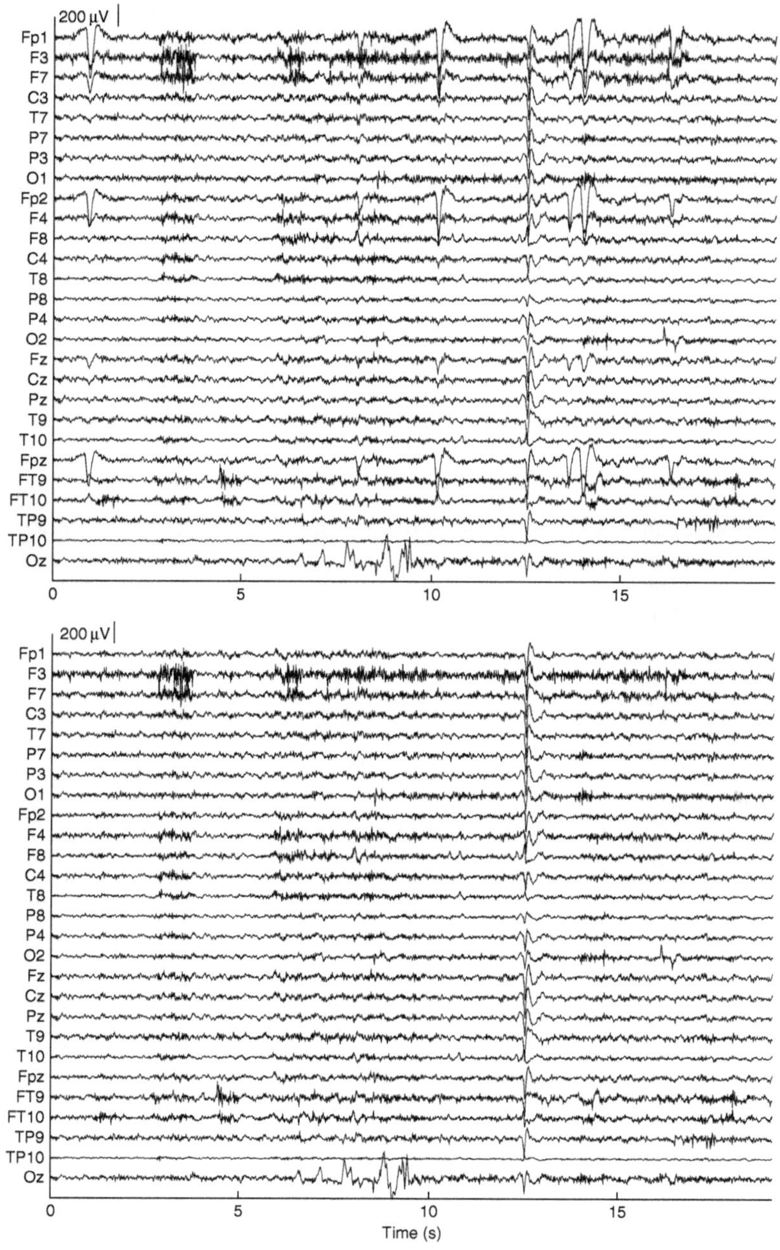

Figure 21 Ocular artifact suppression by exploiting a spatial reference. (Top) The EEG signal with blinks. Signals have been recorded at a sampling frequency of 200 Hz. (Bottom) The EEG fragment after the separation method of [51].

to ocular artifacts, yields the corrected EEG fragment shown in the bottom plot. Since the muscular activity around the 10th second and the spiking activity just before the 15th second have spatial distributions different from that of the ocular artifact, they are not affected by the extraction filter.

5.9. Independent subspaces

Recently, the trend to step forward from ICA techniques toward Independent Subspace Analysis (ISA) has gained increasing attention in the signal processing community. The idea of ISA, originally proposed in Comon [52, p. 125], has been further developed in [53] or [54], among others. This trend goes hand in hand with the dependent component analysis method, that takes prior information about the relation between components to solve the linear mixture system. Left to explore is the influence these novel techniques would have on biomedical signal processing. Since the activity in the three orthogonal axes of the electric heart vector (Section 2.2.4) is temporally dependent, it could reasonably be captured in a three-dimensional measurement subspace, rather than being compressed into a single component. The same observation holds for dependent brain activities (see Section 5.1). Subspace analysis or dependent component analysis could help find coherence between the activities of multiple or spread brain areas.

5.10. Softening the stationarity constraint

If the mixing matrix is constant over a time period, the stationarity of the sources is not mandatory. In fact, when using sample statistics such as moment or cumulants, one actually averages them over the period. What is then important is that these average values should be non-zero, and in particular the underlying source average statistics. Attempts have been made to soften (or entirely discard) the stationary conditions by using, among others, state machines [55] or the ML principle [56]. A similar attempt is made by the algorithm of Phlypo et al. [51] through the combination of short-time and long-term statistics. Another point of view is to exploit independence in the spectral domain. Instead of the ordinary frequency transforms with fixed bases, there is a tendency to use methods that adapt the spectral basis to the data in hand, either through decorrelation [57] or by using HOS as in ICA [58], are drawing increasing interest. Note that many works exploit non-stationarity of the sources when the mixture is constant [56]. In that case, having non-stationary sources is an advantage, not a drawback.

5.11. Revealing more sources than sensor signals

Mixtures having fewer sensors than sources are referred to as underdetermined since 1999 [59]. The difficulty when considering underdetermined mixtures lies in the non-invertibility of the mixing matrix $\mathbf{A}$, since it is no longer full-column rank. As a result, the sources $\mathbf{s}$ cannot be exactly recovered by a linear projection of the observation vector $\mathbf{x}$ onto the columns of $\mathbf{A}$. The problem of extracting the sources in underdetermined mixtures can be addressed by three main approaches: (1) build a non-linear extractor [59], possibly based on a Maximum A Posteriori estimator (well-matched to the case of discrete sources); (2) exploit a possible source sparseness [60,61], whereby only a few sources are assumed to be active simultaneously in the observation window; (3) exploit possible additional diversities by storing the recorded information in a high-dimensional array (a matrix with more than two dimensions, e.g., a cube-like data structure) and performing its decomposition [62]. This last option, and perhaps also the use of sparsity in specific scenarios, would appear as the most promising in biomedical applications. On the other hand, if one is only interested in detecting the number of sources or identifying the mixture (e.g., for source localization), decomposing the cumulant tensor [63] or the characteristic function [64] suffices, and weaker assumptions apply.

6. Summary, Conclusions and Outlook

The extraction of signals of interest from measurements corrupted by noise and interference is a fundamental problem arising in a wide variety of domains. In biomedicine, a successful signal extraction can help the physician diagnose and understand a pathologic condition. The signal extraction techniques presented in this chapter rely on the availability of multiple spatially separated sensors recording simultaneously the biomedical activity under study. Exploiting the spatial dimension brought about by multisensor measurements gives rise to the concept of spatial filtering, and enables powerful signal processing approaches with capabilities reaching far beyond conventional single-channel frequency filters. Specific methods depend on the particular problem in hand, its model and its constraints. Our purpose has been to provide the reader with an overview of some existing signal extraction tools.

Approximating the bioelectrical activity by equivalent current dipoles leads to the definition of signal models in which sources and observations are linearly related. As a result, signal extraction can be carried out by linear processing, namely, the appropriate weighted average of the measured signals. The techniques addressed in this chapter aim at finding the weights or coefficients of the extracting spatial filter by exploiting specific properties of the signals of interest. Such weights are optimal under specific assumptions and extraction criteria. Classical Wiener filtering and spatio-temporal cancellation exploit the uncorrelation between the signal of interest and the interference. Uncorrelation is measured using second-order statistics (covariance). The former method relies on reference signals correlated with the interference but uncorrelated with the desired signal, which demands a careful electrode placement. The latter technique assumes the repetitive character of the interference. BSS is a more general approach capable of dealing with (practically) arbitrary electrode location. The use of statistical independence, a property stronger than uncorrelation and measured using HOS (cumulants), enables the extraction of any type of sources as long as they are non-Gaussian. Even if the relationship between bioelectrical sources and observations is actually more involved than a simple equivalent dipole, these techniques are often able to extract physiologically meaningful and clinically relevant sources.

Each of these signal processing tools has been introduced in the context of a suitable biomedical application, but could also be adapted and applied to the other problems addressed in this chapter. Some comparisons can be found in [6,22,43,65], among other works. Although our focus has been on bioelectrical recordings (ECG, EEG), signals conveying genetic (DNA), chemical, or other kind of biomedical information can also be processed by these techniques. Likewise, the methods are readily applicable, with possible modifications, in other fields such as telecommunications (see, e.g., [39,66–69] and references therein).

Many open problems remain to be tackled in the area of signal extraction in multisensor biomedical recordings. A fundamental question concerns the relationship between the sources estimated by signal extraction methods and the actual sources of underlying physiological activity. As discussed at the end of the previous section, the applicability of existing techniques needs to be extended to more challenging environments such as underdetermined scenarios, and their performance improved by incorporating prior information into the extraction criteria. Multimodal processing is emerging as another promising avenue of research. Signals simultaneously acquired in different modalities (EEG and MEG, EEG and fMRI, PET and SPECT, etc.) can be analyzed sequentially or in parallel to perform mutual validation or to increase the information provided separately by each modality. The potential benefits are many-fold. For instance, the combination of EEG and fMRI can improve both temporal (with respect to fMRI) and spatial (with respect to EEG) resolution, while PET and SPECT can provide information about both anatomical structures and functional behaviour. Recording multimodal measurements, e.g., an EEG in a high magnetic field such as an MRI scanner represents an interesting technical problem in itself. Designing novel signal processing algorithms fully exploiting the information available in the recorded modalities constitutes an exciting research challenge likely to draw the attention of the biomedical signal processing community over the next decades.

ACKNOWLEDGMENTS

This work has been partly supported by contract "DECOTES" (D6-BLAN-0074-01) from Agence National de la Recherche. Most of Figure 5 has been reproduced with permission from the Web edition of Malmivuo and Plonsey [12]. The ECG recording of Figure 9 has been kindly provided by L. De Lathauwer, D. Callaerts, and J. Vandewalle, from K. U. Leuven, Belgium. The ECG recordings of Figures 13 and 14 are courtesy of the Signal Processing Group, Lund University, Sweden. The EEG data of Figures 17 and 21 (top) are due to P. Boon from the Laboratory for Clinical and Experimental Neurology, Ghent University Hospital, Belgium.

REFERENCES

[1] B. Widrow, J. R. Glover, J. M. McCool, et al., Adaptive noise cancelling: principles and applications, Proceedings of the IEEE 63 (12) (1975) 1692–1716.
[2] M. Stridh, L. Sörnmo, Spatiotemporal QRST cancellation techniques for analysis of atrial fibrillation, IEEE Transactions on Biomedical Engineering 48 (1) (2001) 105–111.
[3] P. Comon, Independent component analysis, a new concept? Signal Processing 36 (3) (1994) 287–314, Special Issue on Higher-Order Statistics.
[4] J.-F. Cardoso, Blind signal separation: statistical principles, Proceedings of the IEEE 86 (10) (1998) 2009–2025.
[5] A. Hyvärinen, J. Karhunen, E. Oja, Independent Component Analysis, John Wiley & Sons, New York, 2001.
[6] L. De Lathauwer, B. De Moor, J. Vandewalle, Fetal electrocardiogram extraction by blind source subspace separation, IEEE Transactions on Biomedical Engineering 47 (5) (2000) 567–572, Special Topic Section on Advances in Statistical Signal Processing for Biomedicine.
[7] R. Gulrajani, Bioelectricity and Bioelectromagnetism, John Wiley & Sons, New York, 1998.
[8] H. Hallez, B. Vanrumste, R. Grech, J. Muscat, Y. D'Asseler, W. D. Clercq, A. Vergult, K. P. Camilleri, S. G. Fabri, R. Van de Walle, S. Van Huffel, I. Lemahieu, Review on solving the forward problem in EEG source analysis, Journal of Neuroengineering and Rehabilitation 4 (46), (November 2007) doi:10.1186/1743-0003-4-46, 29 pages.
[9] F. Bijmaa, J. C. de Munck, K. B. Böckerb, H. M. Huizengac, R. M. Heethaard, The coupled dipole model: an integrated model for multiple MEG/EEG data sets, Neuroimage 23 (3) (2004) 890–904.
[10] R. Pascual-Marqui, C. Michel, D. Lehmann, Low resolution electromagnetic tomography: a new method for localizing electrical activity in the brain, International Journal of Psychophysiology 18 (1994) 49–65.
[11] R. M. Rangayyan, Biomedical Signal Analysis. A Case-Study Approach, John Wiley & Sons, Inc., New York, 2002.
[12] J. Malmivuo, R. Plonsey, Bioelectromagnetism: Principles and Applications, Oxford University Press, New York, 1995.
[13] R. Plonsey, Bioelectric Phenomena, McGraw-Hill, New York, 1969.
[14] R. Plonsey, R. C. Barr, Bioelectricity: A Quantitative Approach, 2nd Edition, Kluwer Academic Publishers, New York, 2000.
[15] A. van Oosterom, Beyond the dipole: modeling the genesis of the electrocardiogram, in: Schalij, Janse, van Oosterom, Wellens, van der Wal (Eds.), Einthoven 2002: 100 Years of Electrocardiography, The Einthoven Foundation, Leiden, The Netherlands, 2002, Ch. 2, pp. 7–15.
[16] American Clinical Neurophysiology Society, Guideline 5: Guidelines for standard electrode position nomenclature https://www.acns.org//pdfs/ACFDD46.pdf (2006).
[17] A. Gunji, R. Ishii, W. Chau, R. Kakigi, C. Pantev, Rhythmic brain activities related to singing in humans, NeuroImage 34 (2007) 426–434.
[18] D. J. Amit, Modeling Brain Function, Cambridge University Press, Cambridge, UK, 1992.
[19] P. Robinson, C. Rennie, D. Rowe, S. O'Connor, E. Gordon, Multiscale brain modelling, Philosophical Transactions of the Royal Society of London B 360 (2005) 1043–1050.
[20] G. O. Glentis, K. Berberidis, S. Theodoridis, Efficient least squares adaptive algorithms for FIR transversal filtering, IEEE Signal Processing Magazine 16 (4) (1999) 13–41.
[21] P. Comon, D. T. Pham, An error bound for a noise canceller, IEEE Transactions on Acoustics, Speech and Signal Processing 37 (10) (1989) 1513–1517.
[22] V. Zarzoso, A. K. Nandi, Noninvasive fetal electrocardiogram extraction: blind separation versus adaptive noise cancellation, IEEE Transactions on Biomedical Engineering 48 (1) (2001) 12–18.
[23] P. Langley, J. Bourke, A. Murray, Frequency analysis of atrial fibrillation, IEEE Computer Society Press, Proc. Computers in Cardiology, Vol. 27, Boston, Massachusetts, 2000, 269–272.
[24] M. Stridth, L. Sörnmo, Characterization of atrial fibrillation using the surface ECG: time-dependent spectral properties, IEEE Transactions on Biomedical Engineering 48 (1) (2001) 19–27.
[25] J. Lian, D. Müssig, V. Lang, Computer modeling of ventricular rhythm during atrial fibrillation and ventricular pacing, IEEE Transactions on Biomedical Engineering 53 (8) (2006) 1512–1520.

[26] P. Bonizzi, O. Meste, V. Zarzoso, Atrio-ventricular junction behaviour during atrial fibrillation, IEEE Computer Society Press, Proc. Computers in Cardiology, Vol. 34, Durham, North Carolina, 2007, pp. 561–564.

[27] F. Castells, C. Mora, J. J. Rieta, D. Moratal-Pérez, J. Millet, Estimation of atrial fibrillatory wave from single-lead atrial fibrillation electrocardiograms using principal component analysis concepts, Medical and Biological Engineering and Computing 43 (5) (2005) 557–560.

[28] G. H. Golub, C. F. Van Loan, Matrix Computations, 3rd Edition, The Johns Hopkins University Press, 1996.

[29] F. Castells, P. Laguna, L. Sörnmo, A. Bollmann, J. Millet Roig, Principal component analysis in ECG signal processing, EURASIP Journal on Advances in Signal Processing (2007), doi:10.1155/2007/74580, 21 pages.

[30] O. Meste, N. Serfaty, QRST cancellation using Bayesian estimation for the auricular fibrillation analysis, in: Proceedings of the 27th Annual International Conference of the IEEE Engineering in Medicine and Biology Society, Shanghai, China, 2005, pp. 7083–7086.

[31] J. J. Rieta, F. Castells, C. Sánchez, V. Zarzoso, J. Millet, Atrial activity extraction for atrial fibrillation analysis using blind source separation, IEEE Transactions on Biomedical Engineering 51 (7) (2004) 1176–1186.

[32] F. Castells, J. J. Rieta, J. Millet, V. Zarzoso, Spatiotemporal blind source separation approach to atrial activity estimation in atrial tachyarrhythmias, IEEE Transactions on Biomedical Engineering 52 (2) (2005) 258–267.

[33] J. Slocum, E. Byrom, L. McCarthy, A. Sahakian, S. Swiryn, Computer detection of atrioventricular dissociation from the surface electrocardiogram during wide QRS tachycardia, Circulation 72 (1985) 1028–1036.

[34] G. Beltramello, P. Manganotti, G. Zanette, C. Miniussi, A. Marravita, E. Santorum, A. Polo, C. Marzi, A. Fiaschi, B. Bernardina, Coregistration of EEG and fMRI in rolandic epilepsy with evoked spikes by peripheral tapping stimulation, Electroencephalography and Clinical Neurophysiology 103 (1) (1997) 83.

[35] X. D. Tiége, H. Laufs, S. G. Boyd, W. Harkness, P. J. Allen, C. A. Clark, A. Connelly, J. H. Cross, EEG-fMRI in children with pharmacoresistant focal epilepsy, Epilepsia 48 (2) (2007) 385–389.

[36] E. L. So, Integration of EEG, MRI, and SPECT in localizing the seizure focus for epilepsy surgery, Epilepsia 41 (Suppl 3) (2000) S48–S54.

[37] J. M. Stern, J. Engel Jr., Atlas of EEG Patterns, Lippincott, Williams & Wilkins, Hagerstown, Maryland, 2004.

[38] L. Tang, M. Mantle, P. Ferrari, H. Schiffbauer, H. A. Rowley, N. M. Barbaro, M. S. Berger, T. P. L. Roberts, Consistency of interictal and ictal onset localization using magnetoencephalography in patients with partial epilepsy, Journal of Neurosurgery 98 (4) (2003) 837–845.

[39] P. Comon, C. Jutten, Séparation de sources: concepts de base et analyse en composantes indépendantes, Vol. 1 of Collection IC2, Hermes, 2007, ISBN 978-2-7462-1517-7.

[40] P. McCullagh, Tensor Methods in Statistics, Monographs on Statistics and Applied Probability, Chapman and Hall, London, 1987.

[41] A. Stuart, J. K. Ord, Kendall's Advanced Theory of Statistics, 6th Edition, Vol. I, Edward Arnold, London, 1994.

[42] D. T. Pham, P. Garat, Blind separation of mixture of independent sources through a quasi-maximum likelihood approach, IEEE Transactions on Signal Processing 45 (7) (1997) 1712–1725.

[43] A. Kachenoura, L. Albera, L. Senhadji, P. Comon, ICA: a potential tool for BCI systems, IEEE Signal Processing Magazine 25 (1) (2008), Special Issue on Brain-Computer Interfaces 57–68.

[44] S. Makeig, A. J. Bell, T.-P. Jung, T. J. Sejnowski, Independent component analysis of electroencephalographic data, in: Advances in Neural Information Processing Systems (NIPS) Foundation, Vol. 8, 1996, pp. 145–151.

[45] A. Delorme, S. Makeig, EEGLAB: an open source toolbox for analysis of single trial EEG dynamics, Journal of Neuroscience Methods 134 (2004) 9–21.

[46] A. Adib, E. Moreau, D. Aboutajdine, Source separation contrasts using a reference signal, IEEE Signal Processing Letters 11 (3) (2004) 312–315.

[47] W. Lu, J. C. Rajapakse, Approach and applications of constrained ICA, IEEE Transactions on Neural Networks 16 (1) (2005) 203–212.

[48] A. Hyvärinen, E. Oja, A fast fixed-point algorithm for independent component analysis, Neural Computation 9 (7) (1997) 1483–1492.
[49] N. Ille, P. Berg, M. Scherg, Artifact correction of the ongoing EEG using spatial filters based on artifact and brain signal topographies, Journal of Clinical Neurophysiology 19 (2) (2002) 113–124.
[50] C. W. Hesse, C. J. James, The FastICA algorithm with spatial constraints, IEEE Signal Processing Letters 12 (11) (2005) 792–795.
[51] R. Phlypo, Y. D'Asseler, I. Lemahieu, Removing ocular movement artefacts by a joint smoothened subspace estimator (JSSE), Computational Intelligence and Neuroscience, 2007 doi: 10.1155/2007/75079.
[52] P. Comon, Supervised classification, a probabilistic approach, in: Verleysen (Ed.), ESANN-European Symposium on Artificial Neural Networks, D facto Publ., Brussels, 1995, pp. 111–128, invited.
[53] J.-F. Cardoso, Multidimensional independent component analysis, in: Proc. ICASSP-98, 23rd IEEE International Conference on Acoustics, Speech and Signal Processing, Seattle, 1998, pp. 1941–1944.
[54] F. Theis, Blind signal separation into groups of dependent signals using block diagonalization, in: Proc. ISCAS-2005, IEEE International Symposium on Circuits and Systems, Kobe, Japan, 1995, pp. 5878–5881.
[55] J.-I. Hirayama, S.-I. Maeda, S. Ishii, Markov and semi-Markov switching of source appearances for nonstationary independent component analysis, IEEE Transactions on Neural Networks 18 (5) (2007) 1326–1342.
[56] D.-T. Pham, J.-F. Cardoso, Blind separation of instantaneous mixtures of non stationary sources, IEEE Transactions on Signal Processing 49 (9) (2001) 1837–1848.
[57] R. Phlypo, Y. D'Asseler, I. Lemahieu, Extracting common spectral features by multichannel filtering using circulant matrices, in: Proc. ISSPA-2007, 9th International Symposium on Signal Procession and its Applications, Sharjah, United Arab Emirates, 2007.
[58] M. Davies, C. James, S. Wang, Space-time ICA and EM brain signals, in: M. Davies, C. James, S. Abdallah, M. Plumbley (Eds.), Proc. ICA-2007, 7th International Conference on Independent Component Analysis and Source Separation, Lecture Notes on Computer Science (LNCS 4666), London, UK, 2007, pp. 577–584.
[59] P. Comon, O. Grellier, Non linear inversion of underdetermined mixtures, in: Proc. ICA-99, 1st International Workshop on Independent Component Analysis and Blind Source Separation, Aussois, France, 1999, pp. 461–465.
[60] P. Bofill, M. Zibulevsky, Underdetermined blind source separation using sparse representation, Signal Processing 81 (11) (2001) 2353–2362.
[61] R. Gribonval, Sparse decomposition of stereo signals with matching pursuit and application to blind separation of more than two sources from a stereo mixture, in: Proc. ICASSP-2002, 27th IEEE International Conference on Acoustics, Speech and Signal Processing, Vol. III, Orlando, Florida, 2002, pp. 3057–3060.
[62] N. D. Sidiropoulos, R. Bro, G. B. Giannakis, Parallel factor analysis in sensor array processing, IEEE Transactions on Signal Processing 48 (8) (2000) 2377–2388.
[63] L. Albera, A. Ferreol, P. Comon, P. Chevalier, Blind identification of overcomplete mixtures of sources (BIOME), Linear Algebra and its Applications 391 (2004) 1–30.
[64] P. Comon, M. Rajih, Blind identification of under-determined mixtures based on the characteristic function, Signal Processing 86 (9) (2006) 2271–2281.
[65] P. Langley, M. Stridh, J. J. Rieta, J. Millet, L. Sörnmo, A. Murray, Comparison of atrial signal extraction algorithms in 12-lead ECGs with atrial fibrillation, IEEE Transactions on Biomedical Engineering 53 (2) (2006) 343–346.
[66] N. Delfosse, P. Loubaton, Adaptive blind separation of independent sources: a deflation approach, Signal Processing 45 (1) (1995) 59–83.
[67] L. Castedo, C. Escudero, A. Dapena, A blind signal separation method for multiuser communications, IEEE Transactions on Signal Processing 45 (5) (1997) 1343–1348.
[68] S. Houcke, A. Chevreuil, P. Loubaton, Blind source separation of a mixture of communication sources emitting at various baud-rates, Transactions of IEICE E86-A (3) (2003) 564–572.
[69] M. E. Davies, C. J. James, S. A. Abdallah, M. D. Plumbley (Eds.), Proc. ICA-2007, 7th International Conference on Independent Component Analysis and Signal Separation, Vol. 7 of Lecture Notes in Computer Science (LNCS 4666), Springer, London, UK, 2007.

CHAPTER 4

Fluorescence Lifetime Spectroscopy and Imaging of Visible Fluorescent Proteins

Ankur Jain, Christian Blum, *and* Vinod Subramaniam

Contents

1. Introduction	146
2. Introduction to Fluorescence	146
2.1. Interaction of light with matter	146
2.2. The Jabłoński diagram	147
2.3. Fluorescence parameters	151
2.4. Fluorescence lifetime	151
2.5. Measurement of fluorescence lifetime	153
2.6. Fluorescence anisotropy and polarization	155
2.7. Factors affecting fluorescence	157
3. Fluorophores and Fluorescent Proteins	160
3.1. Green fluorescent protein	161
3.2. Red fluorescent protein	165
4. Applications of VFPs	166
4.1. Lifetime spectroscopy and imaging of VFPs	167
5. Concluding Remarks	170
References	170

Abstract

The excited state lifetime of a fluorophore is a fundamental parameter that reports sensitively on the fluorophore nanoenvironment and on interactions between the fluorophore and other molecules. When coupled with the spatial resolution inherent to imaging techniques, fluorescence lifetime spectroscopy provides information about the spatial distributions of the molecules and the dynamics of related processes. In combination with genetically encodable visible fluorescent protein fluorophores, fluorescence lifetime imaging enables the real-time visualization of dynamic biological interactions in the complex context of the living cell. This chapter provides an introduction to fluorescence spectroscopy with an emphasis on fluorescence lifetimes and to the use of visible fluorescent proteins and fluorescence lifetime microspectroscopy for biophotonics applications.

Keywords: fluorescence, fluorescence lifetime imaging, FRET, GFP, protein localization, photophysics

Advances in Biomedical Engineering
ISBN 978-0-444-53075-2

1. Introduction

Fluorescence spectroscopy is a sensitive and specific tool for biochemical and biophysical studies [1,2]. Rapid developments in instrumentation, light sources, detection methods, and fluorescent reagents have established the prominent role of fluorescence spectroscopy and imaging in biophotonics [3]. The influence of the local nano- and microenvironment on fundamental fluorescence observables such as spectral shape and position, lifetime, and polarization make it a powerful tool for studying molecular structure and dynamics [4–7]. The combination of fluorescence spectroscopy and microscopy with the discovery and development of genetically encodable fluorescent proteins [8,9] has enabled the in vivo visualization [10] of molecular targeting, interactions, and dynamics in intact cells and organisms. The multitude of spectroscopic properties of fluorescence can be further exploited in an imaging mode [11] to obtain information not only about the macromolecules in a spatially resolved manner but also about the immediate molecular nanoenvironment. The information content is significantly enhanced as the fluorescence observables are time resolved [12,13], and provide a means to elucidate and monitor the spatio-temporal distributions and functional states of constituent molecules of the biological systems.

2. Introduction to Fluorescence

The ubiquitous phenomenon of emission of light when molecules revert to the ground state from an electronically excited state created by physical, mechanical, or chemical mechanism is known as luminescence. Luminescence generated by the excitation of a molecule by ultraviolet or visible photons is called photoluminescence, which can be formally divided into two categories, fluorescence and phosphorescence, depending upon the electronic configuration of the excited state and the emission pathway. The category of molecules capable of undergoing electronic transitions that result in fluorescence are known as fluorochromes or fluorophores.

2.1. Interaction of light with matter

Light is an electromagnetic wave consisting of electric and magnetic fields oscillating orthogonal to the direction of propagation. The energy associated with a light wave with frequency ν is given by

$$E = h\nu = \frac{hc}{\lambda} \tag{1}$$

where h is the Planck's constant, c is the speed of light in vacuum, and λ is the wavelength of the incident photon. Because of the electric field associated with the light wave, when a charged particle is placed in the path of a light wave, it

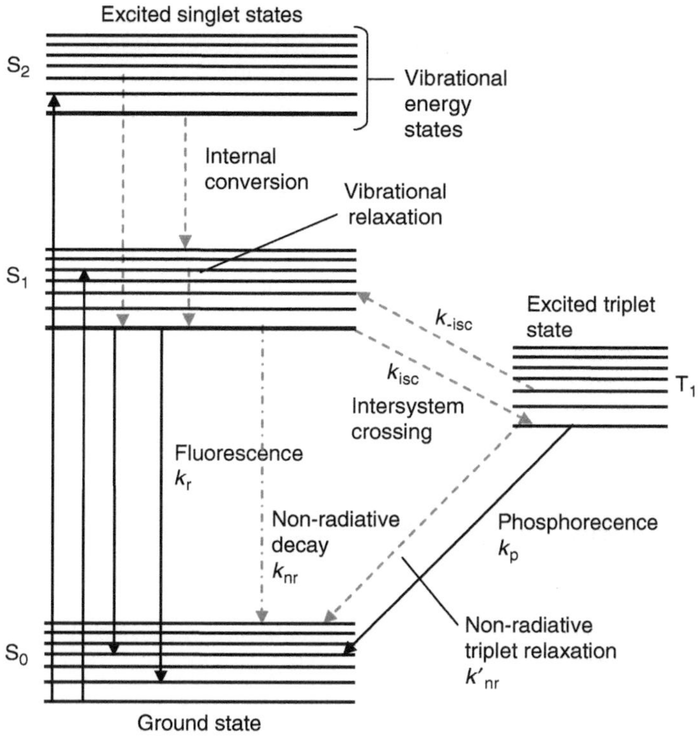

Figure 1 Jabłoński diagram illustrating energy levels and photophysical processes. The different deactivation pathways through which an excited molecule can return to its ground state are labeled with the corresponding rate constants: Fluorescence k_r, non-radiative decay k_{nr}, intersystem crossing k_{isc}, and phosphorescence k_p. Processes such as quenching, energy transfer, and photochemical reactions are not included. The energy levels shown are not to scale.

experiences a force and is capable of absorbing energy from the electric field of the light wave. When an electron in the ground state molecular orbital absorbs this energy, it is promoted to a higher unoccupied molecular orbital, known as the excited state of the electron. Figure 1 depicts a schematic 1D representation of the energies of molecular transitions.

2.2. The Jabłoński diagram

The processes that can occur during an excitation cycle are conveniently illustrated by a Jabłoński diagram, named after the Polish scientist Alexander Jabłoński. In this diagram (Figure 1), the various energy levels of the molecule are depicted by a stack of horizontal lines. The singlet ground, first, second and *n*th electronic excited states are labeled as S_0, S_1, S_2 and S_n, respectively. The excited triplet states are similarly labeled by T_1, T_2, and so on. The vibrational sublevels of each electronic state are denoted by the vibrational quantum numbers 0, 1, 2,... in increasing order of energy. The processes involving the transition of the electron between the energy states are represented by vertical arrows. The fluorescence process can be divided into three major events: absorption, vibrational relaxation, and emission.

2.2.1. Absorbance

The electronic transition accompanying absorption of a photon by a molecule is known as electronic absorption or electronic excitation. The energy required to change the electronic distribution is of the order of a few electron volts for most fluorophores; thus, the corresponding photons lie in the visible or ultraviolet region of the spectrum. The absorption of a photon promotes the molecule from the ground electronic state to a higher vibronic sublevel of an excited singlet state. Absorption mainly occurs from the vibronic ground states of molecules, since, at room temperature, thermal energy is not sufficient to significantly populate the excited vibrational states. The spacing of the vibrational energy levels of the excited states is similar to that of the ground state. Electronic transitions are fast compared with the time scale of nuclear motions so that vibrational levels that correspond to a minimal change in the nuclear coordinates are favored. This approximation that an electronic transition occurs within a stationary nuclear framework is known as the Franck–Condon principle (Figure 2).

The extent to which light of a given wavelength is absorbed by a substance is given by the Beer-Lambert law:

$$A(\lambda) = -\log \frac{I(\lambda)}{I_0(\lambda)} = -\varepsilon(\lambda)Cl \quad (2)$$

where the quantity $A(\lambda)$ represents the absorbance of a given sample, I_0 and I are the intensities of light of wavelength λ incident on and transmitted through the absorber respectively, C is the concentration of absorbing molecules, and l is

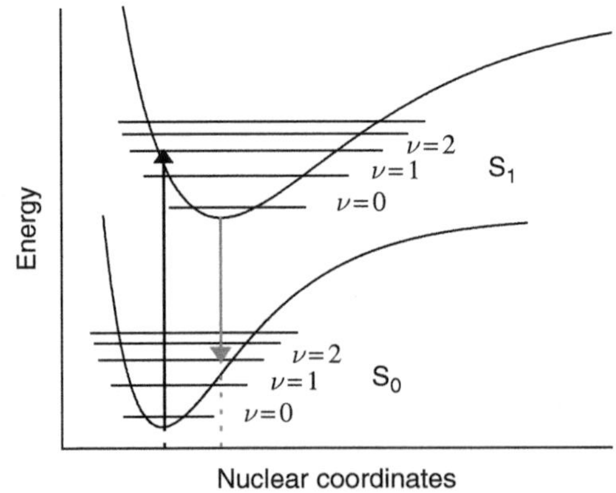

Figure 2 Franck–Condon principle. Since electronic transitions are very fast compared with nuclear motions, vibrational levels are favored when they correspond to a minimal change in the nuclear coordinates. The potential wells are shown favoring transitions between $v = 0$ and $v = 2$.

the path length of light through the sample. $\varepsilon(\lambda)$ is a molecular constant under a given set of physical conditions and solvent, related to the electronic transition dipole moment, and is called the molar extinction coefficient of the sample. A graphical representation of the absorbance of a species as a function of wavelength or frequency is known as the absorption spectrum. Regions of the spectrum where the intensity of absorbed light is high are known as absorption bands. Spectrally broad absorption bands are a consequence of closely spaced vibrational energy levels together with the thermal motion that enables a range of photon energies to cause a particular transition.

2.2.2. Non-radiative relaxation processes

The absorption of a photon leaves a molecule in one of the possible higher vibrational states of one of its electronically excited states. Excess vibrational energy is dissipated as heat by a process termed as *internal conversion* and the molecule relaxes to the ground vibrational level of the first excited singlet state within a few picoseconds, an effect known as *vibrational relaxation*. As the time required for internal conversion is typically much less than that for fluorescence emission, the following transition occurs from the lowest vibrational sublevel of the lowest excited state.

The primary mechanism of relaxation of excited non-emitting molecules is radiationless decay, in which the excess energy is transferred into the vibration, rotation, and translation of the surrounding molecules. The excitation energy is thus dissipated as heat, and the molecule returns to the ground electronic state.

The electrons in the excited singlet state S_1 can also undergo *intersystem crossing*, a non-radiative transition between states of different spin multiplicity. A transition to the triplet state T_1 is usually followed by another radiative (phosphorescence) or non-radiative deactivation. In addition, a molecule in its excited state can non-radiatively transfer its excitation energy to other molecules, either by collision, a process known as dynamic quenching, or by resonance energy transfer.

2.2.3. Radiative relaxation of electronically excited molecules

Molecules that do not possess effective non-radiative decay channels can dissipate their excitation energy upon radiation of a photon. Depending on the electronic nature of the excited state, one can distinguish between fluorescence and phosphorescence.

2.2.3.1. Fluorescence

The processes of internal conversion and vibrational relaxation leave the electronically excited molecule in the lowest vibrational sublevels of the first singlet excited state. If subsequent relaxation to the ground electronic state is accompanied by the emission of a photon, the process is formally known as fluorescence. An excited molecule capable of fluorescence typically stays in the lowest excited singlet state S_1 for a period of several nanoseconds before finally relaxing to the ground state. Internal conversion and vibrational relaxation are responsible for the emitted photon having lower energy than the incident photon. This shift to a longer wavelength is a phenomenon known as the Stokes shift. The Stokes shift is essential for the sensitivity of fluorescence detection because it allows effective separation of the fluorescence emission signal from Rayleigh-scattered excitation light. As emission occurs from the lowest vibrational sublevel of first excited singlet state, the emission spectra are, in general, independent of excitation wavelength.

Emission spectra are usually mirror symmetric to the lowest energy absorption band. This symmetry is a result of relaxation to the vibrationally excited states of the electronic ground state S_0, and the similarities of the vibrational energy levels of S_0 and S_1. According to the Franck–Condon principle, all electronic transitions are vertical (Figure 2) that is, they occur without change in the position of the nuclei. Thus, the various transition probabilities (Franck–Condon factors) are similar and lead to the symmetric nature of the absorption and emission spectra.

2.2.3.2. *Phosphorescence*

The emission of a photon from the excited triplet state T_1 is defined as phosphorescence. The phosphorescence spectrum is generally shifted to longer wavelengths relative to the fluorescence spectrum. Although transitions between states of different spin multiplicity are strictly forbidden, singlet states occasionally have some triplet character due to mixing of the singlet and triplet wavefunctions, thus relaxing this selection rule. Nevertheless, the probability for such a transition remains very small, and as a consequence, the triplet–singlet transition is very long-lived and results in phosphorescence lifetimes on the order of milliseconds to a few seconds.

2.2.3.3. *Delayed fluorescence*

Occasionally, in rigid and viscous media at very low temperature, another long-lived emission band is observed at a frequency corresponding to that of fluorescence but with a decay time similar to phosphorescence. This phenomenon is known as delayed fluorescence and is a consequence of thermal excitation of the molecule from the first excited triplet state back to the first excited singlet state, followed by fluorescence emission.

2.2.4. Timescales of various fluorescence processes

The various relaxation processes of an excited molecule occur on timescales differing by several orders of magnitude; characteristic rate constants and timescales are summarized in Table 1 [1]. The initial absorption of a quantum of energy takes place within a few femtoseconds. The vibrational relaxation takes place in less than a few picoseconds. The final emission of a longer wavelength photon from the first excited singlet state, that is, fluorescence emission, occurs in a relatively longer time period (several nanoseconds), whereas phosphorescence lifetimes can range from several milliseconds to a few seconds. The excited state lifetime is further influenced by processes such as quenching, energy transfer, and intersystem crossing, which provide for alternative decay pathways.

Table 1 Timescales of various fluorescence processes

Transition	Process	Rate constant	Timescale (s)
$S_0 \rightarrow S_1$ or S_n	Absorption (Excitation)		10^{-15}
$S_n \rightarrow S_1$	Internal conversion	k_{ic}	10^{-14} to 10^{-10}
$S_1 \rightarrow S_1$	Vibrational relaxation	k_{vr}	10^{-12} to 10^{-10}
$S_1 \rightarrow S_0$	Fluorescence	k_r	10^{-9} to 10^{-7}
$S_1 \rightarrow T_1$	Intersystem crossing	k_{isc}	10^{-10} to 10^{-8}
$S_1 \rightarrow S_0$	Non-radiative relaxation quenching	k_{nr}, k_q	10^{-7} to 10^{-5}
$T_1 \rightarrow S_0$	Phosphorescence	k_p	10^{-3} to 10^{0}
$T_1 \rightarrow S_0$	Non-radiative relaxation quenching	k'_{nr}, k'_q	10^{-3} to 10^{0}

2.3. Fluorescence parameters

Three fundamental parameters commonly used in describing and comparing the effectiveness of fluorophores are the extinction coefficient (ε), quantum yield (Φ), and the brightness of fluorescence (B). The extinction coefficient is a measure of the amount of impedance a given substance offers to the passage of electromagnetic radiation of a given wavelength. It is related to the absorbance of the given sample by the Beer-Lambert law [Eq. (2)]. The extinction coefficient is the characteristic ability of a fluorophore to absorb incident light of a given wavelength.

The quantum yield is a gauge for measuring the efficiency of fluorescence emission relative to all other pathways of relaxation and is expressed as the dimensionless ratio of the number of photons emitted to the total number of photons absorbed by a fluorophore, or represented in terms of decay rates as

$$\Phi = \frac{k_r}{k_r + k_{nr}} \tag{3}$$

where k_r and k_{nr} are the radiative and non-radiative decay rate constants respectively, as depicted in Figure 1.

The brightness of a fluorophore is proportional to the ability of a substance to absorb light and fluorescence quantum yield, and is given by

$$B = \Phi \cdot \varepsilon \tag{4}$$

Highly fluorescent molecules have high values of both molar extinction coefficient and fluorescence quantum yield, and thus high absorbance and efficient emission.

2.4. Fluorescence lifetime

The fluorescence lifetime is an intrinsic property of fluorescent probes that is extensively used for studying biomolecules, their microenvironment, and their molecular associations [12,13]. The fluorescence lifetime is the characteristic time that a molecule remains in its excited state before returning to the ground state. During the excited state lifetime, a fluorophore can undergo conformational changes, interact with other molecules, and rotate and diffuse through the local environment.

When a sample containing a fluorophore is excited using a pulse of light, an initial population (n_0) of fluorophores in the excited state is created. This population decays with time due to fluorescence emission and non-radiative processes, a process described by the differential equation:

$$\frac{dn(t)}{dt} = -(k_r + k_{nr})n(t) \qquad (5)$$

where $n(t)$ is the number of molecules in the excited state at time t following the excitation. k_r and k_{nr} are the radiative and non-radiative rate constants respectively, as shown in the Figure 1. This results in an exponential decay of the excited state, given by

$$n(t) = n_0 \exp(-t/\tau). \qquad (6)$$

τ is the fluorescence lifetime of the molecule and is expressed as the reciprocal of the total decay rate,

$$\tau = (k_r + k_{nr})^{-1}. \qquad (7)$$

The fluorescence decay following pulsed excitation is schematically depicted in Figure 3. The natural or radiative lifetime of the fluorophore is defined as $\tau_0 = k_r^{-1}$. The fluorescence lifetime and quantum yield are related by the expression

$$\Phi = \frac{k_r}{k_r + k_{nr}} = \frac{\tau}{\tau_0} \qquad (8)$$

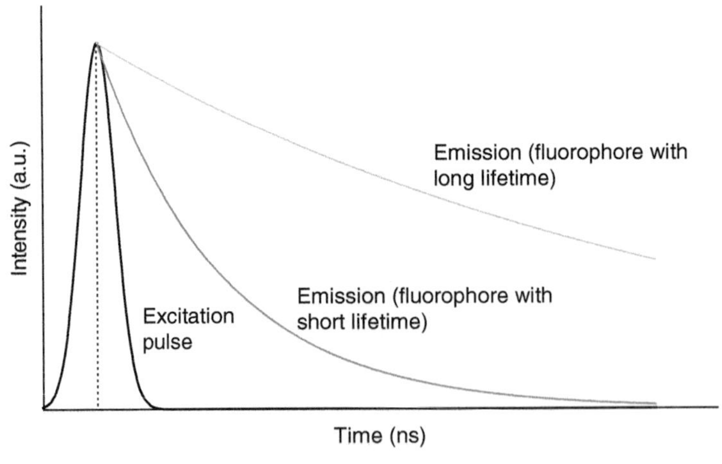

Figure 3 Schematic of fluorescence decay upon pulsed excitation. Fluorophores are excited using a short pulse of light. The emitted fluorescence is measured in a time-resolved manner, forming the basis for time-domain lifetime measurement.

The decay of fluorescence intensity as a function of time in a uniform population of molecules excited with a brief pulse of light is described by an exponential function which follows from Eq. (6):

$$I(t) = I_0 e^{-(t/\tau)} \tag{9}$$

where $I(t)$ is the fluorescence intensity measured at time t.

2.5. Measurement of fluorescence lifetime

Two complementary approaches are widely used for measuring the fluorescence lifetime, the time-domain, and frequency-domain methods.

2.5.1. Time-domain lifetime measurement

In the time-domain method, the sample is excited using a short excitation pulse, and the time-dependent intensity of emission is measured following the excitation pulse (Figure 3). The excited state population decays exponentially (Eq. (6)). In the simplest case, the decay is monoexponential and the lifetime is deduced from the slope of the plot of the logarithm of the intensity versus time. Most fluorophores however display multicomponent decays, and the decay times and associated pre-exponential factors are determined by fitting multiexponential decays to the data. The average lifetime of a fluorophore can in general be calculated by averaging the excited state time over the intensity decay of the fluorophore:

$$\langle t \rangle = \frac{\int_0^\infty tI(t)dt}{\int_0^\infty I(t)dt} = \frac{\int_0^\infty t\exp(-t/\tau)dt}{\int_0^\infty \exp(-t/\tau)dt} \tag{10}$$

Most time-domain fluorimeters are based on the *time-correlated single photon counting* (TCSPC) method [14]. This method relies on the fact that the probability of detecting a photon at time t after the excitation pulse is proportional to the fluorescence intensity at that time. The fluorescence intensity curve is reconstructed after timing and recording the single photons emitted after a large number of excitation pulses. TCSPC systems provide outstanding sensitivity, but data acquisition can be quite time consuming. Another approach is the use of time-gated systems where the photons are detected during a fixed time window following the excitation [15].

An ideal fluorophore exhibits a monoexponential decay of fluorescence as given by Eq. (9). Figure 4 depicts the fluorescence decay of an aqueous solution of rhodamine 6G measured by TCSPC. A monoexponential fit to these data yields a lifetime of ~4.1 ns. However, most biological systems like viable tissues and living cells contain heterogeneous environments and multiple emitting species and thus often yield multiexponential lifetimes.

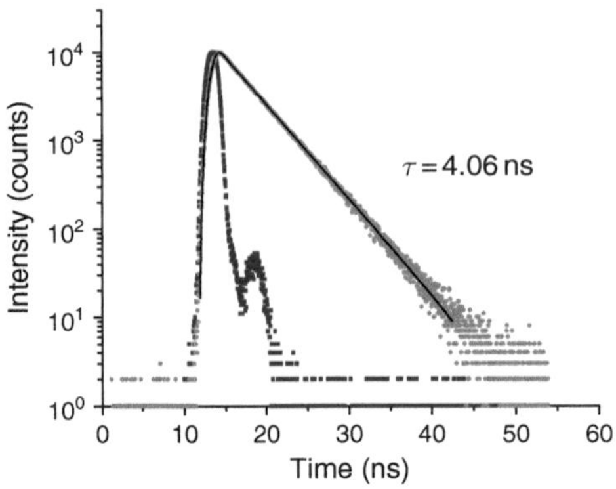

Figure 4 Fluorescence decay of rhodamine 6G obtained using TCSPC. The lifetime is extracted from a monoexponential fit to the experimental data. The measured intensity is shown in light gray dots on a logarithmic scale. The instrument response function is depicted in dark gray.

2.5.2. Frequency-domain lifetime measurement

The alternative method of measuring the decay time is the frequency domain or the phase modulation method [16]. In this case, the sample is excited using a continuous source of excitation that is modulated in intensity at a very high frequency. The resulting fluorescence emission is thus also modulated at the same frequency. However, owing to the finite lifetime of fluorescence, the emission is delayed in time relative to the excitation. This delay is measured as a phase shift (ϕ), which can be used to calculate the decay time, called the phase lifetime (τ_ϕ), given by the equation:

$$\tau_\phi = \omega^{-1} \tan \phi \tag{11}$$

Further, the lifetime leads to a decreased modulation depth of the emission relative to that of the excitation light, as shown in Figure 5. The extent to which

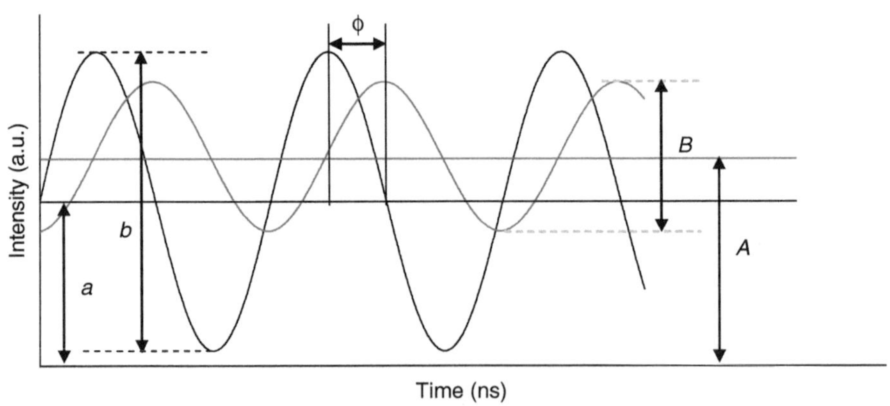

Figure 5 Principle of frequency domain lifetime measurement. The fluorescence lifetime is calculated from the phase shift and demodulation of the emitted light (gray curve) with respect to the sinusoidally modulated excitation light (black curve).

this occurs depends upon the decay time of the fluorophore and the modulation frequency. The demodulation factor, m, is given by

$$m = \frac{B/A}{b/a} \tag{12}$$

where b/a is the modulation of the excitation light and B/A, for the emitted light (Figure 5). The lifetime calculated from the demodulation is called the modulation lifetime (τ_m) and is given by

$$\tau_m = \frac{1}{\omega}\left[\frac{1}{m^2} - 1\right]^{1/2} \tag{13}$$

For monoexponential decays, the phase and modulation lifetimes correspond. A difference in the phase and modulation lifetimes is indicative of multiexponential fluorescence decay or excited state interactions of the fluorophore. Frequency domain lifetime determination is particularly suited for wide-field illumination strategies and spatially-resolved fluorescence lifetime imaging [17].

2.6. Fluorescence anisotropy and polarization

The interaction of the exciting light with the molecule can be described as the interaction of the electric field component of the light with the relevant transition dipole moment of the molecule. For excitation by linearly polarized light, fluorophores whose transition moments are aligned parallel to the electric field vector of the excitation light preferentially absorb photons, a process known as photoselection. This selective excitation results in a partially oriented population of excited fluorophores and thus in a partially polarized emission. The extent of fluorescence polarization is determined by the parameter anisotropy r, given by

$$r = \frac{I_{\parallel} - I_{\perp}}{I_{\parallel} + 2I_{\perp}} \tag{14}$$

where $I_{\parallel}$ and $I_{\perp}$ are the intensities of vertically ($\parallel$) and horizontally ($\perp$) polarized emission, measured relative to the polarization of the excitation light. The limiting anisotropy, r_0, for a randomly oriented molecule upon excitation by linearly polarized light under one photon excitation conditions, is given by

$$r_0 = 0.6\cos^2\beta - 0.2 \tag{15}$$

where β is the angle by which the absorption and emission transition dipole moments of the fluorophore are displaced relative to each other. It follows that the theoretical anisotropy maximum for collinear absorption and emission transition dipole moments is 0.4.

When the emission and excitation transition dipole moments are non-collinear, there will be some depolarization even when the molecules are fixed (Eq. (15)).

Several other phenomena can further decrease the measured anisotropy values, the most common being rotational diffusion during the lifetime of the excited state that results in additional displacement of the emission transition dipole from its original orientation. The rate of rotational diffusion is dependent upon the size of the fluorophore or the macromolecule to which it is bound, and the viscosity of its local environment. The fluorescence anisotropy of a solution, assuming no other processes result in loss of anisotropy, is given by the Perrin equation. This expression relates the observed anisotropy, the fluorescence lifetime of the fluorophore, and its relaxation time:

$$r = \frac{r_0}{1 + (\tau/\phi)} \tag{16}$$

where r is the measured anisotropy, r_0 is the limiting anisotropy of the fluorophore in the absence of rotational diffusion or energy transfer, and τ is the fluorescence lifetime. ϕ is the rotational correlation time of the fluorophore, given by the Stokes equation:

$$\phi = \frac{\eta V}{RT} \tag{17}$$

where η is the viscosity of the environment, V is the volume of the rotating molecule, R is the universal gas constant, and T is the absolute temperature of the solution.

For smaller fluorophores, the rotational correlation times are typically 50–100 ps. As the excited state lifetime is several nanoseconds, fluorophores are able to rotate several times before emission. As a result, the polarized emission is randomized so that the net anisotropy is zero. The absorption, excitation, and fluorescence emission spectra of rhodamine 6G are shown in Figure 6. Free rhodamine in solution exhibits a near zero value of anisotrophy (0.015), which can be attributed to fast rotational diffusion of the

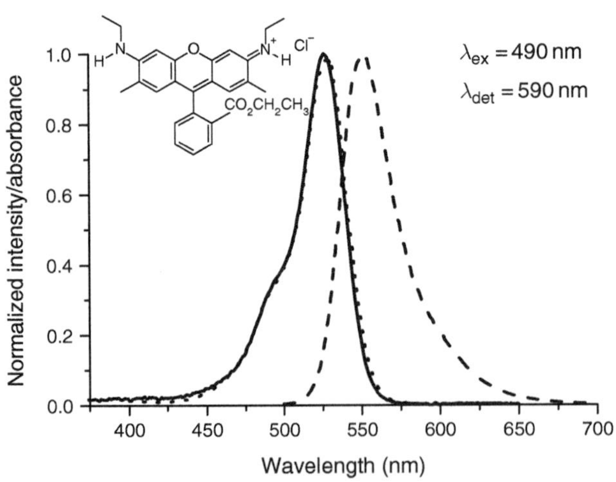

Figure 6 Absorption (solid line), excitation (dot), and emission (dash) spectra for rhodamine 6G. For the emission spectrum, the sample was excited using light with wavelength $\lambda_{ex} = 490$ nm. The excitation spectrum was obtained at detection wavelength $\lambda_{det} = 590$ nm.

molecule leading to depolarization. Association of small-molecule fluorophores with larger macromolecules results in a dramatic increase in rotational correlation time and thus to increased anisotropies. Due to their sensitivity to molecular rotation, molecular weight, shape, and energy transfer, fluorescence anisotropy measurements are extensively utilized to study protein–protein associations [18–22], energy transfer, and quenching [23,24], for probing fluidity of membranes and in immunoassays [25].

2.7. Factors affecting fluorescence

Although the spectral positions, intensities, and contours of fluorescence emission are mainly determined by the molecular electronic structure and molecular geometry, these properties may be strongly influenced by the environment of the fluorophore. Fluorescence spectra and quantum yields are generally more dependent on the environment than absorption spectra and extinction coefficients. The main environmental factors influencing fluorescence properties include the solvent polarity, ionic strength, oxygen concentration, pH of the solution, and temperature. Other processes that affect the fluorescence properties of a fluorophore are quenching and resonance energy transfer. Fluorescence quenching and resonance energy transfer allow fluorescence measurements to detect fluorophore dynamics and hence report sensitively on its microenvironment.

2.7.1. Fluorescence quenching

Quenching is the decrease in the observed fluorescence intensity as a result of interaction of the ground or excited states of a fluorophore with other species in solution. Quenching can be divided into two broad categories: dynamic and static quenching. In dynamic or collisional quenching, interaction of an excited state fluorophore with the quencher results in radiationless deactivation of the fluorophore to the ground state. The efficiency of dynamic quenching is hence sensitively dependent on the concentration of the quenching species. The lifetime of the fluorophore, and as a consequence its quantum yield, decreases with increasing quencher concentration and an additional term, $k_q[Q]$, is added to the denominator of Eqs (3), (7), and (8), to yield

$$\Phi = \frac{k_r}{k_r + k_{nr} + k_q[Q]} \tag{18}$$

where k_q is the quenching rate constant and [Q] is the concentration of the quenching species. The quenching of rhodamine 6G by iodide ions is an example of dynamic quenching and is illustrated in Figure 7(a). The corresponding lifetimes are summarized in Table 2. Figure 7(b) illustrates the dependence of fluorescence lifetime on the quencher concentration.

Static quenching is characterized by the formation of non-fluorescent complexes between the quencher and the fluorophore in its ground state. In static quenching, fluorescent emission is reduced without altering the excited state lifetime or quantum yield, and the two kinds of quenching can be distinguished on the basis of the excited state lifetime of the fluorophore.

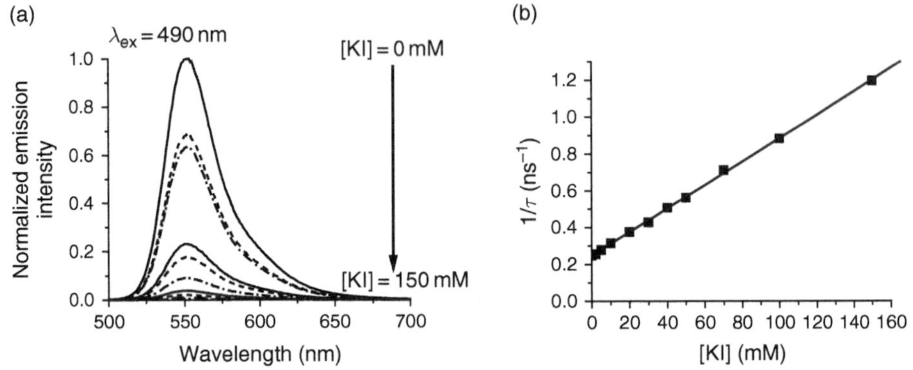

Figure 7 Effect of dynamic quenching by iodide on rhodamine 6G. (a) The observed fluorescence intensity for rhodamine 6G decreases with increasing concentration of KI. (b) Stern–Volmer plot showing the dependence of fluorescence lifetime upon the iodide concentration. The line is the best linear fit to the average lifetime obtained by the frequency domain method (Table 2). KI was used as the source of iodide. The ionic strength of the solution was held constant by addition of KCl.

Table 2 Dynamic quenching of R6G: Lifetime of rhodamine 6G decreases with increasing iodide ion concentration. The ionic strength of the solution is kept constant by adding KCl. The lifetimes were measured using a frequency domain setup

Solution	[KCl] M	[KI] M	τ_ϕ (ns)	τ_m (ns)	τ_{avg} (ns)
S_0	0.200	0.000	4.040	4.026	4.033
S_1	0.198	0.002	3.892	3.941	3.916
S_2	0.195	0.005	3.582	3.598	3.590
S_3	0.190	0.010	3.139	3.199	3.169
S_4	0.180	0.020	2.641	2.688	2.665
S_5	0.170	0.030	2.284	2.378	2.331
S_6	0.160	0.040	1.945	1.994	1.969
S_7	0.150	0.050	1.735	1.905	1.820
S_8	0.130	0.070	1.405	1.404	1.405
S_9	0.100	0.100	1.084	1.179	1.132
S_{10}	0.050	0.150	0.813	0.858	0.836

2.7.2. Fluorescence resonance energy transfer

Fluorescence (or Förster) resonance energy transfer (FRET) [26,27,28] is a non-radiative transfer of the excitation energy from a donor to an acceptor chromophore that involves a distance-dependent interaction between the emission and the absorption transition dipole moments of the donor and acceptor, respectively. The rate of energy transfer depends on the spectral overlap of the donor emission and acceptor absorbance, the donor fluorescence quantum yield, the relative orientation of their transition dipole moments, and the distance between donor and acceptor molecules. The energy transfer rate, k_{ET}, is given by

$$k_{ET} = \frac{1}{\tau_d}\left(\frac{R_0}{r}\right)^6 \tag{19}$$

where τ_d is the decay time of the fluorophore in the absence of an acceptor, and r is the distance between the donor and the acceptor. R_0 is the Förster radius (typically 2 to 9 nm) characterizing the donor/acceptor pair. It is defined as the distance at which the efficiency of resonance energy transfer is 50% and can be estimated as

$$R_0(\text{in nm}) = 979\left(\kappa^2 n^4 \Phi_0 J\right)^{1/6} \tag{20}$$

where n is the refractive index of the medium, Φ_0 is the fluorescence quantum yield of the donor, J is the spectral overlap integral, and κ^2 is the orientation factor. The overlap integral, J, expresses the extent of overlap between the donor emission and the acceptor absorption.

$$J = \int_0^{\infty} F_D(\lambda)\varepsilon_A(\lambda)\lambda^4 d\lambda \tag{21}$$

where $F_D(\lambda)$ is the normalized fluorescence spectrum of the donor and $\varepsilon_A(\lambda)$ is the molar extinction coefficient of the acceptor as a function of wavelength, λ. The rate of energy transfer varies linearly with the overlap integral [29]. A large value of overlap integral ensures high sensitivity for imaging and sensing as well as high selectivity.

Resonance energy transfer provides an additional deactivation pathway for the excited fluorophore and results in reduced excited state lifetime of the donor fluorophore. The Förster distances are comparable to the size of biological macromolecules, and thus, FRET is extensively used as a "spectroscopic ruler" for measuring distances between sites on interacting proteins [30–36].

The magnitude of k_{ET} can be determined from the efficiency of energy transfer, E_T, using the relation

$$k_{ET} = \frac{1}{\tau_d}\left(\frac{E_T}{1 - E_T}\right) \tag{22}$$

and E_T can be determined experimentally by measuring the decrease in the intensity F or the lifetime τ of the donor in the presence of the acceptor,

$$E_T = 1 - \frac{F}{F_d} = 1 - \frac{\tau}{\tau_d}. \tag{23}$$

Hence, the change in observed fluorescence intensity or the excited state lifetime of the donor fluorophore, due to resonance energy transfer, can be used to determine the distance between the donor acceptor pair. However, the distances so estimated are also influenced by the orientation factor, κ^2, which depends on the relative orientation of donor emission transition moment and acceptor absorption transition moment. The value of κ^2, in general, varies from 0 (perpendicular orientation) to 4 (for parallel orientation of transition moments). For fast and freely rotating fluorophores, its value may be isotropically averaged over all possible orientations, and is often taken to be 2/3. However, this assumption may not be valid for immobilized fluorophores. For a given FRET pair, under a given set of physical conditions, the value of κ^2 can be estimated by polarization measurements [37].

3. Fluorophores and Fluorescent Proteins

Fluorophores are typically polyaromatic compounds having a conjugated π-electron system. Fluorophores in biological applications can be broadly divided into two main categories: intrinsic and extrinsic. Intrinsic fluorophores are the ones that occur naturally and include aromatic amino acids, nicotinamide adenine dinucleotide (NADH), flavins, and derivatives of pyridoxal and chlorophyll. Intrinsic protein fluorescence arises from the aromatic amino acids tryptophan, tyrosine, and phenylalanine [5], although practically one observes fluorescence from tryptophan and tyrosine residues only. Protein fluorescence is highly sensitive to the local environment of these residues and is thus widely used to study protein conformational changes, binding to co-factors and ligands, and protein–protein associations [6,7].

However, in most cases, the molecule of interest is either non-fluorescent or a very poor emitter. The lipids and DNA, for example, are essentially devoid of intrinsic fluorescence. Most intrinsic fluorophores require excitation by short wavelength ultraviolet and blue light which is often hazardous for live cells. Further, the brightness and quantum yield of intrinsic fluorophores is, in general, quite low for most practical applications. It is thus often beneficial to label the molecule of interest exogenously by fluorescent molecules with desirable fluorescent properties such as absorption at longer wavelengths, higher quantum yield, and photostability. Such extrinsic fluorophores can be bioconjugated to the molecule of interest by various chemical strategies. However, the use of extrinsic fluorophores for in vivo studies presents experimental disadvantages. Bioconjugation with an exogenous fluorophore may lead to conformational changes and loss in activity of the molecule of interest. Introduction of labeled molecules into cells often requires invasive techniques such as microinjection.

A recent revolution in the use of fluorescence for studying biological systems came with the development and use of naturally fluorescent proteins as fluorescent probes. The jellyfish *Aequorea victoria* produces an intrinsically fluorescent protein known as the green fluorescent protein (GFP) [8,9]. The remarkable feature of GFP is that the chromophore forms spontaneously upon folding of the polypeptide chain, without the need of enzymatic synthesis. Extensive mutagenesis of the *Aequorea* GFP has generated a palette of visible fluorescent proteins with different colors, lifetimes, and photoactivation properties [38]. The use of genetically encoded fluorescent proteins as fluorescent probes provides a non-invasive means to visualize, track, and quantify molecules and events in living cells with high spatial and temporal resolution essential for understanding biological systems.

3.1. Green fluorescent protein

The green fluorescent protein from *A. victoria* was first reported by Shimomura et al. [39,40] in 1962, as a side product after purification of aequorin, a chemiluminescent protein. Emission of blue light by aequorin leads to the excitation of its companion protein GFP, thereby resulting in the characteristic green fluorescence of the species. Proteolysis of denatured GFP and further analysis of the peptide that retained visible absorbance led to the identification of the chromophore [41] as a 4-(*p*-hydroxybenzylidene)imidazolidin-5-one attached to the peptide backbone through the 1- and 2-positions of the ring (Figure 10). Crucial breakthroughs came with the cloning of the gene in 1992 [42], and the demonstrations that expression of the gene in other organisms creates fluorescence [43,44]. Hence, the gene contains all the information necessary for formation of the chromophore and no jellyfish specific enzymes are required.

Wild-type GFP has a major absorption peak at 398 nm and a minor absorption at 475 nm; the molar extinction coefficients at these wavelengths are 25,000 and 9500 respectively [45,46]. Excitation at 398 nm results in an emission maximum at 508 nm while excitation at 475 nm leads to emission with maximum at 503 nm. The GFP chromophore is highly luminescent with a quantum yield of 0.79 [45,46]. The absorption, excitation, and emission spectra of wild-type GFP are shown in Figure 12(a).

3.1.1. Structure of GFP

The green fluorescent protein is a small protein (28 kDa), with a barrel-like structure composed of 11 β-sheets with a central α-helix running up the axis of the cylinder [47] (Figure 8). The chromophore is attached to the α-helix and is buried in the centre of the cylinder or the β-can, and is well insulated from the environment. Thus, GFP fluorescence is not affected in general by quenching and other solvent interactions. The inaccessibility of the chromophore to quenching agents is illustrated in Figure 9, depicting the fluorescence emission spectra of EGFP, an optimized version of GFP, as a function of iodide concentration. As can be seen, the quencher has limited effect upon the observed fluorescence from EGFP. In contrast, Figure 7(a) depicts the emission spectra of rhodamine 6G at equivalent iodide concentrations, where the observed fluorescence intensity falls considerably with increasing iodide concentration. The robust shielding of the EGFP chromophore by the beta-barrel scaffold is also reflected in the excited state lifetimes of EGFP in solutions containing usual quenching agents. While the lifetime of quenched rhodamine decreases sharply with increasing quencher concentration (Table 2), the lifetime of EGFP remains almost constant (3.0 ns).

3.1.2. GFP chromophore

The chromophore of the wild-type GFP consists of Ser65-Tyr66-Gly67 residues [49,50] and is formed by a series of autocatalytic steps as depicted in Figure 11 [8]. First, GFP folds into a nearly native configuration and the imidazolinone is formed by nucleophilic attack of amide of Gly67 on the carboxyl of Ser65, followed by

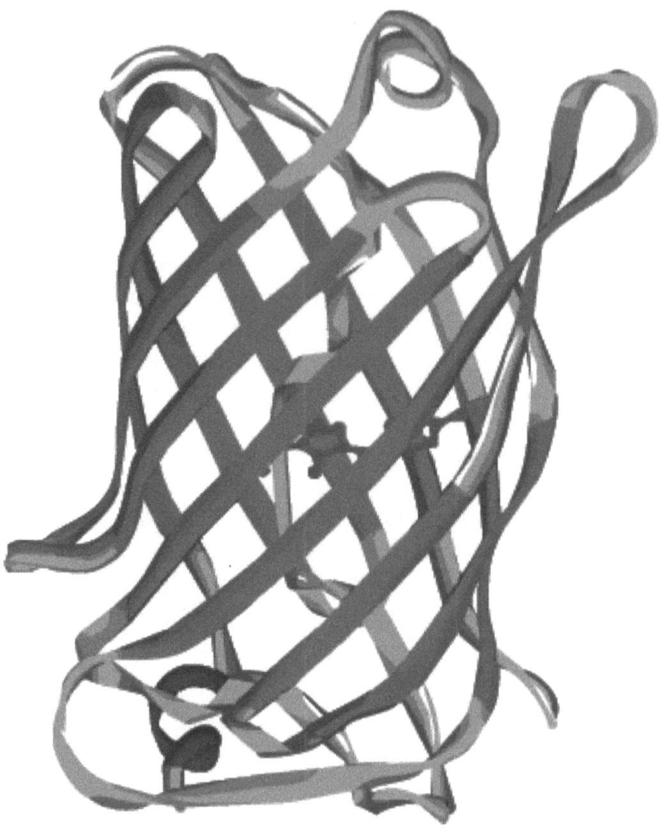

Figure 8 Schematic crystal structure of *Aequorea* GFP showing 11 β strands forming a cylinder with the embedded chromophore shown in ball and stick representation. Structure generated with the Accelrys DS ViewerPro Suite v5.1 and X-ray data from the Protein Data Bank [48].

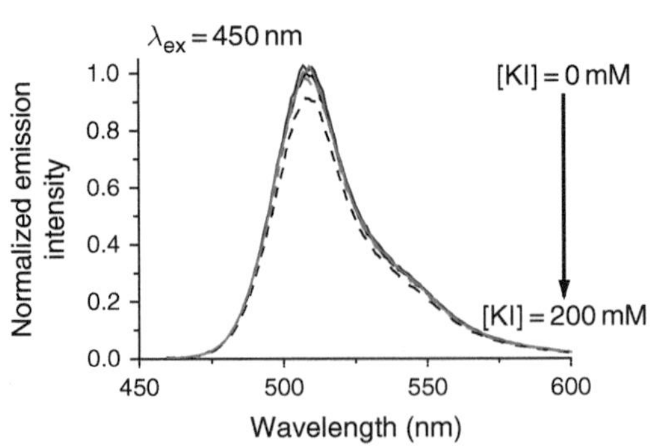

Figure 9 Effect of dynamic quenching by iodide on EGFP. The EGFP chromophore is shielded from the solvent by the protein beta-barrel. Thus, the observed fluorescence intensity remains nearly constant.

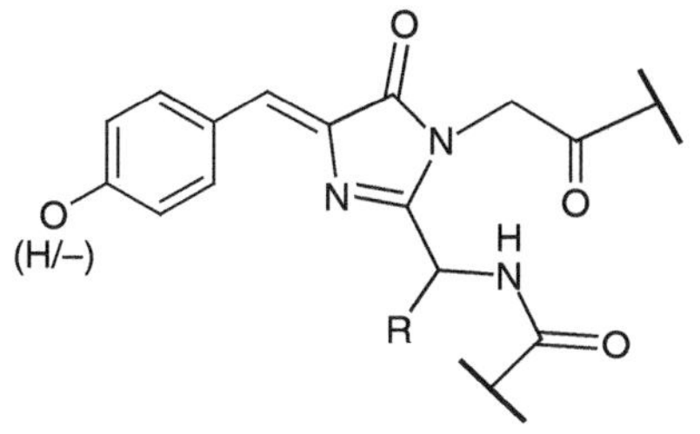

Figure 10 Structure of the chromophore of *Aequoria* green fluorescent protein.

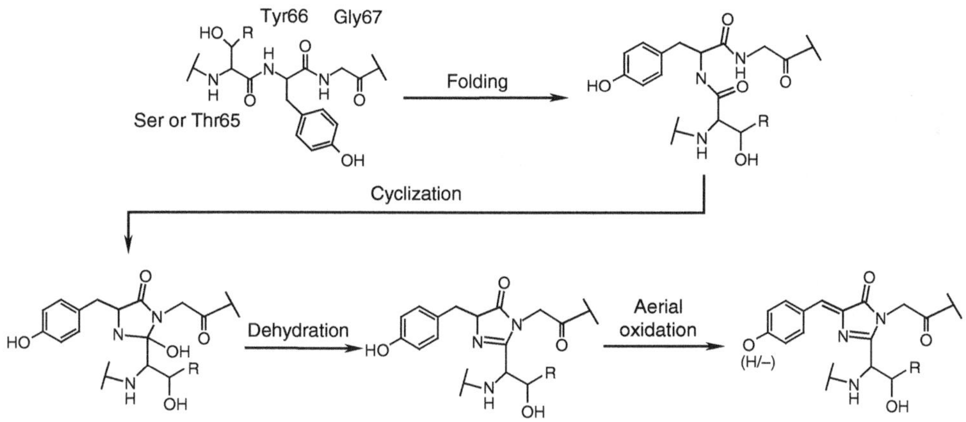

Figure 11 Mechanism of GFP chromophore formation.

dehydration. Further oxidation results in conjugation of the imidazolinone ring with Tyr66, and finally in the formation of the fluorophore.

3.1.3. GFP variants

Mutagenesis studies of GFP have yielded variants with improved fluorescence properties and kinetics of fluorophore formation [8,9,51] that have stimulated the widespread use of GFPs as fluorescent tags [52,53]. Although the wild-type green fluorescent protein produces significant fluorescence and is largely insensitive to changes in environmental conditions like solvent polarity, ionic strength, viscosity, and temperature, its excitation maximum is close to the UV region that can damage living cells. The wild-type GFP also exhibits significant photochromicity. A point mutation $Ser^{65} \rightarrow Thr^{65}$ (S65T) shifts the absorption maximum of the protein to 488 nm and also accelerates the rate of fluorophore formation [54]. Further mutations were made to change the codon usage to improve translation in the mammalian cells, to prevent protein aggregation and improve its brightness and photostability. These various improved characteristics were combined in the GFP variant known as enhanced GFP (EGFP), which is among the most commonly used GFP variants in cell biology [52]. Further improvements led to the development of numerous mutants with differing absorbance and emission spectra allowing simultaneous visualization of distinct fluorescent proteins in a cell. Various spectral

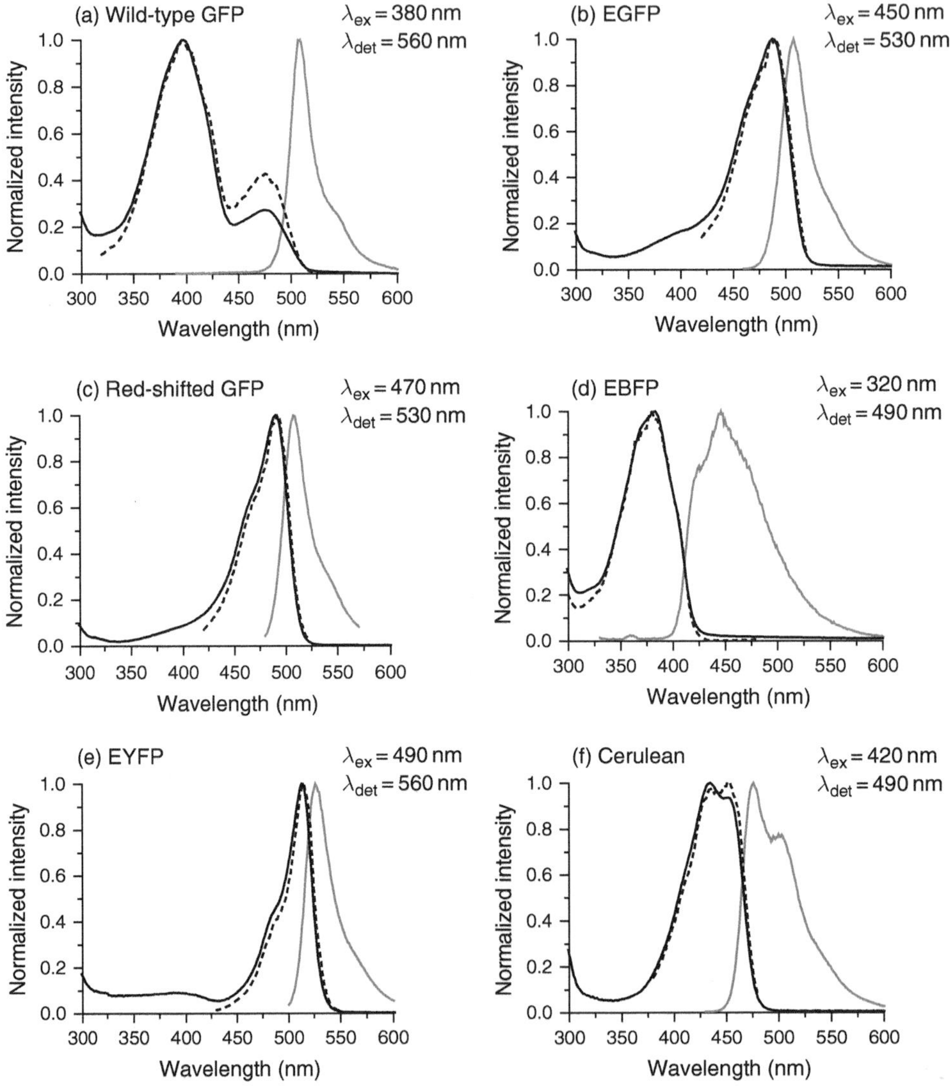

Figure 12 Absorption (black solid), excitation (black dotted) and emission spectra (gray) of VFPs: (a) wild-type GFP, (b) EGFP, (c) red-shifted GFP (RSGFP) (d) EBFP, (e) EYFP, and (f) Cerulean. Protein solutions at a concentration ~1.5 μM were prepared in 100 mM Tris buffer containing 100 mM NaCl at pH 8.5. The fluorescence excitation and emission spectra were determined at room temperature (~22° C) with a Varian Cary Eclipse spectrofluorometer. The excitation wavelength for emission spectra (λ_{ex}) and the detection wavelength for excitation spectra (λ_{det}) are mentioned on the top right corner of the figures (a)–(f).

variants of GFP, emitting in blue, cyan, and yellow (EBFP, ECFP, and EYFP, respectively) were developed and opened new dimensions in fluorescence imaging [55]. The emission spectrum of ECFP overlaps considerably with the absorption spectrum of EYFP, forming an efficient FRET pair [38,55] for determining protein–protein interactions and use in FRET-based biosensors. Further screening has led to the development of improved variants such as Cerulean (2.5 times brighter than ECFP) [56] and Citrine (an improved variant of EYFP) [57]. The absorption, excitation, and emission spectra of various GFP mutants are shown in Figure 12.

3.2. Red fluorescent protein

The fluorescence protein palette was extended to the red wavelength range [58] with the discovery and cloning of fluorescent protein from the reef coral of *Discosoma* genus, commonly known as DsRed. DsRed forms an obligate tetramer [59–61], as shown in Figure 13, with peak excitation at 558 nm and emission at

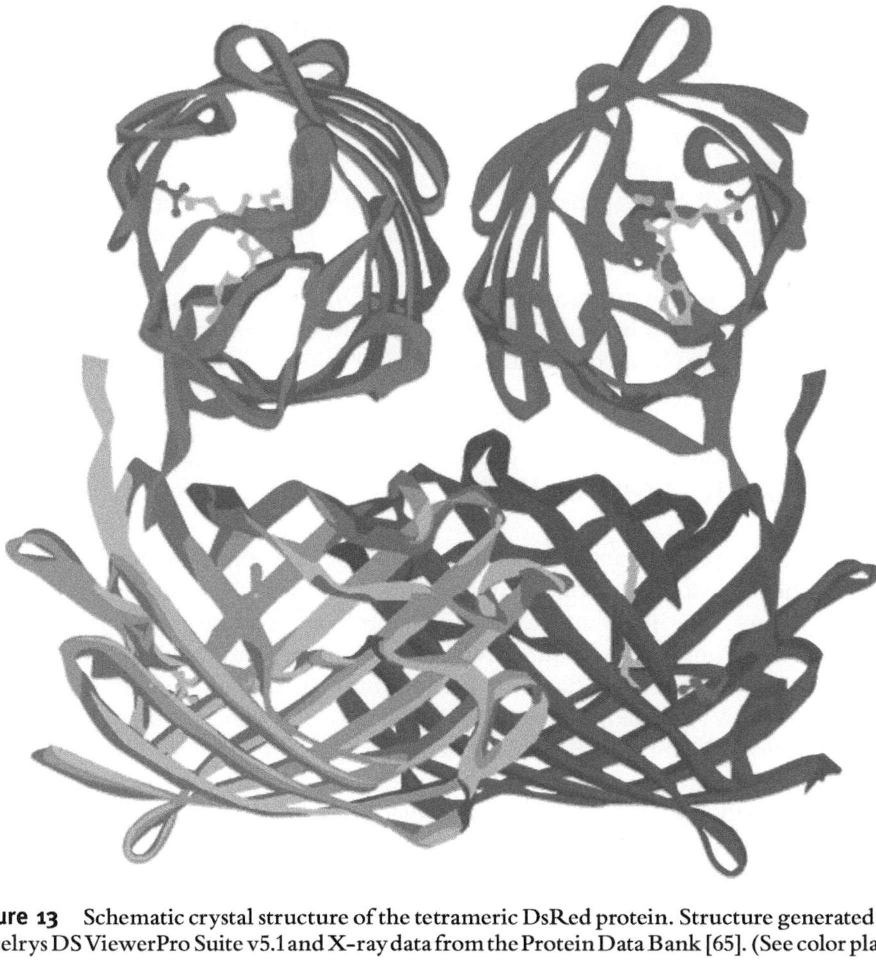

Figure 13 Schematic crystal structure of the tetrameric DsRed protein. Structure generated with Accelrys DS ViewerPro Suite v5.1 and X-ray data from the Protein Data Bank [65]. (See color plate 8).

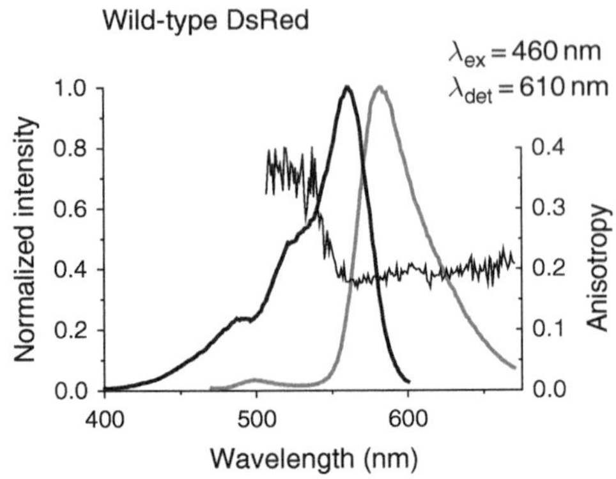

Figure 14 The absorption, excitation, and emission spectra for the tetrameric DsRed protein. The emission anisotropy is depicted on the scale on the right. The two levels in anisotropy indicate the presence of two different chromophores in the protein.

583 nm [62] (Figure 14). The chromophore is formed by a mechanism [63,64] similar to that for the formation of GFP with an additional oxidation reaction at backbone atoms to extend the conjugated π-electron system.

3.2.1. DsRed variants

Mutagenesis of DsRed has generated numerous variants with faster maturation rates and reduced aggregation (DsRed2, DsRed Express) [66]. Monomeric variants of DsRed have also been developed, such as mRFP1 [67], but these have reduced quantum yields and extinction coefficients. Directed evolution and iterative somatic hyper-mutation has led to the development of whole palette of monomeric RFP variants like mHoneydew, mBanana, mOrange, mTangerine, mStrawberry, mCherry [68], mRaspberry, mPlum, mGrape and tandem dimeric tdTomato.

4. Applications of VFPs

Fluorescent proteins act as genetically encodable fluorophores and provide non-invasive probes to study biological systems from individual molecules and cells to whole organisms. The main applications of fluorescent proteins include visualization of target-gene promoter up- and down-regulation, protein labeling, detection of protein–protein interactions, tracking protein movement, and monitoring cellular parameters using VFP-based fluorescent sensors. Several advanced microscopy techniques benefit from the introduction of VFPs to follow dynamics of biological systems in live cells. These techniques include 4D microscopy, fluorescence recovery after photobleaching (FRAP), fluorescence loss in photobleaching (FLIP), fluorescence correlation spectroscopy (FCS), and FRET imaging [28]. In 4D microscopy [69], time-lapse 3D images of fluorescent molecules are collected to provide spatial and temporal information about the changes in distribution of targeted protein as it relates to some complex cellular process being studied. FRAP [70–73] is used to study the mobility and diffusion of fluorescent molecules in living cells. In this technique, a region of interest is selectively photobleached with a high-intensity laser, and the recovery that occurs as fluorescent molecules move into the bleached region is monitored over time with low-intensity laser light. Depending on the protein studied, fluorescence recovery can result from protein diffusion, binding/dissociation, or transport processes. Complementary to the photobleaching technique, the continuity of a cell compartment can be monitored using FLIP [73]. In a FLIP experiment, a fluorescent cell is repeatedly photobleached within a small region while the whole cell is continuously imaged. Regions of the cell that are connected to the area being bleached will gradually lose fluorescence due to lateral movement of mobile proteins into this area. By contrast, the fluorescence in unconnected regions will not be affected. In FCS [74], fluorescent intensity fluctuations for molecules diffusing in and out of a defined focal volume (~1 fL) are measured with high time resolution, and provide information about protein's diffusion coefficient, binding constants, and concentrations. The main application of FRET analysis is in the monitoring of protein interactions using suitable chimeras of proteins of interest. Suitable pair of proteins having sufficient spectral overlap (e.g., ECFP-EYFP) are used to construct the fusion proteins. The biological process of interest induces a conformational change in the fusion construct, which in turn, alters the FRET efficiency. FRET biosensors have been engineered to detect a variety of molecular events such as protein interactions and conformational changes, protein activity levels, concentration of biomolecules and mechanical forces [27,28].

4.1. Lifetime spectroscopy and imaging of VFPs

As previously discussed, fluorescence lifetime is an intrinsic property of the fluorophore and is, in general, not modulated by fluorescence intensity, fluorophore concentration, light path length, scattering, and photobleaching, which renders it an extremely important tool for studying the biological systems where it is extremely difficult to quantify, control, and maintain these parameters. Because of the spectral overlap of emission, it is not always possible to resolve two or more fluorophores. This is illustrated in Figure 12 for enhanced green fluorescent protein (EGFP) and red-shifted green fluorescent protein (RS GFP), two mutants of the *Aequorea* GFP. However, these do differ considerably in their excited state lifetime, and the two proteins can be discriminated on this basis [75]. The fluorescence decay profiles for EGFP and red-shifted GFP obtained by TCSPC are shown in Figure 15.

The fluorescence lifetimes of various fluorescent proteins in solution are summarized in Table 3. These lifetimes were determined in solution using a frequency domain lifetime-imaging instrument consisting of a widefield microscope (Nikon TE2000U) with a Lambert Instruments fluorescence lifetime-imaging attachment (LIFA).

Fluorescence lifetime-imaging microscopy (FLIM) is a robust technique to map the spatial distribution of excited state lifetimes within microscopic images [76]. The lifetime of any given sample is measured at each pixel. FLIM combines the advantages of lifetime spectroscopy with fluorescence microscopy by revealing the spatial distribution of a fluorescent molecule together with information about its microenvironment. Fluorescence lifetime image acquisition is also rapid enough (~ hundreds of milliseconds to seconds) to make measurements in

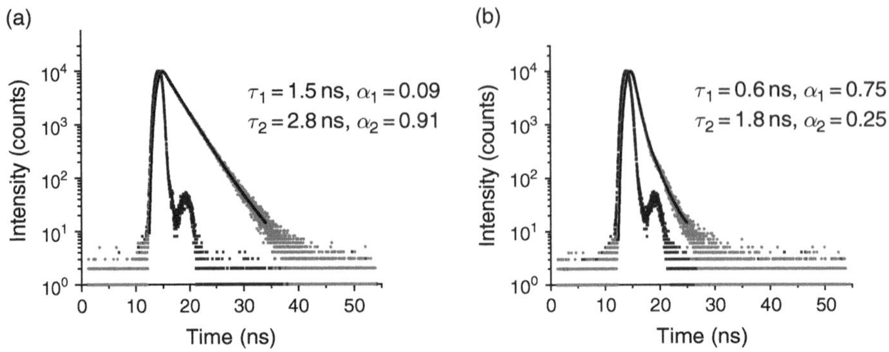

Figure 15 The fluorescence decay profiles of (a) EGFP and (b) red-shifted GFP obtained by TCSPC. The measured intensity is shown in light gray dots on a logarithmic scale. The instrument response function is depicted in dark gray. The decay was fitted with a bi-exponential fit (solid line). τ_1 and τ_2 are the two lifetimes extracted from a two-exponential fit; α_1 and α_2 are the relative amplitudes of the two components.

Table 3 Frequency domain lifetimes of a selection of visible fluorescent proteins

Protein	τ_ϕ (ns)	τ_m (ns)	τ_{avg} (ns)
WT GFP	3.654	3.333	3.494
S65T	3.133	3.015	3.074
EGFP	2.990	3.008	2.999
RS GFP	1.231	1.459	1.345
EYFP	3.005	3.091	3.048
Cerulean	4.223	3.779	4.001
Wild-type DsRed	3.931	3.773	3.852

live cells feasible [11]. Lifetime imaging, together with the use of genetically encoded fluorescent biosensors based upon fluorescent protein fusion, provides access to the protein state maps and a sensitive means to probe the cellular environment. These properties have been exploited to spatially resolve physiological parameters such as pH [77], calcium ion concentration [78], oxygen concentration, molecular association [79], and proteolytic processing [80]. FLIM has been implemented with both time domain and frequency domain lifetime detection. Time-domain fluorescence lifetime imaging can be implemented using a scanning confocal microscope and a TCSPC setup [81] enabling pixel-by-pixel determination of lifetimes, or by using a time-gated ICCD [15] in either a multifocal excitation configuration or a wide field illumination. Frequency domain lifetime determination is particularly suited for wide-field illumination strategies [17].

One of the most important applications of FLIM in live-cell imaging is in FRET-based biosensors. FLIM is independent of the local concentrations of fluorophores and the excitation intensity. Moreover, spectral bleedthrough is not an issue in FLIM because only the donor lifetime is measured. Therefore, it provides a more reliable quantitative estimate of FRET efficiency (Eq. 23) as

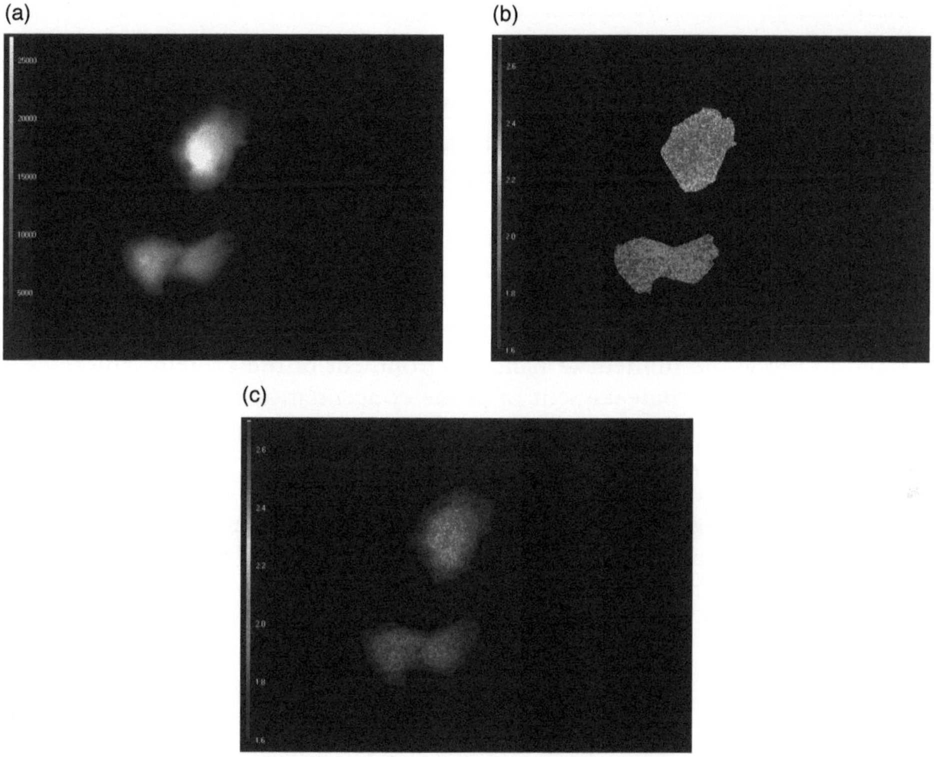

Figure 16 Fluorescence lifetime images of mammalian cells transfected with CFP alone and a CFP–YFP fusion construct. Panel (a) fluorescence intensity image, Panel (b) lifetime image, and Panel (c) overlay of fluorescence intensity and lifetime images. Change in CFP lifetime was selectively monitored by using a CFP band pass filter. The upper cell is transfected with CFP alone and yields a lifetime of 2.5 ns. For the cells expressing the fusion construct, the lifetime is substantially lowered (1.8 ns) due to FRET from CFP (donor) to the YFP (acceptor) molecule. Images courtesy of Lambert Instruments. (See color plate 9).

compared to fluorescence intensity based methods. By measuring the donor lifetime in the presence and the absence of acceptor one can accurately quantify FRET and calculate the distance between the donor and acceptor-labeled proteins. The CFP–YFP pair has been used extensively for constructing fusion proteins for FRET detection. Figure 16 shows images of mammalian cells transfected with either CFP or with a CFP–YFP fusion construct. The lifetimes of CFP are monitored using the Lambert Instruments frequency domain lifetime-imaging instrument. The lifetimes observed for the fusion protein are much shorter than that for the CFP alone. Detection of FRET by lifetime measurements has been used in a variety of applications for determining protein localizations [82,83], protein–protein [84,85] and protein–DNA interactions [86], receptor-antibody binding [87], and immunoassays [88].

5. Concluding Remarks

The discovery and development of fluorescent proteins as molecular tags in the recent years has led to a revolution by allowing complex biochemical processes to be correlated with the functioning of proteins in living cells. The green fluorescent protein from *A. victoria* and its various variants can be fused to virtually any protein of interest to analyze its spatial distribution, biochemical reactions, and distribution in the biological systems. Lifetime spectroscopy and imaging of cells expressing fluorescent protein-tagged fusion constructs provides a non-invasive means to detect and quantify various biochemical processes in taking place in live cells and to probe the immediate nano-environment of the protein. Fluorescence lifetime of a species is independent of probe concentration and light path length: variables difficult to measure and control in live cells. Interactions, proteolytic processing, covalent modifications, and conformational changes can be followed in a live cell by detecting FRET between tagged proteins. Further, recent advances like the development and use of new fluorescent tools such as semiconductor nanocrystals (quantum dots), multi-photon excitation, single molecule detection and development of better fluorescent proteins have opened up newer dimensions for the application of fluorescence lifetime for studying biological systems.

REFERENCES

[1] J.R. Lakowicz, *Principles of Fluorescence Spectroscopy*, Second Edition, Kluwer Academic/Plenum Publishers, New York, 2002.

[2] A. Sharma and S.G. Schulman, *Introduction to Fluorescence Spectroscopy*, John Wiley and Sons, New York, 1999.

[3] M. Hof, V. Fidler and R. Hutterer, in *Fluorescence Spectroscopy in Biology*, Vol. 3 (ed. M. Hof, V. Fidler and R. Hutterer), Springer-Verlag, New York, 2005, pp. 1–29.

[4] A.P. Demchenko, in *Topics in Fluorescence Spectroscopy: Biochemical Applications* (ed. J.R. Lakowicz), Plenum, New York, 1991, pp. 65–112.

[5] A.P. Demchenko, *Ultraviolet Spectroscopy of Proteins*, Springer-Verlag, New York, 1981.

[6] A. Saxena, J.B. Udgaonkar and G. Krishnamoorthy, in *Fluorescence Spectroscopy in Biology*, Vol. 3 (ed. M. Hof, V. Fidler and R. Hutterer), Springer-Verlag, New York, 2005, pp. 163–179.

[7] M.R. Eftink, The use of fluorescence methods to monitor unfolding transitions in proteins, *Biophys. J.*, 1994, **66**, 482–501.

[8] R.Y. Tsien, The green fluorescent protein, *Annu. Rev. Biochem.*, 1998, **67**, 509–544.

[9] M. Zimmer, Green fluorescent protein (GFP): application, structure and related photophysical behavior, *Chem. Rev.*, 2002, **102**, 759–781.

[10] A. Miyawaki, A. Sawano and T. Kogure, Lighting up cells: labelling proteins with fluorophores, *Nat. Cell Biol.*, 2003, **5**, Suppl., S1–S7.

[11] P.I.H. Bastiaens and A. Squire, Fluorescence lifetime imaging microscopy: spatial resolution of biochemical processes in the cell, *Trends Cell Biol.*, 1999, **9**, 48–52.

[12] J.R. Lakowicz, in *Topics in Fluorescence Spectroscopy: Probe Design and Chemical Sensing* (ed. J.R. Lakowicz), Plenum, New York, 1991, pp. 1–20.

[13] K. Suhling, P.M. French and D. Phillips, Time-resolved fluorescence microscopy, *Photochem. Photobiol. Sci.*, 2005, **4**, 13–22

[14] D.J.S. Birch and R.E. Imhof, in *Topics in Fluorescence Spectroscopy: Techniques* (ed. J.R. Lakowicz), Plenum, New York, 1991, pp. 1–96.

[15] H. Schneckenburger, M.H. Gschwend, R. Sailer, H.P. Mock and W.S. Strauss, Time-gated fluorescence microscopy in cellular and molecular biology, *Cell. Mol. Biol.*, 1998, **44**, 795–805.
[16] J.R. Lakowicz and I. Gryczynski, in *Topics in Fluorescence Spectroscopy: Techniques* (ed. J.R. Lakowicz), Plenum, New York, 1991, pp. 293–336.
[17] T.W.J. Gadella Jr, R.M. Clegg and T.M. Jovin, Fluorescence lifetime imaging microscopy: pixel-by-pixel analysis of phase-modulation data, *Bioimaging*, 1994, **2**, 139–159.
[18] R.F. Steiner, in *Topics in Fluorescence Spectroscopy: Principles* (ed. J.R. Lakowicz), Plenum, New York, 1991, pp. 1–52.
[19] D.M. Jameson and S.E. Seifried, Quantification of protein-protein interactions using fluorescence polarization, *Methods*, 1999, **19**, 222–233.
[20] V. LeTilly and C.A. Royer, Fluorescence anisotropy assays implicate protein-protein interactions in regulating trp repressor DNA binding, *Biochemistry*, 1993, **32**, 7753–7758.
[21] R.V. Weatherman and L.L. Kiessling, Fluorescence anisotropy assays reveal affinities of *C*- and *O*-glycosides for concanavalin A[1], *J. Org. Chem.*, 1996, **61**, 534–538.
[22] D.M. Jameson and W.H. Sawyer, Fluorescence anisotropy applied to biomolecular interactions, *Meth. Enzymol.*, 1995, **246**, 283–300.
[23] J. Larsen, J. Andersson, T. Polívka, J. Sly, M.J. Crossley, V. Sundström and E. Åkesson, Energy transfer and conformational dynamics in Zn-porphyrin dendrimers, *Chem. Phys. Lett.*, 2005, **403**, 205–210.
[24] Q.H. Xu, S. Wang, D. Korystov, A. Mikhailovsky, G.C. Bazan, D. Moses and A.J. Heeger, The fluorescence resonance energy transfer (FRET) gate: a time-resolved study, *Proc. Natl. Acad. Sci. U.S.A.*, 2005, **102**, 530–535.
[25] M.S. Nasir and M.E. Jolley, Fluorescence polarization: an analytical tool for immunoassay and drug discovery, *Comb. Chem. High Throughput Screen*, 1999, **2**, 177–190.
[26] P. Wu and L. Brand, Resonance energy transfer: methods and applications, *Anal. Biochem.*, 1994, **218**, 1–13.
[27] E.A. Jares-Erijman and T.M. Jovin, FRET imaging, *Nat. Biotechnol.*, 2003, **21**, 1387–1395.
[28] A. Peyker, O. Rocks, and P.I. Bastiaens, Imaging activation of two Ras isoforms simultaneously in a single cell, *Chembiochem*, 2005, **6**, 78–85.
[29] R.P. Haugland, J. Yguerabide and L. Stryer, Dependence of the kinetics of singlet-singlet energy transfer on spectral overlap, *Proc. Natl. Acad. Sci. U.S.A.*, 1969, **63**, 23–30.
[30] L. Stryer, Fluorescence energy transfer as a spectroscopic ruler, *Annu. Rev. Biochem.*, 1978, **47**, 819–846.
[31] C.G. dos Remedios, M. Miki and J.A. Barden, Fluorescence resonance energy transfer measurements of distances in actin and myosin. A critical evaluation, *J. Muscle Res. Cell. Motil.*, 1987, **8**, 97–117.
[32] P.R. Selvin, The renaissance of fluorescence resonance energy transfer, *Nat. Struct. Biol.*, 2000, **7**, 730–734.
[33] P.R. Selvin, Fluorescence resonance energy transfer, *Meth. Enzymol.*, 1995, **246**, 300–334.
[34] C.G. dos Remedios and P.D. Moens, Fluorescence resonance energy transfer spectroscopy is a reliable "ruler" for measuring structural changes in proteins. Dispelling the problem of the unknown orientation factor, *J. Struct. Biol.*, 1995, **115**, 175–185.
[35] C. Berney and G. Danuser, FRET or no FRET: a quantitative comparison, *Biophys. J.*, 2003, **84**, 3992–4010.
[36] A. Periasamy, Fluorescence resonance energy transfer microscopy: a mini review, *J. Biomed. Opt.*, 2001, **6**, 287–291.
[37] R.E. Dale, J. Eisinger and W.E. Blumberg, The orientational freedom of molecular probes. The orientation factor in intramolecular energy transfer, *Biophys. J.*, 1979, **26**, 161–193.
[38] J. Zhang, R.E. Campbell, A.Y. Ting and R.Y. Tsien, Creating new fluorescent probes for cell biology, *Nat. Rev. Mol. Cell Biol.*, 2002, **3**, 906-918.
[39] O. Shimomura, F.H. Johnson and Y. Saiga, Extraction, purification and properties of aequorin, a bioluminescent protein from the luminous hydromedusan *Aequorea*, *J. Cell. Comp. Physiol.*, 1962, **59**, 223–239.
[40] O. Shimomura, The discovery of aequorin and green fluorescent protein, *J. Microsc.*, 2005, **217**, 1–15.

[41] O. Shimomura, Structure of the chromophore of *Aequorea* green fluorescent protein, *FEBS Lett.*, 1979, **104**, 220–222.
[42] D.C. Prasher, V.K. Eckenrode, W.W. Ward, F.G. Prendergast and M.J. Cormier, Primary structure of the *Aequorea victoria* green-fluorescent protein, *Gene*, 1992, **111**, 229–233.
[43] M. Chalfie, Y. Tu, G. Euskirchen, W.W. Ward and D.C. Prasher, Green fluorescent protein as a marker for gene expression, *Science*, 1994, **263**, 802–805.
[44] S. Inouye and F.I. Tsuji, *Aequorea* green fluorescent protein: expression of the gene and fluorescence characteristics of the recombinant protein, *FEBS Lett.*, 1994, **341**, 277–280.
[45] G.H. Patterson, S.M. Knobel, W.D. Sharif, S.R. Kain and D.W. Piston, Use of the green fluorescent protein and its mutants in quantitative fluorescence microscopy, *Biophys. J.*, 1997, **73**, 2782–2790.
[46] W.W. Ward, in *Green Fluorescent Protein: Properties, Applications, and Protocols*, Second edition (ed. M. Chalfie and S.R. Kain), John Wiley & Sons, New York, 2005, pp. 39–65.
[47] M. Ormö, A.B. Cubitt, K. Kallio, L.A. Gross, R.Y. Tsien and S.J. Remington, Crystal structure of the *Aequorea victoria* green fluorescent protein, *Science*, 1996, **273**, 1392–1395.
[48] F. Yang, L. Moss and G. Phillips, The molecular structure of green fluorescent protein, *Nat. Biotechnol.*, 1996, **14**, 1246–1251.
[49] B.G. Reid and G.C. Flynn, Chromophore formation in green fluorescent protein, *Biochemistry*, 1997, **36**, 6786–6791.
[50] R. Heim, D.C. Prasher and R.Y. Tsien, Wavelength mutations and posttranslational autoxidation of green fluorescent protein, *Proc. Natl. Acad. Sci. U.S.A.*, 1994, **91**, 12501–12504.
[51] A.B. Cubitt, R. Heim, S.R. Adams, A.E. Boyd, L.A. Gross and R.Y. Tsien, Understanding, improving and using green fluorescent proteins, *Trends Biochem. Sci.*, 1995, **20**, 448–455.
[52] J.A. Schmid and H. Neumeier, Evolutions in science triggered by green fluorescent protein (GFP), *Chembiochem*, 2005, **6**, 1149–1156.
[53] J. Lippincott-Schwartz and G.H. Patterson, Development and use of fluorescent protein markers in living cells, *Science*, 2003, **300**, 87–91.
[54] R. Heim, A.B. Cubitt and R.Y. Tsien, Improved green fluorescence, *Nature*, 1995, **373**, 663–664.
[55] P. van Roessel and A.H. Brand, Imaging into the future: visualizing gene expression and protein interactions with fluorescent proteins, *Nat. Cell Biol.*, 2002, **4**, E15–E20.
[56] M.A. Rizzo, G.H. Springer, B. Granada and D.W. Piston, An improved cyan fluorescent protein variant useful for FRET, *Nat. Biotechnol.*, 2004, **22**, 445–449.
[57] O. Griesbeck, G.S. Baird, R.E. Campbell, D.A. Zacharias and R.Y. Tsien, Reducing the environmental sensitivity of yellow fluorescent protein, *J. Biol. Chem.*, 2001, **276**, 29188–29194.
[58] M.V. Matz, A.F. Fradkov, Y.A. Labas, A.P. Savitsky, A.G. Zaraisky, M.L. Markelov and S.A. Lukyanov, Fluorescent proteins from nonbioluminescent Anthozoa species, *Nat. Biotechnol.*, 1999, **17**, 969–973.
[59] G.S. Baird, D.A. Zacharias and R.Y. Tsien, Biochemistry, mutagenesis, and oligomerization of DsRed, a red fluorescent protein from coral, *Proc. Natl. Acad. Sci. U.S.A.*, 2000, **97**, 11984–11989.
[60] A. Sacchetti, V. Subramaniam, T.M. Jovin and S. Alberti, Oligomerization of DsRed is required for the generation of a functional red fluorescent chromophore, *FEBS Lett.*, 2002, **525**, 13–19.
[61] D. Yarbrough, R.M. Wachter, K. Kallio, M.V. Matz and S.J. Remington, Refined crystal structure of DsRed, a red fluorescent protein from coral, at 2.0-Å resolution, *Proc. Natl. Acad. Sci. U.S.A.*, 2001, **98**, 462–467.
[62] A.A. Heikal, S.T. Hess, G.S. Baird, R.Y. Tsien and W.W. Webb, Molecular spectroscopy and dynamics of intrinsically fluorescent proteins: coral red (dsRed) and yellow (Citrine), *Proc. Natl. Acad. Sci. U.S.A.*, 2000, **97**, 11996–12001.
[63] V.V. Verkhusha, D.M. Chudakov, N.G. Gurskaya, S. Lukyanov and K.A. Lukyanov, Common pathway for the red chromophore formation in fluorescent proteins and chromoproteins, *Chem. Biol.*, 2004, **11**, 845–854.
[64] L.A. Gross, G.S. Baird, R.C. Hoffman, K.K. Baldridge and R.Y. Tsien, The structure of the chromophore within DsRed, a red fluorescent protein from coral, *Proc. Natl. Acad. Sci. U.S.A.*, 2000, **97**, 11990–11995.

[65] M.A. Wall, M. Socolich and R. Ranganathan, The structural basis for red fluorescence in the tetrameric GFP homolog DsRed, *Nat. Struct. Biol.*, 2000, **7**, 1133–1138.
[66] B.J. Bevis and B.S. Glick, Rapidly maturing variants of the Discosoma red fluorescent protein (DsRed), *Nat. Biotechnol.*, 2002, **20**, 83–87.
[67] R.E. Campbell, O. Tour, A.E. Palmer, P.A. Steinbach, G.S. Baird, D.A. Zacharias and R.Y. Tsien, A monomeric red fluorescent protein, *Proc. Natl. Acad. Sci. U.S.A.*, 2002, **99**, 7877–7882.
[68] N.C. Shaner, R.E. Campbell, P.A. Steinbach, B.N. Giepmans, A.E. Palmer and R.Y. Tsien, Improved monomeric red, orange and yellow fluorescent proteins derived from *Discosoma* sp. red fluorescent protein, *Nat. Biotechnol.*, 2004, **22**, 1567–1572.
[69] D. Gerlich, J. Beaudouin, M. Gebhard, J. Ellenberg and R. Eils, Four-dimensional imaging and quantitative reconstruction to analyse complex spatiotemporal processes in live cells, *Nat. Cell Biol.*, 2001, **3**, 852–855.
[70] D. Axelrod, D.E. Koppel, J. Schlessinger, E. Elson and W.W. Webb, Mobility measurement by analysis of fluorescence photobleaching recovery kinetics, *Biophys. J.*, 1976, **16**, 1055–1069.
[71] R.D. Phair and T. Misteli, Kinetic modeling approaches to *in vivo* imaging, *Nat. Rev. Mol. Cell Biol.*, 2001, **2**, 898–907.
[72] E.A. Reits and J.J. Neefjes, From fixed to FRAP: measuring protein mobility and activity in living cells, *Nat. Cell Biol.*, 2001, **3**, E145–E147.
[73] J. Lippincott-Schwartz, E. Snapp and A. Kenworthy, Studying protein dynamics in living cells, *Nat. Rev. Mol. Cell Biol.*, 2001, **2**, 444–456.
[74] P. Schwille, Fluorescence correlation spectroscopy and its potential for intracellular applications, *Cell. Biochem. Biophys.*, 2001, **34**, 383–408.
[75] A. Volkmer, V. Subramaniam, D.J.S. Birch and T.M. Jovin, One- and two-photon excited fluorescence lifetimes and anisotropy decays of green fluorescent proteins, *Biophys. J.*, 2000, **78**, 1589–1598.
[76] E.C. van Munster and T.W.J. Gadella, Fluorescence lifetime imaging microscopy (FLIM), *Adv. Biochem. Eng. Biotechnol.*, 2005, **95**, 143–175.
[77] R. Sanders, A. Draaijer, H.C. Gerritsen, P.M. Houpt and Y.K. Levine, Quantitative pH imaging in cells using confocal fluorescence lifetime imaging microscopy, *Anal. Biochem.*, 1995, **227**, 302–308.
[78] J.R. Lakowicz, H. Szmacinski, K. Nowaczyk, W.J. Lederer, M.S. Kirby and M.L. Johnson, Fluorescence lifetime imaging of intracellular calcium in COS cells using Quin-2, *Cell Calcium*, 1994, **15**, 7–27.
[79] P.I.H. Bastiaens and T.M. Jovin, Microspectroscopic imaging tracks the intracellular processing of a signal transduction protein: fluorescent-labeled protein kinase CβI, *Proc. Natl. Acad. Sci. U.S.A.*, 1996, **93**, 8407–8412.
[80] T.W. Gadella Jr. and T.M. Jovin, Oligomerization of epidermal growth factor receptors on A431 cells studied by time-resolved fluorescence imaging microscopy. A stereochemical model for tyrosine kinase receptor activation, *J. Cell Biol.*, 1995, **129**, 1543–1558.
[81] W. Becker, A. Bergmann, M.A. Hink, K. Konig, K. Benndorf and C. Biskup, Fluorescence lifetime imaging by time-correlated single-photon counting, *Microsc. Res. Tech.*, 2004, **63**, 58–66.
[82] J.W. Legg, C.A. Lewis, M. Parsons, T. Ng and C.M. Isacke, A novel PKC-regulated mechanism controls CD44 ezrin association and directional cell motility, *Nat. Cell Biol.*, 2002, **4**, 399–407.
[83] P.J. Verveer, F.S. Wouters, A.R. Reynolds and P.I. Bastiaens, Quantitative imaging of lateral ErbB1 receptor signal propagation in the plasma membrane, *Science*, 2000, **290**, 1567–1570.
[84] F. S. Wouters and P. I. Bastiaens, Fluorescence lifetime imaging of receptor tyrosine kinase activity in cells, *Curr. Biol.*, 1999, **9**, 1127–1130.
[85] A.G. Harpur, F.S. Wouters and P.I. Bastiaens, Imaging FRET between spectrally similar GFP molecules in single cells, *Nat. Biotechnol.*, 2001, **19**, 167–169.
[86] F.G.E. Cremazy, E.M. Manders, P.I. Bastiaens, G. Kramer, G.L. Hager, E.B. van Munster, P.J. Verschure, T.W.J. Gadella Jr. and R. van Driel, Imaging in situ protein-DNA interactions in the cell nucleus using FRET–FLIM, *Exp. Cell Res.*, 2005, **309**, 390–396.

[87] T. Ng, M. Parsons, W.E. Hughes, J. Monypenny, D. Zicha, A. Gautreau, M. Arpin, S. Gschmeissner, P.J. Verveer, P.I. Bastiaens and P.J. Parker, Ezrin is a downstream effector of trafficking PKC-integrin complexes involved in the control of cell motility, *EMBO J.*, 2001, **20**, 2723–2741.
[88] W. Zwart, A. Griekspoor, C. Kuijl, M. Marsman, J. van Rheenen, H. Janssen, J. Calafat, M. van Ham, L. Janssen, M. van Lith, K. Jalink and J. Neefjes, Spatial separation of HLA-DM/HLA-DR interactions within MIIC and phagosome-induced immune escape, *Immunity*, 2005, **22**, 221–233.

CHAPTER 5

Monte Carlo Simulations in Nuclear Medicine Imaging

Steven Staelens *and* Irène Buvat

Contents

1. Introduction 176
2. Nuclear Medicine Imaging 176
   2.1. Single photon imaging 177
   2.2. Positron emission tomography 178
   2.3. Emission tomography in small animal imaging 179
   2.4. Reconstruction 179
3. The MC Method 180
   3.1. Random numbers 180
   3.2. Sampling methods 181
   3.3. Photon transport modeling 182
   3.4. Scoring 183
4. Relevance of Accurate MC Simulations in Nuclear Medicine 184
   4.1. Studying detector design 184
   4.2. Analysing quantification issues 184
   4.3. Correction methods for image degradations 185
   4.4. Detection tasks using MC simulations 186
   4.5. Applications in other domains 186
5. Available MC Simulators 187
6. Gate 188
   6.1. Basic features 188
   6.2. GATE: Time management 192
   6.3. GATE: Digitization 193
7. Efficiency-Accuracy Trade-Off 194
   7.1. Accuracy and validation 194
   7.2. Calculation time 194
8. Case Studies 195
   8.1. Case study I: TOF-PET 195
   8.2. Case study II: Assessment of PVE correction 196
   8.3. Case study III: MC-based reconstruction 197
9. Future Prospects 200
10. Conclusion 200

Acknowledgments 201

References 201

Advances in Biomedical Engineering
ISBN 978-0-444-53075-2

Abstract

The purpose of this chapter is to serve as an educational review regarding Monte Carlo (MC) simulations in Nuclear Medicine imaging complementary to reviews in specific scientific journals. After the basics of Nuclear Medicine are briefly described, the principles of MC simulations for modeling particle transport are presented. The major fields of applications of MC simulations in Nuclear Medicine imaging are discussed. To better illustrate the principles of a Monte Carlo simulator in Nuclear Medicine imaging, a recent but widely used simulation tool, GATE, is taken as an example. Finally, three case studies are presented as current applications representative of the use of MC simulations in Nuclear Medicine imaging.

Keywords: Monte Carlo, simulation, emission tomography, SPECT, PET, tomographic reconstruction, GATE

1. INTRODUCTION

Monte Carlo (MC) simulations are a modeling tool that is increasingly used in Nuclear Medicine imaging, namely in Single Photon Emission Computed Tomography (SPECT) and in Positron Emission Tomography (PET), mostly for the optimization of detector design and of acquisition and processing protocols, and also as part of the image formation process itself. In this chapter, we first recall the basics of Nuclear Medicine imaging to set the context. We then present the general principles of MC simulations when applied to modeling particle transport, and we explain why accurate MC modeling is of foremost importance for various applications in emission tomography. After briefly presenting the state-of-the-art regarding MC simulation tools appropriate for emission tomography modeling, we focus on a recent and widely used tool, GATE, to illustrate the major components of a simulator dedicated to SPECT and PET modeling, and to explain the way the simulation of a SPECT or PET acquisition can be designed. Lastly, three typical examples representative of the current use of MC simulations in emission tomography are presented. Future prospects regarding MC simulations in the field of Nuclear Medicine imaging are briefly discussed.

2. NUCLEAR MEDICINE IMAGING

Nuclear medicine imaging involves the use of radioactively labeled pharmaceuticals to diagnose and assess disease in the human body. The radiopharmaceutical is intravenously injected, inhaled or ingested: the choice of pharmaceutical and route of administration depends on the disease under investigation. Position-sensitive detectors are used to detect the radiation emitted from the pharmaceutical, and so, in principle, both the temporal and spatial distribution of the pharmaceutical within the body can be determined. Historically, Nuclear Medicine imaging was the first modality for functional imaging, i.e., giving information regarding the way organs work, as opposed to anatomical imaging, giving information regarding the structures of the organs. Nowadays, although functional imaging is also achievable using magnetic resonance imaging, ultrasound, or even X-ray computed tomography, Nuclear Medicine still plays a major role in the realm of functional imaging given its exquisite sensitivity which makes it possible to target molecular processes. Modern Nuclear Medicine imaging primarily consists of two main branches: single photon imaging and positron annihilation photon imaging, in which two annihilation photons are detected simultaneously [1,2].

2.1. Single photon imaging

Single photon imaging requires at minimum just one detector fixed in one position to obtain a two-dimensional (2D) projection of a three-dimensional (3D) radiopharmaceutical distribution. A typical imaging system suitable for this task is the gamma camera (Figure 1). A typical gamma camera consists of a collimator that

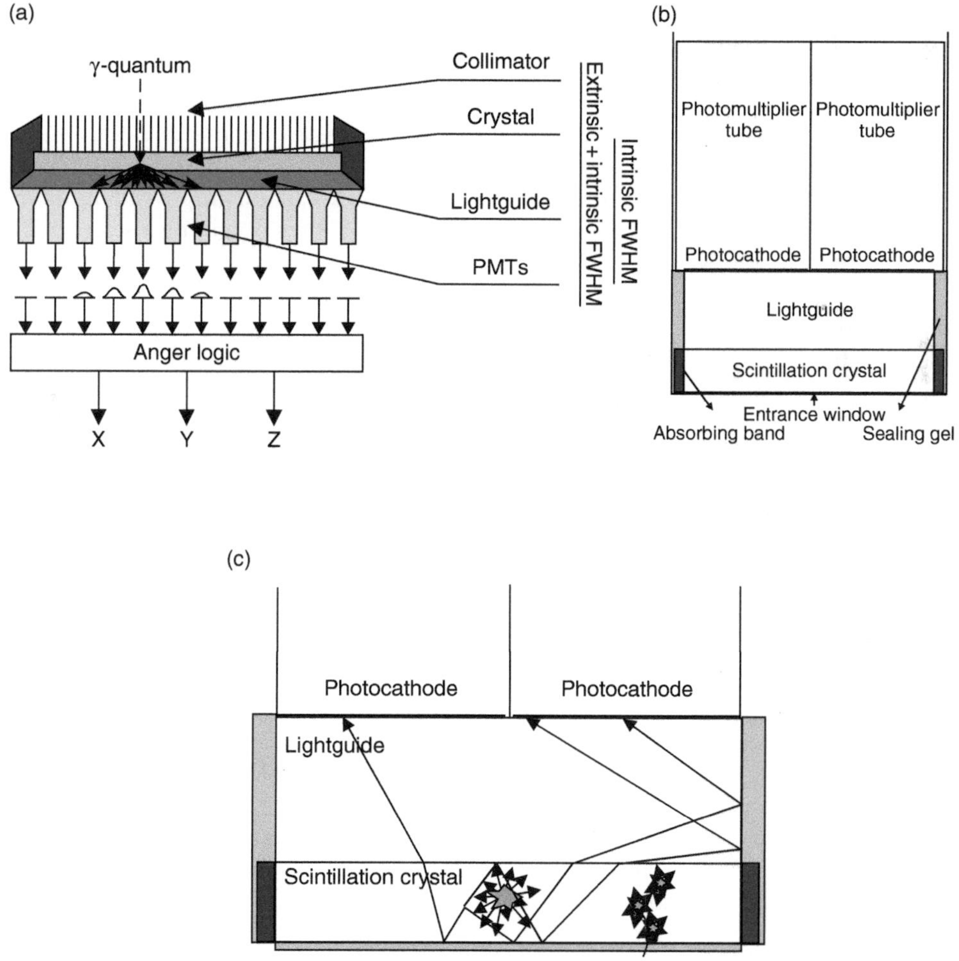

Figure 1 Gamma Camera: (a) technical scheme, (b) detailed view, (c) optical photon transport.

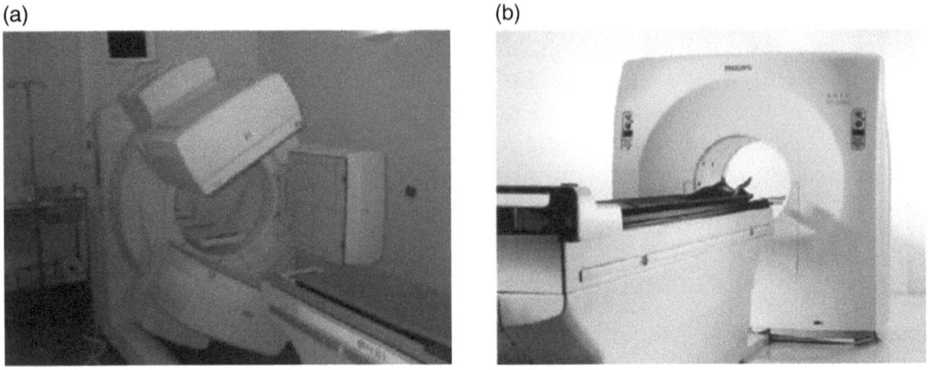

Figure 2 (a) SPECT: Philips Irix, (b) PETcamera: Philips Allegro.

limits the angle of incidence of the detected photons on the detector to a specific direction, for instance around 90° for a parallel hole collimator. The deposited energy is converted by a scintillating crystal to visible light which travels through the crystal and the light guide towards a set of photomultipliers (PMTs) containing a photocathode which converts the optical photons (Figure 1(c)) into electrons that are transported over dynodes towards an Anger logic readout for position and energy signal calculation. Single projection imaging can be improved upon by simply rotating the gamma camera around the patient, thus obtaining a series of 2D projections. These can be used to retrieve depth information, i.e., the 3D distribution of the radiopharmaceutical, which is known as SPECT. The quality of the information (i.e. the time-dependent reconstructed radiopharmaceutical images) obtained from SPECT depends considerably on the imaging time, camera sensitivity, and position resolution. Consequently, the temporal and spatial resolutions of SPECT can be improved through the simultaneous use of more than one detector, and most systems currently consist of two or three heads mounted on a single gantry (Figure 2(a)). The radionuclide mostly used in SPECT is Technetium 99m (^{99m}Tc) which has a half life of 6.03 h and which emits photons of 140.5 keV. SPECT application fields in order of importance are cardiac imaging, bone scanning, neurology, and renal and liver function imaging.

2.2. Positron emission tomography

Pharmaceuticals can also be labeled with neutron-deficient isotopes which are positron emitters. A number of elements that are fundamentally used by the human body can be positron emitters, allowing more scope for radiopharmaceutical design than is possible with single photon emitters. However, production costs are higher since a cyclotron is needed to generate these instable radionuclides. Fluorine 18 (^{18}F) is mostly used to label the fluorodeoxyglucose (FDG) compound, which is a glucose analogue and thus suitable for the visualization of the glucose metabolism. The emitted positron annihilates with an electron giving rise to two virtually antiparallel 511 keV photons, corresponding to the conversion of the rest mass of the two particles into energy. The primary advantage of detecting both annihilation photons in time coincidence is the electronic collimation. Detection of two photons is sufficient to determine the line on which the annihilation took place and so no physical collimation is required, allowing higher sensitivity than in single photon imaging. Tomographs appropriate for Positron Emission Tomography (PET) (Figure 2(b)) typically consist of a series of ring detectors, having a large number of separate blocks coupled to PMTs. Depending on the required sensitivity, the individual detectors can be in coincidence only with other opposing detectors in the same ring using lead septa, which is called PET in 2D mode, or also with detectors in other rings (3D mode). The 3D mode increases the sensitivity but also increases the image degradation resulting from the detection of scattered photons and random coincidences. PET application fields in order of importance are oncology, neurology, and cardiology.

2.3. Emission tomography in small animal imaging

The past 4–5 years were a major breakthrough for small animal imaging, which has received considerable attention in the field of medical imaging. Dedicated high-resolution small animal-imaging systems have recently emerged as important new tools for cancer research. These new imaging systems permit researchers to non-invasively screen animals for mutations or pathologies and to monitor disease progression and response to therapy. Several research groups have extended the use of imagers from human to small animal applications. The first and most obvious challenge to small animal-imaging technology derives from the magnitude of the difference between the physical size of human subjects for which clinical systems have been developed and the laboratory rat or mouse (70 kg, 300 g, and 30 g, respectively). To address the same biological questions in mice that can currently be investigated in humans, small animal systems must have similar ratios of volumetric spatial resolution to the volume of the object of interest. This suggests a reconstructed spatial resolution smaller than 1 mm. In the past decade, the development of small animal SPECT-imaging technologies has progressed along two main paths: firstly, the adaptation of clinical systems incorporating novel collimator designs [3] and, secondly, the development of dedicated small animal scanners based on compact high-resolution detectors [4]. In small animal PET, the very demanding spatial resolution requirement calls for new approaches in both front-end detector materials and overall system design. In recent years, a number of new scintillators and photodetectors have been explored [5]. Increased light per gamma ray interaction, faster rise and decay times, greater stopping power, and improved energy resolution are the desired characteristics. Improvements in these characteristics enable detectors to be divided into smaller elements, thus increasing resolution and minimizing dead time losses.

2.4. Reconstruction

A key ingredient for SPECT and PET is image reconstruction, consisting in retrieving the 3D spatial distribution of the radiopharmaceutical from the projection data acquired under different angles. Tomographic reconstruction can be performed in several ways [6]. The reconstruction problem can be solved analytically or iteratively. In the latter case, a standard algebraic approach can be used such as the Algebraic Reconstruction Technique (ART) [7], or, if the acquisition process can be assumed to be statistical, a statistical reconstruction technique can be preferred, such as Maximum Likelihood Expectation Maximization (ML-EM) [8]. This latter technique is frequently used in emission tomography as it accounts for the Poisson nature of the acquired data. The main advantage of incorporating a statistical model into the reconstruction process is that this yields a smaller variance in the reconstructed images. Iterative reconstruction techniques also enable the modeling of image-degrading effects in the reconstruction algorithm, so that these are automatically compensated for during image reconstruction [9].

3. The MC Method

MC methods are statistical simulation methods wherein a statistical simulation is any simulation that uses a sequence of random numbers. In order to do MC calculations, some a priori information about the occuring physics processes in the simulations is needed; this information is expressed in the form of probability density functions (pdfs). When simulating photon interactions, the partial and total cross-section data (based on the material constitution) represent such information used to calculate the path length and the type of interaction. Afterwards, these pdfs are sampled by predefined rules using randomly generated numbers. The energy of a photon can be dissipated along its path or the photon can penetrate all scattering and attenuating media to reach the detector where a new pdf-sampling decides whether it should be accounted for in the scoring region or whether it should be discarded [10].

3.1. Random numbers

Random numbers are key important when modeling a physical system by a statistical model. Every random number generator has to deliver uncorrelated, uniform and reproducible sequences with a very long period in a short amount of time. Some algorithms are able to generate repeating sequences of C integers which are, to a fairly good approximation, randomly distributed in the range 0 to $C-1$ where C is a large integer, for instance mod(2^k) with k the integer word size of the computer. An example of such an algorithm is the linear congruental algorithm. The formula linking the nth and $(n+1)$th integers in the sequence is

$$I_{n+1} = (AI_n + B)\mathrm{mod}(C), \tag{1}$$

where A and B are constants. The first number of the series, the seed value, must be randomly changed, for instance by triggering a value from a call to the system clock.

3.2. Sampling methods

To obtain a stochastic variable that follows a particular pdf, $f(x)$, three different sampling methods can be used.

3.2.1. The distribution function method

A cumulative distribution function $F(x) = P(X \leq x)$ is constructed from the integral of $f(x)$ over the interval $[a, x]$ according to

$$F(x) = \int_{a}^{x} f(x')dx'. \tag{2}$$

The variable x is then sampled by replacing $F(x)$ in Eq. (2) with a uniformly distributed random number in the range of [0, 1] and solving for x. This sampling method is used if the inverse of $F(x)$ can be easily calculated.

3.2.2. The rejection method

Possibly there can arise some mathematical difficulties in calculating the inverse of the cumulative distribution function F. In such case, the rejection method can offer a solution. Let $f(x)$ be defined in the range $[a, b]$ and let $r(x)$ be the rejection function $\forall\, x \in [a, b] \Rightarrow 0 < r(x) \leq 1$. Here $r(x)$ is given as $f^*(x)$, being the pdf after normalization to unity. Then, a uniform distributed value of x within the range $[a, b]$ can be sampled from the relation

$$x = a + R_1(b - a), \tag{3}$$

where R_1 is a random number in [0, 1]. A second random number R_2 in [0, 1] is then drawn to decide whether the sampled x should be accepted by checking whether

$$R_2 < r(x) = f^*(x). \tag{4}$$

If this relation is fulfilled, then x is accepted as a properly distributed stochastic value, otherwise x needs to be resampled.

3.2.3. Mixed methods

When the previous two methods are not applicable, a mixture of the two methods above is used. The $f(x)$ is written as the product of two probability distribution functions $m(x).r(x)$ with $r(x)$ acting as the rejection function defined in the previous section. One determines an x value using the distribution function method on $m(x)$ and applies the rejection method with that x to $r(x)$ [11].

3.2.4. Non-analog sampling

An MC simulation involving the exact pdfs may be impractical from a computational point of view, requiring unrealistic computation time to achieve reliable results. One remedy consists in biasing the sampling. When simulating photon transport, this can be done using stratification, i.e., by sampling more from regions that highly contribute to the results, or using techniques such as splitting or Russian Roulette in which particles are weighted, or by implementing forced detection, i.e., by forcing particles to interact so that they contribute to the simulation result [12].

3.3. Photon transport modeling

Compton incoherent scattering and photoelectric effect are the two most important interaction mechanisms in Nuclear Medicine simulations. The total photoelectric cross section for a given energy, E, is calculated using discretized libraries by means of interpolation. The incident photon is absorbed and an electron is emitted in the same direction as the incident photon. The kinetic energy of the electron is defined as the difference between the energy of the original photon and the binding energy of the electron. The subshell of emission is again randomly sampled, thereby using cross-section data for all subshells. When simulating the Compton scattering of a photon from an atomic electron, an empirical cross-section formula is used, and sampling of the final state is done by a variant of the mixed MC method as noted in Section 3.2.3. The following illustrates how this is implemented for GEANT4 [13], a particle tracking code which forms the basis of the MC simulator to be discussed in Section 6. The quantum mechanical Klein–Nishina differential cross section per atom is [14]:

$$f(x) = \pi r_e^2 \frac{m_e c^2}{E_0} Z \left[\frac{1}{x} + x\right] \left[\frac{1 - x\sin^2\theta}{1 + x^2}\right], \tag{5}$$

where

r_e = classical electron radius
$m_e C^2$ = electron mass
E_0 = energy of the incident photon
E_1 = energy of the scattered photon
$x = \frac{E_1}{E_0}$.

Assuming an elastic collision, the scattering angle θ is defined by the Compton formula:

$$E_1 = E_0 \frac{m_e c^2}{m_e c^2 + E_0(1 - \cos\theta)}. \tag{6}$$

The value of x corresponding to the minimum photon energy (backscatter) is given by

$$a = \frac{m_e c^2}{m_e c^2 + 2E_0}, \tag{7}$$

hence $x \in [a, 1]$. One may state that

$$f(x) \approx \left[\frac{1}{x} + x\right] \left[\frac{1 - x\sin^2\theta}{1 + x^2}\right] = m(x) \cdot r(x) = [\alpha_1 m_1(x) + \alpha_2 m_2(x)] \cdot r(x), \quad (8)$$

where

$$\begin{aligned} \alpha_1 &= \ln\left(\frac{1}{a}\right) \\ m_1(x) &= \frac{1}{\alpha_1 x} \\ \alpha_2 &= \frac{(1 - a^2)}{2} \\ m_2(x) &= \frac{x}{\alpha_2}. \end{aligned} \quad (9)$$

m_1 and m_2 are probability density functions defined on the interval $[a, 1]$, and $r(x)$ is set to

$$r(x) = \left[\frac{1 - x\sin^2\theta}{1 + x^2}\right], \quad (10)$$

being the rejection function $\forall\ x \in [a, 1] \Rightarrow 0 < r(x) \leq 1$. Given a set of three random numbers R_0, R_1, R_2 uniformly distributed on the interval [0, 1], the sampling procedure for x is the following:

(1) decide whether to sample from $m_1(x)$ or $m_2(x)$: if $R_1 < \frac{\alpha_1}{(\alpha_1 + \alpha_2)}$ select $m_1(x)$ otherwise select $m_2(x)$,
(2) sample x from the distributions to m_1 or m_2:

for m_1: $x = a^{R_1}$
for m_2: $x^2 = a^2 + (1 - a^2)R_1$,

(3) calculate $\sin^2\theta = t(t - 1)$ where $t \equiv (1 - \cos\theta) = \frac{m_e c^2 (1 - x)}{E_0 x}$,
(4) test the rejection function: if $r(x) \geq R_2$ accept x, otherwise go to (1).

The polar angle θ is deduced from the sampled x value and in the azimuthal direction the angular distributions of both the scattered photon and the recoil electron are considered to be isotropic [13,14].

3.4. Scoring

A selection of relevant physical quantities such as energy, detection location and photon interactions must be accumulated into tallies or scores. Also, an estimate of the statistical error as a function of the number of trials (and other quantities) must be determined.

4. Relevance of Accurate MC Simulations in Nuclear Medicine

MC modeling is the only possible approach for all applications where measurements are not feasible or where analytic models are not available due to the complex nature of the problem [11,15]. In addition, such modeling is a practical approach in nuclear medical imaging in four important application fields [10,16], presented hereafter.

4.1. Studying detector design

MC methods can assist in the development of new collimator and detector designs, evaluation of new electronics, aso. MC simulations have been extensively used to analyze the performance of new long bore parallel collimators as well as rotating slit collimators [17], fan beam [18], cone beam and pinhole collimators [19]. For the latter, simulations have been performed with various aperture span angles, different hole sizes and various materials to evaluate the penetration [20]. MC methods also play an important role in new system design, for instance in the research field of solid-state detectors with improved energy resolution and low-noise electronics [21–23]. Similarly, in the field of PET, MC techniques have been used to study the performance of interplane septa with varying constitution, thickness and geometry, to compare single to true coincidence event ratios in single-slice, multi-slice and open collimator 3D configurations and to assess the effect of collimation on scatter fraction (e.g., [24–26]). MC simulations of detector responses and efficiencies are of key importance since the scintillation crystal is the critical component in emission tomography [27]. Simulations also prove their usefulness in Time-of-Flight PET (TOF) design since they can be used to test several detector crystals and to simulate the influence of timing resolution on variance reduction (e.g., [28–30]). MC methods can also assist in the design of new detectors with depth-of-interaction (DOI) information by simulating multilayer crystals and by estimating the gain in reconstructed resolution through incorporation of DOI information [31,32].

4.2. Analysing quantification issues

The presence of the scatter, attenuation and partial volume effects in Nuclear Medicine images limits the accuracy of activity estimates (i.e., what is called quantification). Each of these degradations has a particular impact [33–35]. Scatter does not produce major artefacts comparable to those caused by attenuation but reduces contrast by introducing a low-frequency blur in the image. Moreover, the impact of scatter and attenuation generally depends on the photon energy, camera energy resolution and energy window settings, as well as the object size and shape, detector geometry and the source distribution. Many of these parameters are non-stationary which implies a potential difficulty when developing proper scatter and attenuation correction techniques. MC calculations have been found to be powerful tools for an in depth analysis since the user has the ability to separate the detected photons according to their interaction history: primary (i.e., unscattered) events, scattered events, contribution of downscattered events, scatter in patient or detector, etc. [36]. MC modeling thus enables a detailed investigation of the spatial and energy distribution of Compton scatter: energy spectra, point-spread functions and scatter fractions can be simulated [37]. This is impossible to perform using present experimental techniques, even with very good energy resolution detectors. Another important application field for MC simulations in quantification is resolution recovery. For instance in SPECT, distance dependent collimator response can be subdivided into a geometric component, collimator scatter and septal penetration. MC simulations can be used to determine the relative contribution of each of these components in specific imaging configurations [38].

In PET, the penetration of annihilation photons into the detector material before interaction is a statistical process which leads to significant displacement and anisotropy of the point spread function [39,40]. Compensation for crystal penetration is thus one of the most important issue in order to recover the spatial resolution in PET. Theoretical models for that anisotrope spatial resolution can only be verified by MC simulations since those DOI-dependent experiments are very hard to perform with current devices.

Finally, MC analysis has been proven useful in evaluating motion artefacts. If the simulator is capable of using explicit timing information to synchronize decay with detector and patient movement then degradations such as cardiac or respiratory motions can be studied [41].

4.3. Correction methods for image degradations

The image-degrading factors (as described in the previous section) can have a large impact on quantitative accuracy and on clinical diagnosis and ideally all have to be corrected for to achieve optimal clinical imaging. Traditionally, this was done by preprocessing the projection data or by postprocessing the reconstructed images. MC-simulated datasets are often used to evaluate the accuracy of these correction methods. For such evaluation studies, the simulated datasets must have the appropriate statistical properties and include all detector specific imperfections. For instance, when evaluating a scatter correction method, MC simulations can be used to generate scatter-free projections. Images reconstructed from scatter-free projections constitute the gold standard for any scatter correction method. Nowadays, MC simulations often form the basis of the aforementioned correction techniques since iterative reconstruction algorithms are able to correct for image degradations by simulating the effect during the reconstruction. The simulations can be based on MC methods or on a combination of MC and analytical methods. These SPECT simulators face a considerable challenge; they have to balance accuracy against efficiency. A large variety of simulators have been proposed [42–46]. Other groups achieve similar goals by simulating and store the system matrix [47,48]. Reconstruction methods based on MC simulators have been shown to improve contrast-to-noise [49–51] and lesion detectability [52–54]. Finally, MC-generated datasets are also used in evaluation or comparison of different reconstruction schemes, for instance to evaluate the robustness of an algorithm to parameter variations.

4.4. Detection tasks using MC simulations

Receiver Operating Characteristic (ROC) analysis is considered as the most reliable method for evaluating the diagnostic ability of medical imaging techniques. It provides a measure of the diagnostic performance of an imaging modality by plotting the sensitivity versus the specificity for a wide and continuous range of decision criteria. These studies require many images (typically hundreds) so that statistical analysis of the detection performance can be performed. Observer studies based on MC-based projection data can answer a large range of problems from the simple comparison between analytical and statistical reconstruction [55] to complex studies on the benefits of anatomical a priori information in the reconstruction [56]. Using MC simulations to generate the datasets is therefore only feasible if efficient codes are available. Also, sharing these datasets encourages and facilitates evaluation studies. In addition, evaluating data-processing tools developed by different groups worldwide using the same datasets removes the evaluation biases introduced by the use of different data. Therefore, some efforts have recently been accomplished to propose publicly available databases of MC-simulated data dedicated to evaluation studies [57,58].

4.5. Applications in other domains

In the Nuclear Medicine field, MC simulations are also widely used in dosimetry and radiotherapy research. One application is the individual treatment planning in radionuclide or radioimmunotherapy [59] where prospective dose estimates are made in the latter by using a tracer activity of a radiolabeled antibody to obtain biodistribution information prior to administration of a large therapeutic activity (e.g., [60]). The MC simulations return the dose distribution calculated from the quantitative images of the tracer distribution. The clinical applicability of individual treatment planning algorithms will depend on the time required to generate those absorbed dose estimates for each individual.

MC simulations are also used for the evaluation of external beam radiotherapy setups [61]. A complete simulation model of the accelerator head, including scatter foil or X-ray target and flattening filter, multileaf collimator, support plates, and so on, is necessary for this purpose. In that context, hybrid devices such as SPECT/CT and PET/CT might allow for a better delineation of the Planned Target Volume (PTV) on CT using the functional information provided by the registered SPECT or PET images, thus fundamentally increasing the accuracy. Precise MC simulations should be used in the virtual treatment-planning system to maintain the most precise dose estimates [62] to further improve the treatment plan.

Another field of interest is to perform a MC-based dosimetry for brachytherapy applications [63] to provide detailed descriptions of local doses. Hadrontherapy is also a MC-assisted research area [64] where simulations can reproduce the beam-induced β^+ emissions of the irradiated target nuclei along the beam path, to assist the design of in-beam PET systems with high sensitivity and high spatial resolution.

5. AVAILABLE MC SIMULATORS

Some MC simulation codes dedicated to the simulation of SPECT and PET have been developed, such as SimSET [65] enabling SPECT and PET modeling, SIMIND [66] and SimSPECT [67] for SPECT, PETSIM [68], Eidolon [69] and SORTEO [70] for PET.

Such dedicated packages are very powerful for their specific design goal but are often not detailed and flexible enough to enable realistic simulations of emission tomography detector geometries. Moreover, most of them do not account for time explicitly, which limits their use for modeling time-dependent processes such as decay, tracer kinetics, patient and bed motion, dead time or detector orbits. To ensure high flexibility in simulation design, general purpose nuclear physics codes such as EGS4 [71], GEANT4 [13,72], or MCNP [73] have also been used for SPECT and PET simulations. The main advantages of these general purpose codes are that they are widely used and extensively tested, they can be regarded as long-term existent as well as supported, and they are continuously evolving and therefore use the best of current hardware and software capabilities. There are fewer limitations on their possible applications than for dedicated codes, and fewer simplifying assumptions are made. For example, the processes taking place in the collimator can be thoroughly simulated whereas dedicated codes often use a parametric model. Also, time-dependent processes can be simulated and their non-specific design makes them appropriate to implement non-conventional SPECT and PET cameras [16]. In the end of the 1990s, 14 codes were used for MC simulations in SPECT and PET, among which 10 were not used by another group than those who developed the code, and 4 were publicly released or available from the authors. In the early 2000, 15 codes were used among which 8 were home-made and 7 were publicly released or available from authors [74]. This suggests that up to recently, there was no code that was considered as a standard for MC simulations in emission tomography. Based on this observation, a new code, GATE [75–77], has recently been designed as an upper layer of the GEANT4 nuclear physics code [13,72] tuned for simulating SPECT and PET acquisitions. GATE thus takes advantage of all GEANT4 features, including well-validated physics models, basic event-timing information, geometry modeling and visualization tools. It is written in an object-oriented language that ensures high modularity. In addition, GATE uses a scripting language making it easy for the user to design a SPECT or PET simulation. GATE has been developed as a collaboration effort of about 10 laboratories worldwide (the OpenGATE collaboration), and was publicly released in May 2004. It is now shared by a large research community, while long-term support and maintenance is still ensured by the OpenGATE collaboration. During 2006, already more than 650 individuals had subscribed to the GATE community and at the 2005 IEEE Medical Imaging Conference (the most relevant conference in this domain), GATE and/or GEANT4 were used as tool of excellence in 33% of all papers involving MC simulations while this number was 13% for SimSET and 7% for SIMIND, respectively [74]. It is therefore expected that GATE will become the code most frequently used for SPECT and PET simulations. Given this explicit tendency, we will describe GATE hereafter as an example of a simulation platform.

6. GATE

6.1. Basic features

6.1.1. Software architecture

GATE is a C++ design, enabling a modular structure built on three fundamental layers: (1) the core layer defines the basic mechanisms available in GATE for geometry definition, time management, source definition, digitization, and data output, (2) the application layer is composed of classes derived from the core layer classes to model-specific objects or processes and finally (3) the user layer. The latter allows the user to simulate a setup through the use of scripting. Indeed, the functionality provided by each class is available through script commands, so that the end-user of GATE does not have to manipulate any C++ language. Therefore, a complete Nuclear Medicine experiment can be defined using the GATE script language, including the object geometry, the radioactive sources, the camera geometry, the detector electronics, the physics processes, the passing of time, the kinetic parameters, and the output format. Some important modeling instructions will be demonstrated hereafter using examples of scripts, while discussing the implemented functionality.

6.1.2. Defining the geometric features of an acquisition

All geometric features involved in modeling a SPECT or PET acquisition are defined using basics elements known as "logical volumes." A logical volume is defined by its name, shape, size, and material composition. Repeaters can be used to replicate and place logical volumes at multiple positions and orientations. Repeaters are elementary geometrical transformations such as rotations and translations applied in succession. When logical volumes are placed at specific positions, they form "physical volumes."

```
#Multiply head to create triple headed SPECT system
/gate/SPECThead/repeaters/insert ring
/gate/SPECThead/ring/setRepeatNumber 3
/gate/SPECThead/ring/setAngularPitch 120. deg
```

More complex structures can be created by combining various types of repeaters (e.g., ring, linear, quadrant and cubic array), as shown in Figure 3 which demonstrates how the construction of a folded lead collimator for SPECT simulations was performed by repeating a hexagon on a rectangular array. This array was then filled with a translated duplicate, creating the correct distribution of air holes.

Accessing a material database file is the primary method for assigning material properties to volumes. This file contains all material parameters required by GEANT4 to calculate the interaction cross sections. The following example shows how breast material is obtained by defining the fraction of constituting elements.

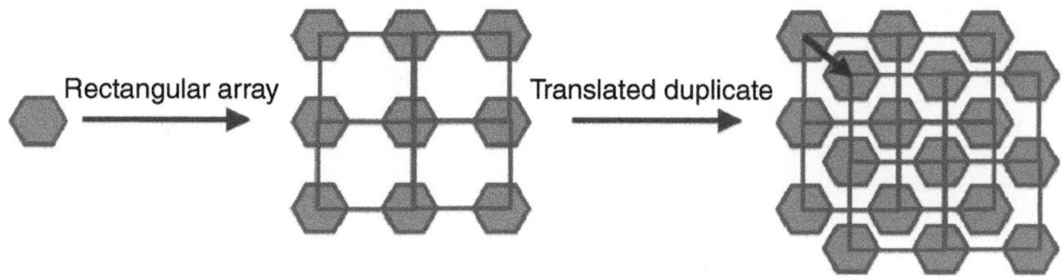

Figure 3 Detail of a collimator modeling process.

```
#[Elements]
Hydrogen:    S=H  ;  Z=1   ;     A=   1.01  g/mole
Carbon:      S=C  ;  Z=6   ;     A=  12.01  g/mole
Nitrogen:    S=N  ;  Z=7.  ;     A=  14.01  g/mole
Oxygen:      S=O  ;  Z=8.  ;     A=  16.00  g/mole
Sodium:      S=Na ;  Z=11. ;     A=  22.99  g/mole
Phosphor:    S=P  ;  Z=15. ;     A=  30.97  g/mole
Sulfur:      S=S  ;  Z=16. ;     A=  32.066 g/mole
Chlorine:    S=Cl ;  Z=17. ;     A=  35.45  g/mole

#[Mixture]
Breast: d=1.020 g/cm3 ; n=8
+el: name=Oxygen;      f=0.5270
+el: name=Carbon;      f=0.3320
+el: name=Hydrogen ;   f=0.1060
+el: name=Nitrogen;    f=0.0300
+el: name=Sulfur ;     f=0.0020
+el: name=Sodium ;     f=0.0010
+el: name=Phosphor;    f=0.0010
+el: name=Chlorine ;   f=0.0010
```

When modeling a scanner using combinations of volumes, specific guidelines with respect to the geometrical hierarchy must be followed. Most PET scanners are built following comparable concepts: one or more rings, each ring consisting of several blocks of scintillating material, each block being subdivided in crystal parts, etc. For SPECT, the detector mostly consists in a gamma camera with a continuous or a pixelated crystal and a collimator. Most of these geometrical concepts are common to many different imaging systems. To facilitate the modeling of detectors, predefined global *systems* are used. From the user's point of view, the main property of a *system* is that its geometric hierarchy is automatically accounted for by the corresponding data output formats. ASCII and ROOT [78] output files are available for all *systems* and can easily be transformed to a list mode file for reconstruction purposes.

Examples of a modeled SPECT and PET scanner using these predefined systems are shown in Figure 4.

6.1.3. Defining radioactive sources

The spatial distribution of a radioactive source can be specified using five 2D shapes, i.e. circle, annulus, ellipse, square, and rectangle, and four 3D shapes, i.e.

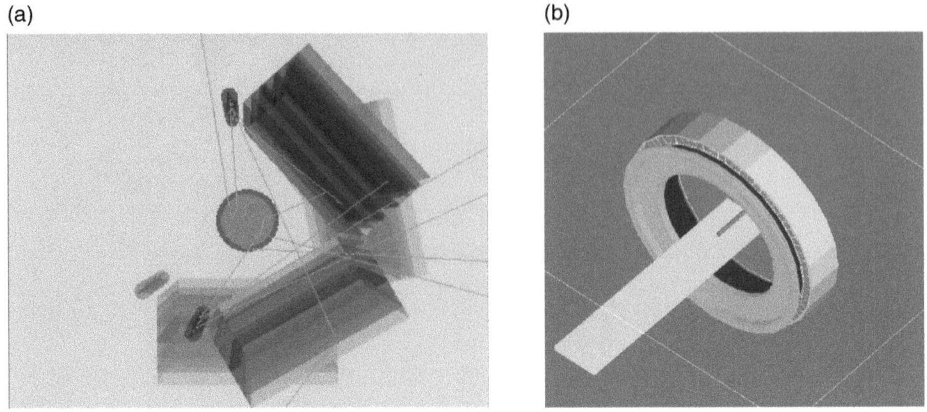

Figure 4 Example scanner types modeled in GATE: (a) SPECT system (IRIX-Beacon), (b) PETsystem (ECAT HR+).

sphere, ellipsoid, cylinder, and parallelepiped. Examples of simple analytical sources are shown in Figure 5. Moreover, the angular distribution of the decay particles can be tuned by using azimuthal and polar angles, making it possible to emit particles in different angular spans. In PET, the user can force the annihilation photons to be emitted back-to-back to significantly decrease the simulation time.

Radioactive sources can also be described using images, what is called a voxelized description. Using such voxelized descriptions, any inhomogeneous, anthropomorphic source or patient image can be used to model realistic acquisitions [79]. As an example, simulated projection data of the Hoffman hardware phantom (voxelized source) [80] with and without a voxelized attenuation geometry is shown in Figure 6.

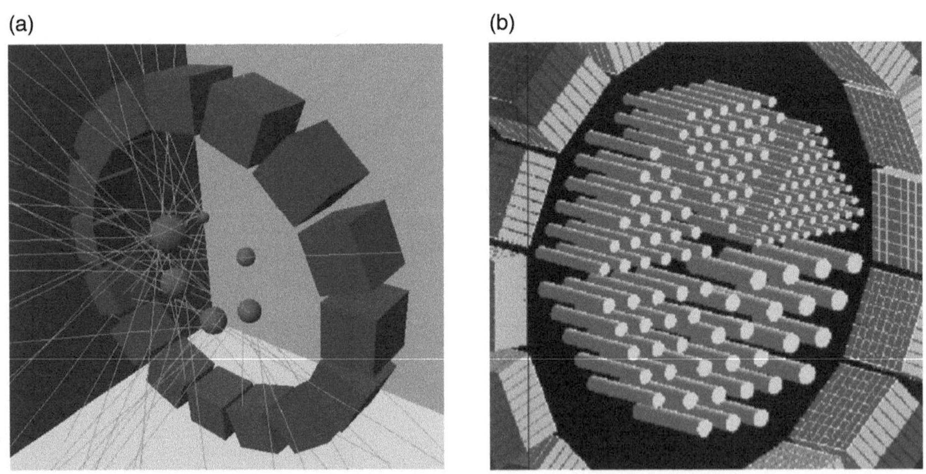

Figure 5 Example of analytical sources: (a) simple test object, (b) Derenzo phantom.

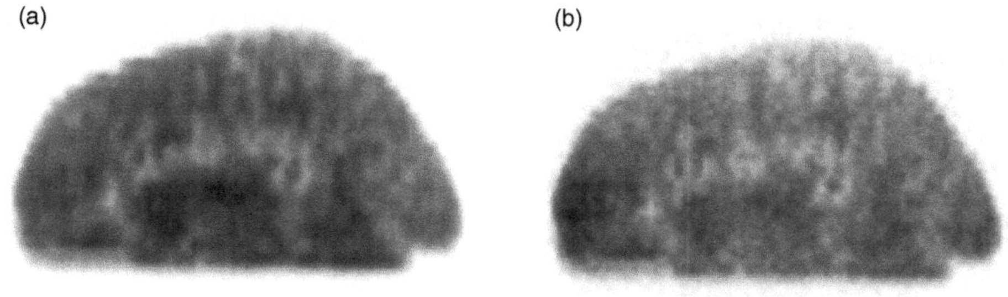

Figure 6 Simulated projection data of the Hoffman brain phantom: (a) without attenuation; (b) with a voxelized attenuation geometry based on the same phantom.

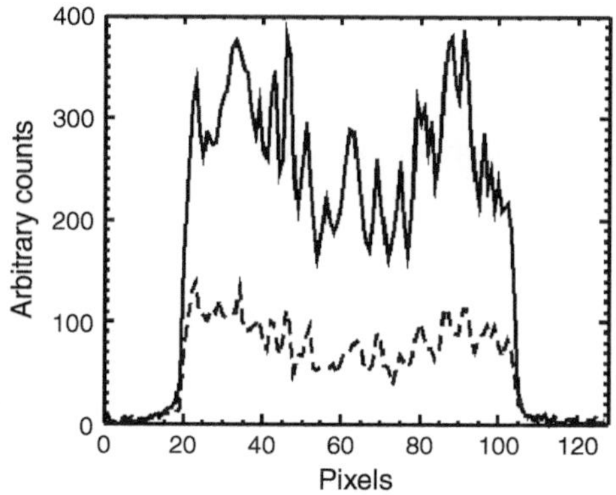

Figure 7 Midway horizontal profiles of the simulated sagittal projections of the digital Hoffman voxelized brain phantom with (dashed line) and without (full line) the voxelized attenuation geometry.

Both projection data are compared by plotting two midway horizontal profiles of the sagittal projections as is shown in Figure 7. The voxelized nature of the results is obvious, as well as the appropriate attenuation effect.

Although quite realistic anthropomorphic phantoms are available for human and for small animal studies, the impact of MC simulations in SPECT and PET would be increased by the development of a wider variety of numerical anthropomorphic phantoms. These should include humans with a wide variety of body habitus and motions, and small animal models with breathing and cardiac motions (see also Section 6.2). Several GATE development efforts are currently ongoing in this direction [81].

6.1.4. Physics

The electromagnetic interactions used in GATE are derived from GEANT4. The physics package manages electrons, positrons, γ-rays, X-rays, optical photons, muons, hadrons, and ions. When using the low-energy extension, the modeling of photons and electrons is extended and verified down to 250 eV and also includes Rayleigh scattering.

The non-collinearity of two annihilation photons for PET applications is not taken into account by GEANT4. Because this effect can be important in PET, an additional module has been incorporated in GATE that models the angular distribution of these photons based upon a Gaussian angular distribution in water (0.58° FWHM) [82]. On the other hand, the positron range is intrinsically modeled.

6.2. GATE: Time management

A major and recent advance in MC simulations is the current possibility to model time-dependent phenomena. This makes it possible to realistically model radioactive decay, dynamic biodistributions, physiological motions such as respiratory and cardiac motions, rotation of the scanner, TOF-PET, dead time, count rate, random coincidences, and event pile-up. At least three codes have been specifically designed so that time-dependent phenomena can be more easily handled. PET-SORTEO [70] can model the evolution of activity concentration in time but it cannot easily model respiratory or cardiac motion. SimSET has been recently extended to keep track of time-of-flight [83], but it cannot easily handle simulations of time-activity curves in different physiological regions, nor detector motion.

To the best of our knowledge, the only code that can model any time-dependent effect is GATE. In GATE, time is explicitly kept track of during the simulation by the implementation of a virtual clock that synchronizes all time-dependent processes. The following script example shows the movement of the gantry at a speed of 1° per second as well as a translation of the patient support table. The rotation of the gantry and the translation of the table are kept synchronized which enables the simulation of whole body acquisitions.

```
#Movement of the gantry
/gate/SPECThead/moves/insert orbiting
/gate/SPECThead/orbiting/setSpeed 1 deg/s
/gate/SPECThead/orbiting/setPoint1 0 0 0 cm
/gate/SPECThead/orbiting/setPoint2 0 0 1 cm
#Movement of the patient support
/gate/Table/moves/insert translation
/gate/Table/translation/setSpeed 0 0 1. cm/s
/gate/application/setTimeSlice 1. s
/gate/application/setTimeStart 0. s
/gate/application/setTimeStop 360. s
/gate/application/startDAQ
```

In GATE, the start (0 s in the previous example) and stop (6 min here) times of the acquisition must be provided by the user, as well as the sampling interval (1 s in the example). In such a "time slice," the geometry is kept at rest and the update is performed at the time slice transition. However, the sources are allowed to decay within the time slice and the particle transport proceeds.

6.3. GATE: Digitization

Parts of the scanner geometry are designated as sensitive detectors, and particle interactions are recorded and scored within these regions. Firstly, *hits* are generated from interactions that occur inside the detector parts of the scanner (e.g., the crystal). The data contained in these *hits* include the energy that has been deposited, the interaction positions, the origin of the particle, the type of interaction, the volume name in which the interaction took place, and the time information. Secondly, similar sensitive regions are used to detect and count the Compton and Rayleigh interactions occurring within the scanner's field-of-view (FOV). For example, the number of scattering interactions that occurred in the light guide, the isolation, the shielding, and the phantom or patient can be recorded.

The digitizer consists of a chain of processing modules (Figure 8) that takes a list of *hits* from the sensitive detectors and transforms them into *pulses* referred to as singles. The processing modules of the digitizer can be controlled using the GATE-scripting language. The key elements of this chain are now briefly described. Firstly, the adder sums the deposited energy of all interactions in a sensitive detector (*hits*) to yield a *pulse*. The position of the *pulse* is calculated from the energy-weighted centroid of the *hit* positions, and the time of the *pulse* is set to that of the first *hit* within the volume. If a particle interacts in several sensitive detectors, for instance after crystal scatter or after crystal penetration, the hit adder will generate a list of *pulses*, one for each sensitive detector. The second instruction of the digitizer (readout) implements the readout segmentation of the scanner. The position of the *pulse* is set to that of the *pulse* from the adder that has the most energy (winner-takes-all paradigm).

The energy response instruction applies a blurring to the energy of the *pulse*. The spatial response instruction applies a Gaussian blur of the position for SPECT. For PET, spatial resolution is calculated by the pulse reader, which simulates the intrinsic spatial resolution of the detector. Furthermore, an energy discrimination is performed to set an acceptance window (thresholder electronics). Also, both paralyzable (the detector is dead for a period t after every event, even if the event arrives during the dead period of the previous event) and non-paralyzable dead times (the detector is dead for a period t after every event) can be modeled on an event-by-event basis.

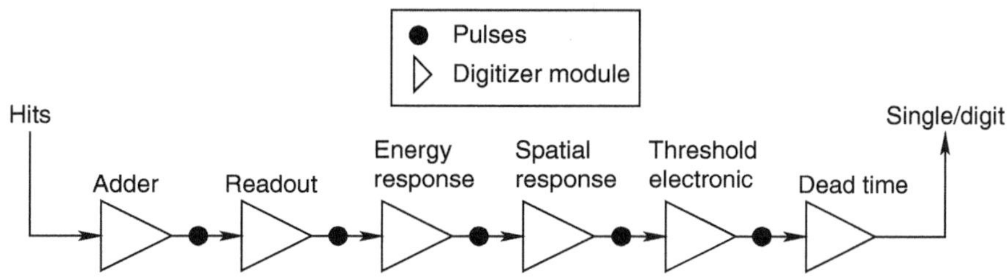

Figure 8 The digitizer is organized as a chain of several modules that processes the *hits* to yield a single, which represents a physical observable.

Other user-defined modules can be added individually to model more specific properties, such as the time resolution. Moreover for PET, at the end of a digitizer chain, a sorter can be added to find pairs of singles that are in coincidence (whenever the time interval between the singles is less than a user-defined coincidence window). Using the event number and the Compton flag, randoms and scatter coincidences can be differentiated from true coincidences. Multiple coincidences and auto-coincidences are also taken into account.

The optimization of the digitizer chain parameters to reproduce the behaviour of a specific scanner is very time-consuming. This is best done by comparing the results from different sets of digitizer parameters using the same series of *hits* with those of the real tomography when available. To perform this optimization, GATE offers an operating mode named DigiGATE. In this mode, *hits* are read from a data file generated by GATE and fed directly into the digitizer chain. All conditions are kept identical in the simulations including time dependencies.

7. EFFICIENCY-ACCURACY TRADE-OFF

7.1. Accuracy and validation

To model a broad range of detector geometries, there are currently two options: either use a generic simulator, such as GEANT4, EGS4 or Penelope, or use GATE, which is currently the code able to accommodate the largest number of detector geometries. For instance, the spherical shape of the Hi-Rez PET tomograph could not be accurately modeled using SimSET, but only using GATE (or another generic code) [84].

The flexibility of GATE is further illustrated by the number of commercial tomographs that have been modeled in GATE, both in SPECT (e.g., AXIS [37], IRIX [85], DST-Xli [86] and Millenium VG Hawkeye [87]), in PET (e.g., ECAT Exact HR+ [88], Advance [89], Allegro [90], ECAT HRRT [91], Hi-Rez [84]) and in small animal imaging (e.g., MicroPET P4 [92], microPET Focus 220 [93], ClearPET [94]). Moreover, GATE is the code used the most for studying prototypes (e.g., [95,96]).

How well simulations can predict the physical response of the tomograph is usually verified by comparing the simulated and empirical values of some parameters, which can be experimentally measured and that characterize the physical performances [97]. The modeling of a system is then considered to be validated if it accurately reproduces the response of the real system. The parameters of interest that are used most often to assess the validity of a model are the spatial resolution, energy spectra, scatter fractions, sensitivity, and count rates obtained in specific configurations, for instance, using the NEMA phantoms [98].

7.2. Calculation time

An important issue of MC simulations is the calculation burden. The computing time for a typical PET benchmark on a 1-GHz basis averages around 850 generated and tracked events per second and 16 simulated coincidence detections per second. For SPECT, this value amounts 420 generated and tracked events per second and 0.83 detections per second or 1.2 second per detection. For simulating clinically realistic acquisitions, many months of computation time on one CPU are accordingly needed. This computational issue is most cumbersome for SPECT since the total number of detected counts in SPECT is less than 0.02% of the generated events because the collimator in front of the crystal stops most of the incoming photons. This effect is most pronounced in high-resolution collimator variants. Secondly, every photon is tracked through every object of the experiment, and a typical Low-Energy High-Resolution (LEHR) collimator for instance consists of 161120 individual air holes. Several approaches are investigated to solve this computational issue both for SPECT and PET. Firstly, variance reduction techniques such as stratification, importance sampling and forced detection can be used [12]. Also, hybrid methods can be proposed that combine MC simulations in the object coupled to analytical or tabulated models for the detector [99,100]. Lastly, efforts are conducted to submit simulations to a cluster of computers by distribution software or by intrinsically parallelizing the code [101].

8. Case Studies

Three case studies will now be discussed to illustrate the relevance of MC simulations in emission tomography. The first case study will demonstrate the value of using MC simulations in detector design for the optimization of TOF-PET. The second case study will show an example of the use of MC simulation for the assessment of the quantitative accuracy of correction techniques in clinical data. The third case study will illustrate the use of MC simulations in image reconstruction to calculate the system matrix in SPECT.

8.1. Case study I: TOF-PET

During the 1980s, different PET systems with TOF information were developed [102–109]. The TOF information leads to more accurate imaging by reducing the noise in image data, resulting in higher image quality, shorter imaging times, and lower dose to the patient. Due to technical limitations of the TOF-capable scintillators, these systems could not compete at that time with non-TOF BGO PET scanners. Recently, new scintillators with high stopping power and high light output (e.g., LSO and $LaBr_3$) that are fast enough for TOF PET have become available. Therefore, different TOF-PET systems are under development and the first commercial system is available since June 2006 [110].

In PET, the time difference measurement between both photons is used to determine whether two single photons will be stored as a coincident event. Two photons coming from a single annihilation arrive at slightly different times depending on their origin in the FOV. For a camera with a ring diameter of 90 cm, the maximum time difference for photons to reach the detectors is a few nanoseconds and therefore accuracy of time measurement has to be only in the range of 1–2 ns. In TOF-PET, the time difference is determined as accurately as possible and the difference is stored and used in the reconstruction. For TOF-PET, it is necessary to use a MC simulator with very accurate time tracking (accuracy of ps) of both photons. To get a good prediction of the system performance, it is also important to have an accurate model of the scintillators' rise and decay time. GATE has an explicit modeling of time and is therefore suitable for simulation of TOF-PET scanners.

The GATE MC simulation tool has been used to model the University of Pennsylvania prototype $LaBr_3$ TOF-PET scanner [111]. The scanner consists of 648 crystals ($4 \times 4 \times 30\,mm^3$) in a 90-cm diameter ring. Sixty rings cover a 25-cm axial length for a maximum axial acceptance angle of $\pm 15°$. The slice thickness is 4.3 mm. Based on measurements on the scanner, the energy resolution was set to 6%. The spatial resolution is determined by the crystal size (4×4 mm) but also by the fact that photopeak detection is often the result of multiple Compton interactions in $LaBr_3$. The singles timing resolution was simulated at 150, 300, and 600 ps; this corresponds to coincidence timing resolutions of 212, 424, and 848 ps (or 32, 64, and 128 mm spatial equivalent, respectively). The TOF information decreases the impact of axial rebinning and lessens the needs for fine transverse angular sampling. When TOF information is ignored during reconstruction (single slice rebinning), the axial resolution worsens dramatically with radial distance due to axial rebinning errors. When TOF information is used during rebinning, the axial resolution is fairly constant across the transverse FOV. In addition, the rebinning error is reduced with improved timing resolution. For the simulated system the axial resolution is about 10–11 mm for 848 ps timing resolution, 6–7.5 mm for 424 ps, and 5–5.5 mm for 212 ps. 3D listmode MLEM reconstruction results in an axial resolution of about 4 mm, independent of timing resolution. Degraded tangential resolution is anticipated when the data are mashed into a limited number of transverse angles. With 848 ps timing resolution, even using 36 angles results in only a slight loss of tangential resolution. With 424 ps timing resolution, as few as 18 angles are needed, and with 212 ps resolution, 9 angles leads to only a small (<1 mm) loss of tangential resolution.

The aforementioned simulations were used to develop a more efficient way of reconstructing TOF-PET data without introduction of resolution loss [112]. The MC method made it possible to investigate the relationship between timing resolution and angular sampling requirements in terms of rebinning and mashing without having to build the systems and their electronics for performing real measurements.

8.2. Case study II: Assessment of PVE correction

In SPECT scans of the dopaminergic system, measurements of striatal uptake are useful for diagnosis and patient follow-up. Most radiopharmaceuticals appropriate for studying the presynaptic transporter binding or the postsynaptic dopamine D2 receptor status are labeled with ^{123}I. Common indications are early diagnosis of Parkinson's disease and evaluation of Parkinsonian syndromes. A more recent application is the differential diagnosis between dementia with Lewy bodies (DLB) and Alzheimer's disease (AD) in demented patients. DLB is the second most common type of degenerative dementia in the elderly. As mentioned in Section 4.2, MC simulations of patient data are extremely useful for assessing the accuracy with which the radiopharmaceutical uptake can be quantified, because they allow to determine absolute biases. In the study taken here as an example [113], 23 patients diagnosed with DLB (10 patients) or AD (13 patients) were simulated. For these patients, processing of MC-simulated data showed that without PVE correction, the binding potential (BP), which characterizes the radiopharmaceutical uptake in the striata (hence the density of dopaminergic transporters), was underestimated by 72.4% on average in the caudate nuclei and by 57.3% on average in the putamen. Similar to what was observed on acquired patient data, PVE correction greatly affected BP values. When combined with scatter and attenuation corrections, it yielded an averaged overestimation of BP estimates of only 13.4% in the caudate nuclei and of 16.1% in the putamen. The mean absolute difference between true BP and measured BP was no more than 0.9 in the caudate nuclei and 0.6 in the putamen.

Accurate quantitative assessment of BP would be of great interest for detection of pre-symptomatic cases of neurodegenerative diseases, both for a better understanding of the degenerative process and for evaluation of neuroprotective treatments. However, pre-symptomatic patients are hard to recruit so using MC simulations is an appealing solution to get the data that would be acquired in such patients, and compare the performance of different processing protocols in detecting subtle changes in striatal uptake. In [113], SPECT scans were simulated with striatal uptake values mimicking pre-symptomatic cases of DLB. For this specifically (difficult) patient population, it was shown that DLB could be differentiated from AD only when PVE correction was combined with attenuation and scatter corrections (Figure 9).

In summary, MC simulations allowed these authors to conclude that only a complete processing scheme including scatter, attenuation, and PVE corrections could accurately estimate BP from SPECT scans of the dopaminergic system, and it became clear that the most important correction for accurate estimates of BP was PVE correction. MC simulations allowed the authors to test their quantification methods while being sure that no segmentation or registration errors occurred. They could demonstrate that PVE correction improved the accuracy of classification of patients when pre-symptomatic DLB cases were considered. These results suggest that the accuracy with which preclinical cases are detected may be increased when applying a complete reconstruction scheme, including scatter, attenuation, and most important, PVE corrections.

8.3. Case study III: MC-based reconstruction

The concept of using MC simulations to estimate the 3D system involved in SPECT tomographic reconstruction was proposed early [114–116] but could not be applied in fully 3D at that time, due to impractical storage and computation time.

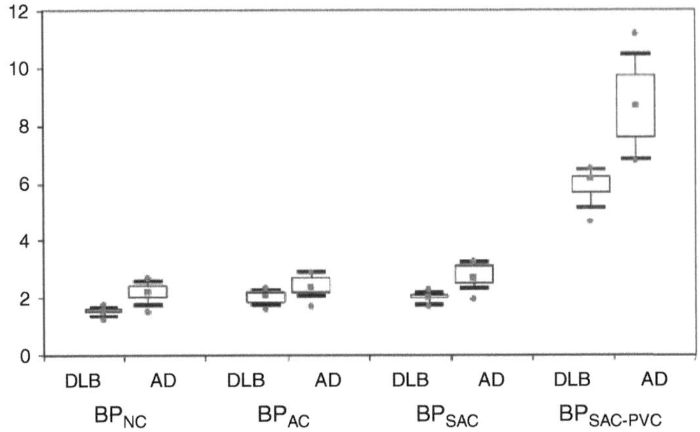

Figure 9 Distribution of BP averaged over the two putamen for simulated patients assigned as pre-symptomatic DLB or AD. Box and whisker plots without any correction (BP_{NC}), with attenuation correction (BP_{AC}), with attenuation and scatter corrections (BP_{SAC}), and with all corrections ($BP_{SAC-PVC}$) are shown.

Renewed efforts have been performed recently to investigate the feasibility of simulating and storing the 3D system matrix (see [48] for a complete review), and then invert it for SPECT reconstruction. Among these, Beekman and colleagues have developed a computationally efficient approach, involving the use of a dual matrix block-iterative expectation maximization algorithm, which avoids the need for massive matrix storage and which is about two orders of magnitude faster than MLEM reconstruction [117].

The dual matrix reconstruction uses a different transition matrix for forward projection (calculating the projection data given the 3D isotope distribution) and back projection (calculating the 3D isotope distribution given the projections). The forward projection step models attenuation, detector blurring, and scatter, whereas the back-projection models only attenuation and detector blurring, which significantly reduces calculation times. For forward projection, scatter is modeled by using a fast MC simulator, based on Convolution-based Forced Detection (CFD) [99], which combines stochastic photon transport calculation in the patient with an analytic (instead of a stochastic) detector modeling.

Using such a fast MC simulator for the projection step makes it possible to perform a MC simulation at each iteration of the reconstruction algorithm, thus avoiding the need for massive system matrix storage. The hybrid simulations (i.e. mixing MC and analytical modeling) involved in the dual matrix-CFD reconstruction approach converge several orders of magnitude faster to low noise projections than those carried out with brute force MC simulators, making the reconstruction computationally practical. Typical reconstruction times of less than 5 min have been reported for a cardiac perfusion SPECT study [118].

Xiao et al. used this framework for ^{99m}Tc cardiac perfusion reconstruction and found that the MC-based reconstruction yields a better contrast-to-noise ratio

(CNR) than if traditional window-based scatter correction is applied [119]. The MC-based reconstruction framework also produces images with larger quantitative differences between defect and non-defect cases and shows less sensitivity to anatomical variations than traditional methods. This is illustrated by Figure 10 where CNR curves for different reconstruction methods are compared. For all the phantom configurations, the images reconstructed by the traditional algorithms had lower overall contrast when compared at equal noise and have higher noise when compared at equal contrast then the images reconstructed by the MC-based algorithm. The contrast achieved with the MC-based approach was approximately 10–20% higher when measured at equal noise level than the contrast obtained with the other two methods while the noise was approximately 14% lower when compared at equal contrast.

This dual matrix-CFD reconstruction framework has been generalized for medium- and high-energy isotopes. Indeed, for such isotopes, the Gaussian modeling of the collimator response function involved in conventional CFD is not accurate enough given that a significant number of photons penetrate the collimator septa, inducing a non-Gaussian collimator response. An efficient simulation method for calculating the effects of septal penetration in SPECT projections using only one Septal Penetration Point Spread Function (SP-PSF) was developed. The latter method has been incorporated in the aforementioned dual matrix CFD framework and accordingly results in a hybrid framework that is now generally applicable for a wide range of isotopes and collimators [120].

Similar work about calculating the matrix system for tomographic reconstruction using an MC approach has also been recently performed in the context of PET (see for instance [121,122]).

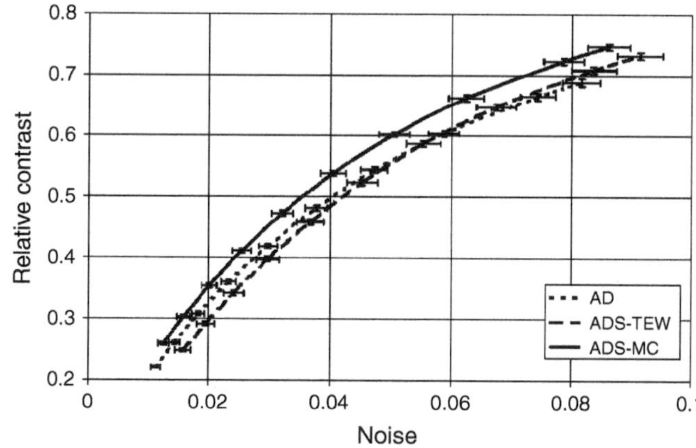

Figure 10 Average defect contrast as a function of average noise in the myocardium for four different phantom configurations for different reconstruction algorithms. AD models attenuation and detector blurring while ADS also corrects for scatter by a triple energy window correction (TEW) or by simulations (MC).

9. Future Prospects

Future developments in MC simulations in the context of Nuclear Medicine will probably include further integration of current MC tools to provide more accurate and powerful diagnosis and treatment systems [123].

A first integration level would consist in bridging the gap between MC calculations used for SPECT and PET modeling and for dose calculation. Nowadays, MC simulations are becoming part of the imaging process itself, when being used for calculating the system matrix involved in tomographic reconstruction, as previously discussed in the third case study. A further extension will be to also calculate the patient-specific dose distributions using this MC approach, which can give accurate dose deposit estimates following, e.g., internal radiation [124]. Currently, when performed with the MC approach, dose calculations usually rely on different MC codes than those used for modeling SPECT or PET scans. Using a single simulation framework for both tomographic reconstruction and dose calculation will facilitate the feasibility (e.g., single coordinate system, common object description, consistent sampling) and the assessment of patient-specific dosimetry.

A second integration level concerns the MC modeling of dual modality imaging system. In the past years, the Nuclear Medicine instrumentation market has shifted toward dual modality imagers: SPECT-CT and PET-CT. SPECT and PET MC modeling are currently very well mastered. MC modeling of CT devices has also been described (see e.g. [125]). Integrating a MC model for the CT part of these hybrid imagers in the same MC simulation framework as that used for SPECT and PET modeling would enable a more detailed investigation of dual modality techniques, such as CT-based SPECT or PET attenuation correction, or CT-based SPECT and PET partial volume correction.

Finally, MC simulations remain computationally demanding methods, while large database of simulated data are ideally needed for evaluation purposes. Therefore, further progress is anticipated in implementing variance reduction techniques. Also, some laboratories endow significant funds in large computer clusters in order to perform detection tasks based on simulated datasets. Forthcoming work will be dedicated to developing distribution software for datagrid deployments or to parallellization solutions. It can also be expected that more efforts will go to the collection of simulated data to create online repositories, that would enable subscribers to test their image correction and processing techniques without the need to extensively simulate or acquire the appropriate datasets.

All these expected developments should make MC simulations even more present in the realm of emission tomography in the future.

10. Conclusion

In the end of 1990s, MC simulations in emission tomography were mostly used by research labs to assess reconstruction and correction techniques, and to study specific aspects of the imaging system response. It is expected that MC simulations will eventually find their way to the clinics, as contributing to the SPECT or PET imaging process, and as provider of extremely realistic and still perfectly controlled datasets for evaluation purpose. The role of MC simulations for optimizing detector design will also increase, as simulations are becoming more and more flexible and accurate, allowing for detailed investigation of a broad range of detector configurations. Overall, MC simulations will become an indispensable tool for all involved in emission tomography.

ACKNOWLEDGMENTS

This work of S.S was supported by the Fund for Scientific Research Flanders, the work of I.B was supported by INSERM.

REFERENCES

[1] J. Sorenson, M. Phelps, Physics in Nuclear Medicine, W.B. Saunders Company, 1987.
[2] Z. Cho, J. Jones, M. Singh, Foundations of Medical Imaging, John Wiley & sons Inc., 1993.
[3] F. Beekman, F. van der Have, B. Vastenhouw, A. van der Linden, P. van Rijk, J. Burbach, M. Smidt, U-SPECT-I: a novel system for submillimeter-resolution tomography with radiolabelled molecules in mice, Journal of Nuclear Medicine, 46 (2005) 1194–1200.
[4] L. Furenlid, D. Wilson, C. Yi-Chun, K. Hyunki, P. Pietraski, M. Crawford, H. Barrett, FastSPECT II: a second-generation high-resolution dynamic SPECT imager, IEEE Transactions on Nuclear Science, 49 (2004) 116–123.
[5] A. Chatziioannou, Molecular imaging of small animals with dedicated PET tomographs, European Journal of Nuclear Medicine, 29 (2002) 98–114.
[6] F. Natterer, F. Wubbeling, Mathematical Methods in Image Reconstruction, Cambridge University Press, 2001.
[7] R. Gordon, A tutorial on ART, IEEE Transactions on Nuclear Science, 21 (1974) 78–93.
[8] L. A. Shepp, Y. Vardi, Maximum likelihood reconstruction for emission tomography, IEEE Transactions on Medical Imaging, 1 (1982) 113–122.
[9] B. Tsui, E. Frey, X. Zhao, D. Lalush, R. Johnston, W. McCartney, The importance and implementation of accurate 3D compensation methods for quantitative SPECT, Physics in Medicine and Biology, 39 (1994) 509–530.
[10] H. Zaidi, Relevance of accurate Monte Carlo modeling in nuclear medical imaging, Medical Physics, 26 (1999) 574–608.
[11] M. Ljungberg, S. Strand, M. King, Monte Carlo Simulations in Nuclear Medicine: Applications in Diagnostic Imaging, Institute of Physics, 1998.
[12] D. Haynor, R. Harrison, T. Lewellen, The use of importance sampling techniques to improve the efficiency of photon tracking in emission tomography simulations, Medical Physics, 18 (1991) 990–1001.
[13] S. Agostinelli, J. Allison, K. Amako, J. Apostolakis, H. Araujo, P. Arce, M. Asai, D. Axen, S. Banerjee, G. Barrand, F. Behner, L. Bellagamba, J. Boudreau, L. Broglia, A. Brunengo, H. Burkhardt, S. Chauvie, J. Chuma, R. Chytracek, G. Cooperman, G. Cosmo, P. Degtyarenko, A. Dell'AcquaG. Depaola, D. Dietrich, R. Enami, A. Feliciello, C. Ferguson, H. Fesefeldt, G. Folger, F. Foppiano, A. Forti, S. Garelli, S. Giani, R. Giannitrapani, D. Gibin, J. J. G. Cadenas, I. Gonzalez, G. G. Abril, G. Greeniaus, W. Greiner, V. Grichine, A. Grossheim, S. Guatelli, P. Gumplinger, R. Hamatsu, K. Hashimoto, H. Hasui, A. Heikkinen, A. Howard,

V. Ivanchenko, A. Johnson, F. W. Jones, J. Kallenbach, N. Kanaya, M. Kawabata, Y. Kawabata, M. Kawaguti, S. Kelner, P. Kent, A. Kimura, T. Kodama, R. Kokoulin, M. Kossov, H. Kurashige, E. Lamanna, T. Lampen, V. Lara, V. Lefebure, F. Lei, M. Liendl, W. Lockman, F. Longo, S. Magni, M. Maire, E. Medernach, K. Minamimoto, P. M. de Freitas, Y. Morita, K. Murakami, M. Nagamatu, R. Nartallo, P. Nieminen, T. Nishimura, K. Ohtsubo, M. Okamura, S. O'Neale, Y. Oohata, K. Paech, J. Perl, A. Pfeiffer, M. G. Pia, F. Ranjard, A. Rybin, S. Sadilov, E. Di Salvo, G. Santin, T. Sasaki, N. Savvas, Y. Sawada, S. Scherer, S. Seil, V. Sirotenko, D. Smith, N. Starkov, H. Stoecker, J. Sulkimo, M. Takahata, S. Tanaka, E. Tcherniaev, E. S. Tehrani, M. Tropeano, P. Truscott, H. Uno, L. Urban, P. Urban, M. Verderi, A. Walkden, W. Wander, H. Weber, J. P. Wellisch, T. Wenaus, D. C. Williams, D. Wright, T. Yamada, H. Yoshida, D. Zschiesche, Geant4: a simulation toolkit, Nuclear Methods and Instruments in Physics Research Section A, 506 (2003) 250–303.

[14] Geant4, http://geant4.web.cern.ch/geant4/UserDocumentation/.

[15] P. Andreo, Monte Carlo techniques in medical radiation physics, Physics in Medicine and Biology, 36 (1991) 861–920.

[16] I. Buvat, I. Castiglioni, Monte Carlo simulations in SPET and PET, Quarterly Journal of Nuclear Medicine, 46 (2002) 48–61.

[17] S. Webb, D. Binnie, M. Flower, R. Ott, Monte Carlo modeling of the performace of a rotating slit collimator for improved planar gamma camera imaging, Physics in Medicine and Biology, 37 (1992) 1095–1108.

[18] A. Cot, J. Sempau, D. Pareto, S. Bullich, J. Pavia, F. Calvino, D. Ros, Evaluation of the geometric, scatter, and septal penetration components in fan-beam collimators using Monte Carlo simulation, IEEE Transactions on Nuclear Science, 49 (2002) 12–16.

[19] T. Song, Y. Choi, Y. Chung, J. Jung, Y. Choe, K. Lee, S. Kim, B. Kim, Optimization of pinhole collimator for small animal SPECT using Monte Carlo simulation, IEEE Transactions on Nuclear Science, 50 (2003) 327–332.

[20] F. van der Have, F. Beekman, Photon penetration and scatter in micro-pinhole imaging: a Monte Carlo investigation, Physics in Medicine and Biology, 49 (2004) 1369–1386.

[21] D. Gagnon, G. Zeng, M. Links, J. Griesmer, C. Valentino, Design considerations for a new solid state gammacamera: SOLSTICE, in: Proceedings of the IEEE Medical Imaging Conference Norfolk, USA, 2001, pp. 1156–1160.

[22] J. Griesmer, B. Kline, J. Grosholz, K. Parnham, D. Gagnon, Performance evaluation of a new CZT detector for nuclear medicine: SOLSTICE, in: Proceedings of the IEEE Medical Imaging Conference Norfolk, USA, 2001, pp. 1050–1054.

[23] S. Staelens, M. Koole, S. Vandenberghe, Y. D'Asseler, I. Lemahieu, R. Van de Walle, The geometric transfer function for a slat collimator mounted on a strip detector, IEEE Transactions on Nuclear Science, 52 (2005) 708–714.

[24] C. Groiselle, Y. D'Asseler, J. Kolthammer, C. Matthews, S. Glick, A Monte Carlo simulation study to evaluate septal spacing using triple-head hybrid PET imaging, IEEE Transactions on Nuclear Science, 50 (2003) 1339–1346.

[25] C. Thompson, The effect of collimation on singles rates in multi-slice PET, IEEE Transactions on Nuclear Science, 35 (1988) 598–602.

[26] S. Surti, J. Karp, G. Muehllehner, Image quality assessment of LaBr3-based whole-body 3D PET scanners: a Monte Carlo evaluation, Physics in Medicine and Biology, 49 (2004) 4593–4610.

[27] S. Surti, J. Karp, A count-rate model for PET scanners using pixelated Angerlogic detectors with different scintillators, Physics in Medicine and Biology, 50 (2005) 5697–5715.

[28] A. Bice, T. Lewellen, R. Miyaoka, R. Harrison, D. Haynor, K. Pollard, C. Hanson, S. Gillispie, Monte Carlo simulations of BaF_2 detectors used in time of flight positron emission tomography, IEEE Transactions on Nuclear Science, 37 (1990) 696–701.

[29] S. Surti, J. Karp, G. Muehllehner, P. Raby, Investigation of lanthanum scintillators for 3-D PET, IEEE Transactions on Nuclear Science, 50 (2003) 348–354.

[30] W. Moses, Time of flight in PET revisited, IEEE Transactions on Nuclear Science, 50 (2003) 1325–1330.

[31] W. Moses, S. Derenzo, Design studies for a PET detector module using a PIN photodiode to measure depth of interaction, IEEE Transactions on Nuclear Science, 41 (1994) 1441–1445.

[32] C. Lartizien, A. Reilhac, N. Costes, M. Janier, D. Sappey-Marinier, Monte Carlo simulation-based design study of a LSO-LuAP small animal PET system, IEEE Transactions on Nuclear Science, 50 (2003) 1433–1438.
[33] G. El Fakhri, I. Buvat, M. Pelegrini, H. Benali, P. Almeida, B. Bendriem, A. Todd-Pokropek, R. Di Paolo, Respective roles of scatter, attenuation, depth-dependent collimator response and finite spatial resolution in cardiac SPECT quantitation: a Monte Carlo study, European Journal of Nuclear Medicine, 26 (1999) 437–446.
[34] G. El Fakhri, I. Buvat, H. Benali, A. Todd-Pokropek, R. Di Paolo, Relative impact of scatter, collimator response, attenuation, and finite spatial resolution corrections in cardiac SPECT, Journal of Nuclear Medicine, 41 (2000) 1400–1408.
[35] M. Soret, P. Koulibaly, J. Darcourt, S. Hapdey, I. Buvat, Quantitative accuracy of dopaminergic neurotransmission imaging using 123I SPECT, Journal of Nuclear Medicine, 44 (2003) 1184–1193.
[36] H. Zaidi, K. Koral, Scatter modeling and compensation in emission tomography, European Journal of Nuclear Medicine, 31 (2004) 761–782.
[37] S. Staelens, D. Strul, S. Vandenberghe, G. Santin, M. Koole, Y. D'Asseler, I. Lemahieu, R. Van deWalle, Monte Carlo simulations of a scintillation camera using GATE: validation and application modeling, Physics in Medicine and Biology, 48 (2003) 3021–3042.
[38] A. Cot, J. Sempau, D. Pareto, S. Bullich, J. Pavia, F. Calvino, D. Ros, Study of the point spread function (PSF) for 123I SPECT imaging using Monte Carlo simulation, Physics in Medicine and Biology, 49 (2004) 3125–3136.
[39] H. Huesman, E. Salmeron, J. Baker, Compensation for crystal penetration in high resolution positron tomography, IEEE Transactions on Nuclear Science, 36 (1989) 1100–1107.
[40] S. Staelens, Y. D'Asseler, S. Vandenberghe, M. Koole, I. Lemahieu, R. Van de Walle, A three-dimensional theoretical model incorporating spatial detection uncertainty in continuous detector PET, Physics in Medicine and Biology, 49 (2004) 2337–2351.
[41] F. Lamare, M. Ledesma Carbayo, G. Kontaxakis, A. Santos, A. Turzo, Y. Bizais, C. Cheze-Le Rest, D. Visvikis, Incorporation of elastic transformations in list-mode based reconstruction for respiratory motion correction in PET, in: Proceedings of the IEEE Medical Imaging Conference, Rome, Italy, 2004, pp. M03–247.
[42] F. J. Beekman, M. A. Viergever, Fast SPECT simulation including object shape dependent scatter, IEEE Transactions on Medical Imaging, 14 (1995) 271–282.
[43] F. J. Beekman, J. M. den Harder, M. A. Viergever, P. P. van Rijk, SPECT scatter modeling in non-uniform attenuating objects, Physics in Medicine and Biology, 42 (1997) 1133–1142.
[44] E. C. Frey, B. M. W. Tsui, A new method for modelling the spatially variant object dependent scatter response function in SPECT, in: Proceedings of the IEEE Medical Imaging Conference, Anaheim, USA, 1996, pp. 1082–1086.
[45] S. R. Meikle, B. F. Hutton, D. L. Bailey, A transmission dependent method for scatter correction in SPECT, Journal of Nuclear Medicine, 23 (1994) 360–367.
[46] S. H. M. Walrand, L. R. van Elmbt, S. Pauwels, Quantification in SPECT using an effective model of the scattering, Physics in Medicine and Biology, 39 (1994) 719–734.
[47] C. E. Floyd, R. J. Jaszczak, K. L. Greer, R. E. Coleman, Inverse Monte Carlo as a unified reconstruction algorithm for ECT, Journal of Nuclear Medicine, 27 (1986) 1577–1585.
[48] D. Lazaro, Z. El Bitar, V. Breton, D. Hill, I. Buvat, Fully 3D Monte Carlo reconstruction in SPECT: a feasibility study, Physics in Medicine and Biology, 50 (2005) 3739–3754.
[49] F. J. Beekman, C. Kamphuis, M. A. Viergever, Improved SPECT quantitation using fully three-dimensional iterative spatially variant scatter response compensation, IEEE Transactions on Medical Imaging, 15 (1996) 491–499.
[50] F. J. Beekman, C. Kamphuis, E. C. Frey, Scatter compensation methods in 3D iterative SPECT reconstruction: a simulation study, Physics in Medicine and Biology, 42 (1997) 1619–1632.
[51] D. Kadrmas, E. Frey, B. Tsui, Application of reconstruction-based scatter compensation to Thallium-201 SPECT: implementations for reduced reconstructed noise, IEEE Transactions on Medical Imaging, 17 (1998) 325–333.
[52] T. Farncombe, H. Gifford, M. Narayanan, H. Pretorius, E. Frey, M. King, Assessment of scatter compensation strategies for ^{67}Ga SPECT using numerical observers and human LROC studies, Journal of Nuclear Medicine, 45 (2004) 802–812.

[53] E. C. Frey, K. Gilland, B. M. W. Tsui, Application of task-based measures of image quality to optimization and evaluation of three-dimensional reconstruction-based scatter compensation in myocardial perfusion SPECT, IEEE Transactions on Medical Imaging, 21 (2002) 1040–1050.
[54] M. Narayanan, H. Pretorius, S. Dahlberg, J. Leppo, N. Botkin, J. Krasnow, W. Berndt, E. Frey, M. King, Evaluation of scatter compensation strategies and their impact on human detection performance Tc-99m myocardial perfusion imaging, IEEE Transactions on Nuclear Science, 50 (2003) 1522–1527.
[55] G. Tourassi, E. Carey, M. Munley, Improved lesion detection in SPECT using MLEM reconstruction, IEEE Transactions on Nuclear Science, 38 (1991) 780–783.
[56] P. Bruyant, H. Gifford, G. Gindi, M. King, Numerical observer study of MAPOSEM regularization methods with anatomical priors for lesion detection in Ga-67 images, IEEE Transactions on Nuclear Science, 51 (2004) 193–197.
[57] I. Castiglioni, I. Buvat, G. Rizzo, M. Gilardi, J. Feuardent, F. Fazio, A publicly accessible Monte Carlo database for validation purposes in emission tomography, European Journal of Nuclear Medicine, 32 (2005) 1234–1239.
[58] A. Reilhac, G. Batan, C. Michel, C. Grova, J. Tohka, D. Collins, N. Costes, A. Evans, PET-SORTEO: validation and development of database of simulated PET volumes, IEEE Transactions on Nuclear Science, 52 (2005) 1321–1328.
[59] A. Syme, C. Kirkby, T. Riauka, B. Fallone, S. McQuarrie, Monte Carlo investigation of single cell beta dosimetry for intraperitoneal radionuclide therapy, Physics in Medicine and Biology, 49 (2004) 1959–1972.
[60] Y. Dewaraja, S. Wilderman, M. Ljungberg, K. Koral, K. Zasadny, M. Kaminiski, Accurate dosimetry in 131I radionuclide therapy using patient-specific, 3-dimensional methods for SPECT reconstruction and absorbed dose calculation, Journal of Nuclear Medicine, 46 (2005) 840–849.
[61] B. Fraass, J. Smathers, J. Deye, Summary and recommendations of a national cancer institute workshop on issues limiting the clinical use of Monte Carlo dose calculation algorithms for megavoltage external beam radiation therapy, Medical Physics, 30 (2003) 3206–3216.
[62] O. Chibani, X. Li, Monte Carlo dose calculations in homogeneous media and at interfaces: a comparison between GEPTS, EGSnrc, MCNP, and measurements, Medical Physics, 29 (2003) 835–847.
[63] B. Reniers, F. Verhaegen, S. Vynckier, The radial dose function of low-energy brachytherapy seeds in different solid phantoms: comparison between calculations with the EGSnrc and MCNP4C Monte Carlo codes and measurements, Physics in Medicine and Biology, 49 (2004) 1569–1582.
[64] S. Garelli, S. Giordano, G. Piemontese, S. Squaricia, Interactive FLUKA: a World Wide Web version for a simulation code in protontherapy, Journal de Chimie Physique et de Physico-Chimie Biologique, 95 (1998) 767–771.
[65] R. Harrison, S. Vannoy, D. Haynor, S. Gillipsie, M. Kaplan, T. Lewellen, Preliminary experience with the photon generator module of a public-domain simulation system for emission tomography, in: Proceedings of the IEEE Medical Imaging Conference, San Fransico, USA, 1993, pp. 1154–1158.
[66] M. Ljungberg, S. Strand, A Monte Carlo program for the simulation of scintillation camera characteristics, Computer Methods and Programs in Biomedicine, 29 (1989) 257–272.
[67] J. Yanch, A. Dobrzeniecki, C. Ramanathan, R. Behrman, Physically realistic Monte Carlo simulation of source, collimator and tomographic data acquisition for emission computed tomography, Physics in Medicine and Biology, 37 (1992) 853–870.
[68] C. Thomson, J. Cantu, Y. Picard, PETSIM: Monte Carlo program simulation of all sensitivity and resolution parameters of cylindrical positron imaging systems, Physics in Medicine and Biology, 37 (1992) 731–749.
[69] H. Zaidi, A. Hermann Scheurer, C. Morel, An object-oriented Monte-Carlo simulator for 3D cylindrical positron tomographs, Computer Methods and Programs in Biomedicine, 58 (1999) 133–145.
[70] A. Reilhac, C. Lartizien, N. Costes, S. Sans, C. Comtat, R. Gunn, A. Evans, PET-SORTEO: a Monte Carlo-based simulator with high count rate capabilities, IEEE Transactions on Nuclear Science, 51 (2004) 46–52.

[71] A. Bielajew, H. Hirayama, W. Nelson, D. Rogers, History, overview and recent improvements of EGS4, Tech. rep., National Research Council, Ottawa, Canada, (1994).
[72] Geant4, http://wwwasd.web.cern.ch/wwwasd/geant4/geant4.html.
[73] J. Briesmeister, MCNP – a general Monte Carlo N-particle transport code, Tech. rep., Los Alamos National Laboratory, Los Alamos, NM, USA, (1993).
[74] I. Buvat, D. Lazaro, Monte Carlo simulations in emission tomography and GATE: an overview. Nuclear Methods and Instruments in Physics Research 569 (2006) 323–329.
[75] D. Strul, G. Santin, D. Lazaro, V. Breton, C. Morel, GATE (Geant4 Application for Tomographic Emission): a PET/SPECT general-purpose simulation platform, Nuclear Physics B, 125C (2003) 75–79.
[76] G. Santin, D. Strul, D. Lazaro, L. Simon, M. Krieguer, M. Vieira Martins, V. Breton, C. Morel, GATE, a Geant4-based simulation platform for PET and SPECT integrating movement and time management, IEEE Transactions on Nuclear Science, 50 (2003) 1516–1521.
[77] S. Jan, G. Santin, D. Strul, S. Staelens, K. Assié, D. Autret, D. Avner, R. Barbier, M. Bardiès, P. Bloomfield, D. Brasse, V. Breton, P. Bruyndonckx, I. Buvat, A. Chatziioannou, Y. Choi, Y. Chung, C. Comtat, D. Donnarieix, L. Ferrer, S. Glick, C. Groiselle, S. Kerhoas-Cavata, A. Kirov, V. Kohli, M. Koole, M. Krieguer, J. van der Laan, F. Lamare, G. Largeron, C. Lartizien, D. Lazaro, M. Maas, L. Maigne, F. Mayet, F. Melot, S. Nehmeh, E. Pennacchio, J. Perez, U. Pietrzyk, F. Rannou, M. Rey, D. Schaart, R. Schmidtlein, L. Simon, T. Song, J. Vieira, D. Visvikis, R. Van de Walle, E. Wieers, C. Morel, GATE: a simulation toolkit for PET and SPECT, Physics in Medicine and Biology, 49 (2004) 4543–4561.
[78] R. Brun, F. Rademakers, ROOT, an object oriented data analysis framework, Nuclear Methods and Instruments in Physics Research Section A, 389 (1997) 81–86.
[79] S. Staelens, G. Santin, D. Strul, M. Koole, S. Vandenberghe, Y. D'Asseler, V. Breton, C. Morel, I. Lemahieu, R. Van de Walle, Monte Carlo simulations of Interfile based emission and attenuation maps for clinical applications, Journal of Nuclear Medicine, 44 (2003) 1048.
[80] E. Hoffman, P. Cutler, W. Digby, J. Mazziota, Three dimensional phantom to simulate cerebral blood flow and metabolic images for PET, IEEE Transactions on Nuclear Science, 37 (1990) 616–620.
[81] R. Taschereau, P. Chow, A. Chatziioannou, Monte Carlo simulations of dose from microCT imaging procedures in a realistic mouse phantom, Medical Physics, 33 (2006) 216–224.
[82] S. Jan, Simulateur Monte Carlo et caméra à xénon liquide pour la Tomographie à Emission de Positons, Ph.D. thesis, Université Joseph Fourier, Grenoble, France (2002).
[83] R. Harrison, S. Gillispie, A. Alessio, P. Kinahan, T. Lewellen, The effects of object size, attenuation, scatter, and random coincidences on signal to noise ratio in simulations of time-of-flight positron emission tomography, in: Proceedings of the IEEE Medical Imaging Conference Puerto Rico, 2005, pp. 1900–1904.
[84] C. Michel, L. Eriksson, H. Rothfuss, B. Bendriem, D. I. Lazaro, Buvat, Influence of crystal material on the performance of the HiRez 3D PET scanner: a Monte-Carlo study, in: IEEE Medical Imaging Symposium Conference Record, 4, (2006), p. 2007.
[85] S. Staelens, G. Santin, S. Vandenberghe, D. Strul, M. Koole, Y. D'Asseler, I. Lemahieu, R. Van de Walle, Transmission imaging with a moving point source: influence of crystal thickness and collimator type, IEEE Transactions on Nuclear Science, 52 (2005) 166–173.
[86] K. Assie, I. Gardin, P. Vera, I. Buvat, Validation of the Monte Carlo simulator GATE for Indium111 imaging, Physics in Medicine and Biology, 50 (2005) 3113–3125.
[87] D. Autret, A. Bitar, L. Ferrer, A. Lisbona, M. Bardies, Monte Carlo modeling of gamma cameras for I-131 imaging in targeted radiotherapy, Cancer Biotherapy and Radio Pharmaceuticals, 20 (2005) 77–84.
[88] S. Jan, C. Comtat, D. Strul, G. Santin, R. Trebossen, Monte Carlo simulation for the ECAT EXACT HR+ system using GATE, IEEE Transactions on Nuclear Science, 52 (2005) 627–633.
[89] C. Schmidtlein, A. Kirov, S. Nehmeh, Y. Erdi, J. Humm, H. Amols, L. Bidaut, A. Ganin, C. Stearns, D. McDaniel, K. Hamacher, Validation of GATE Monte Carlo simulations of the GE Advance/Discovery LS PET scanners, Medical Physics, 33 (2006) 198–208.

[90] F. Lamare, A. Turzo, Y. Bizais, C. Le Rest, D. Visvikis, Validation of a Monte Carlo simulation of the Philips Allegro/GEMINI PET systems using GATE, Physics in Medicine and Biology, 51 (2006) 943–962.
[91] F. Bataille, C. Comtat, S. Jan, R. Trebossen, Monte Carlo simulation for the ECAT HRRT using GATE, in: Proceedings of the IEEE Medical Imaging Conference Rome, Italy, 2004, pp. 2570–2574.
[92] S. Jan, A. Chatziioannou, C. Comtat, D. Strul, G. Santin, R. Trbossen, Monte Carlo simulation for the microPET P4 system using GATE, Molecular Imaging and Biology, 5 (2003) , 138.
[93] S. Jan, A. Desbree, F. Pain, D. Guez, C. Comtat, H. Gurden, S. Kerhoas, P. Laniece, F. Lefebvre, R. Mastrippolito, R. Trebossen, Monte Carlo simulation of the microPET FOCUS system for small rodents imaging applications, in: Proceedings of the IEEE Medical Imaging Conference Puerto Rico, 2005, pp. 1653–1657.
[94] M. Rey, J. Vieira, J. Mosset, B. Moulin Sallanon, P. Millet, J. Loude, C. Morel, Measured and simulated specifications of the Lausanne ClearPET scanner demonstrator, in: Proceedings of the IEEE Medical Imaging Conference Puerto Rico, 2005, pp. 2070–2073.
[95] F. Rannou, V. Kohli, D. Prout, A. Chatziioannou, Investigation of OPET peformance using GATE, a Geant4-based simulation software, IEEE Transactions on Nuclear Science, 51 (2004) 2713–2717.
[96] Y. Chung, Y. Choi, G. Cho, Y. Choe, K. Lee, B. Kim, Optimization of dual layer phoswich detector consisting of LSO and LuYAP for small animal PET, IEEE Transactions on Nuclear Science, 52 (2005) 217–221.
[97] I. Buvat, I. Castiglioni, J. Feuardent, M. Gilardi, Unified description and validation of Monte Carlo simulators in PET, Physics in Medicine and Biology, 50 (2005) 329–346.
[98] National electric manufacturers association, Rosslyn, VA, NEMA standards publication NU 2-2001: Performance measurements of positron emission tomographs (2001).
[99] H. W. de Jong, E. T. P. Slijpen, F. J. Beekman, Acceleration of Monte Carlo SPECT simulation using convolution-based forced detection, IEEE Transactions on Nuclear Science, 48 (2001) 58–64.
[100] X. Song, W. Segars, Y. Du, B. Tsui, E. Frey, Fast modeling of the collimator-detector response in Monte Carlo simulation of SPECT imaging using the angular response function, Physics in Medicine and Biology, 50 (2005) 1791–1804.
[101] J. De Beenhouwer, D. Kruecker, S. Staelens, L. Ferrer, A. Chatziioannou, F. Rannou, Distributed computing platform for PET and SPECT simulations with GATE, in: Proceedings of the IEEE Medical Imaging ConferencePuerto Rico, 2005, pp. 2437–2440.
[102] R. Allemand, C. Gresset, J. Vacher, Potential advantages of a cesium fluoride scintillator for time-of-flight positron camera, Journal of Nuclear Medicine, 21 (1980) 153–155.
[103] N. Mullani, D. Ficke, R. Hartz, J. Markham, G. Wong, System design of a fast PET scanner utilizing time-of-flight, IEEE Transactions on Nuclear Science NS-28, 104–107.
[104] M. Ter-Pogossian, D. Ficke, M. Yamamoto, J. Hood, Super PETT I: a positron emission tomograph utilizing photon time-of-flight information, IEEE Transactions on Medical Imaging MI-1 (1982), 179–192.
[105] R. Gariod, R. Allemand, E. Cormoreche, M. Laval, M. Moszynski, The LETI positron tomograph architecture and time of flight improvements, in: Proceedings of IEEE Workshop on Time-of-Flight Emission Tomography Washington University, St. Louis, USA, 1982, pp. 25–29.
[106] N. Mullani, W. Wong, R. Hartz, E. Philippe, K. Yerian, J. Gaeta, K. Gould, Preliminary results obtained with TOFPET-I – a whole-body time of flight positron emission tomograph, Journal of Nuclear Medicine, 24 (1983) 11–12.
[107] T. Lewellen, A. Bice, R. Harrison, M. Pencke, J. Link, Performance measurements of the SP3000/UW time-of-flight positron emission tomograph, IEEE Transactions on Nuclear Science, 35 (1988) 665–669.
[108] B. Mazoyer, R. Trebossen, C. Schoukroun, B. Verrey, A. Syrota, J. Vacher, P. Lemasson, O. Monnet, A. Bouvier, J. L. Lecomte, Physical characteristics of TTV03, a new high spatial resolution time-of-flight positron tomograph, IEEE Transactions on Nuclear Science, 37 (1990) 778–782.

[109] W. Wong, N. Mullani, E. Philippe, R. Hartz, D. Bristow, K. Yerian, J. Gaeta, N. Ketharnavaz, Performance characteristics of the University of Texas TOFPET-I PET camera, Journal of Nuclear Medicine, 25 (1984) 46–47.
[110] Philips, http://www.medical.philips.com/main/products/pet/.
[111] J. Karp, A. Kuhn, A. Perkins, S. Surti, M. Werner, M. E. Daube-Witherspoon, L. Popescu, S. Vandenberghe, G. Muehllehner, Characterization of TOF PET scanner based on lanthanum bromide, in: Proceedings of the IEEE Medical Imaging Conference Puerto Rico, 2005, pp. 5.
[112] S. Vandenberghe, M. Daube-Witherspoon, R. Lewitt, J. Karp, Fast reconstruction of 3d time-of-flight PET data by axial rebinning and transverse mashing, Physics in Medicine and Biology, 51 (2006) 1603–1621.
[113] M. Soret, P. Koulibaly, J. Darcourt, I. Buvat, Partial volume effect correction in SPECT for striatal uptake measurements in patients with neurodegenerative diseases: impact upon patient classification, European Journal of Nuclear Medicine, 33 (2006) 1062–1072.
[114] C. Floyd, R. Jaszczak, K. Greer, R. Coleman, Inverse Monte Carlo: a unified reconstruction algorithm for ECT, Journal of Nuclear Medicine, 27 (1996) 1577–1585.
[115] E. Veklerov, J. Llacer, E. Hoffman, MLE reconstruction of a brain phantom using a Monte Carlo transition matrix and a statistical stopping rule, IEEE Transactions on Nuclear Science, 35 (1988) 603–607.
[116] J. Bowsher, C. Floyd, Treatment of Compton scatter in maximum-likelihood, expectation-maximization reconstruction of SPECT images, Journal of Nuclear Medicine, 32 (1991) 1285–1291.
[117] F. Beekman, H. W. de Jong, S. van Geloven, Efficient fully 3D iterative SPECT reconstruction with Monte Carlo based scatter correction, IEEE Transactions on Medical Imaging, 21 (2002) 867–877.
[118] T. C. de Wit, J. Xiao, F. J. Beekman, Monte-Carlo based statistical SPECT reconstruction: influence of number of photon tracks, IEEE Transactions on Nuclear Science, 52 (2005) 1365–1369.
[119] J. Xiao, T. De Wit, S. Staelens, F. Beekman, Quantitative evaluation of 3D Monte-Carlo based scatter correction for Tc-99m cardiac perfusion SPECT: a phantom study, Journal of Nuclear Medicine, 47 (10) (2006) 1662–1669.
[120] S. Staelens, T. de Wit, F. Beekman, Ultra-fast SPECT simulation with combined convolution based forced detection and efficient septal penetration modeling, Journal of Nuclear Medicine, 47 (2006) 198.
[121] M. Rafecas, G. Bning, B. Pichler, E. Lorenz, M. Schwaiger, S. Ziegler, Effect of noise in the probability matrix used for statistical reconstruction of PET, IEEE Transactions on Nuclear Science, 51 (2004) 149–156.
[122] S. Vandenberghe, S. Staelens, C. Byrne, E. Soares, I. Lemahieu, S. Glick, Reconstruction of 2D PET data with Monte Carlo generated natural pixels, Physics in Medicine and Biology, 51 (2006) 3105–3125.
[123] J. Lehmann, C. Hartmann Siantar, D. Wessol, C. Wemple, D. Nigg, J. Cogliati, T. Daly, M. Descalle, T. Flickinger, D. Pletcher, G. DeNardo, Monte Carlo treatment planning for molecular targeted radiotherapy within the MINERVA system, Physics in Medicine and Biology, 50 (2005) 947–958.
[124] E. Furhang, C. Chui, K. Kolbert, S. Larson, G. Sgouros, Monte Carlo dosimetry method for patient-specific internal emitter therapy, Medical Physics, 24 (1997) 1163–1172.
[125] A. Colijn, W. Zbijewski, A. Sasov, F. Beekman, Experimental validation of a rapid Monte Carlo based micro-CT simulator, Physics in Medicine and Biology, 49 (2004) 4321–4333.

CHAPTER 6

Biomedical Visualization

Chris R. Johnson *and* Xavier Tricoche

Contents

1. Introduction 210
2. Scalar Field Visualization 211
   2.1. Direct volume rendering 211
   2.2. Isosurface extraction 220
   2.3. Time-dependent scalar field visualization 222
3. Vector Field Visualization 223
   3.1. Vector field methods in scientific visualization 224
   3.2. Streamline-based techniques 225
   3.3. Stream surfaces 226
   3.4. Texture representations 229
   3.5. Topology 232
4. Tensor Field Visualization 234
   4.1. Anisotropy and tensor invariants 235
   4.2. Color coding of major eigenvector orientation 236
   4.3. Tensor glyphs 236
   4.4. Fiber tractography 239
   4.5. Volume rendering 241
   4.6. White matter segmentation using tensor invariants 244
5. Multi-field Visualization 245
6. Error and Uncertainty Visualization 250
7. Visualization Software 254
   7.1. SCIRun/BioPSE visualization tools 255
   7.2. map3d 258
8. Summary and Conclusion 263
Acknowledgments 264
References 264

Abstract

Computers have become indispensable to the research and the clinical practice of biomedicine. Their widespread use now permits the manipulation of massive amounts of measured data, as well as, the study of sophisticated models through numerical simulations. The rapidly growing size and complexity of the resulting information from experiments or simulations creates a challenging demand on the tools and techniques needed to derive knowledge and insight from the data. Scientific visualization offers a very powerful approach to tackle this data analysis challenge by creating visual representations that convey salient

Advances in Biomedical Engineering
ISBN 978-0-444-53075-2

properties of large data sets and permit their effective interpretation and analysis. This chapter presents an overview of the techniques devised by scientific visualization research to address the specific needs of biomedical applications.

Keywords: biomedical visualization, scalar field visualization, vector field visualization, tensor field visualization, uncertainty visualization, visualization software *PACS*

1. INTRODUCTION

Computers are now extensively used throughout science, engineering, and medicine. Advances in computational geometric modeling, imaging, and simulation allow researchers to build and test models of increasing complexity and thus to generate unprecedented amounts of data. As noted in the NIH-NSF Visualization Research Challenges report [1], to effectively understand and make use of the vast amounts of information being produced is one of the greatest scientific challenges of the twenty-first Century. Visualization, namely helping researchers explore measured or simulated data to gain insight into structures and relationships within the data, will be critical in achieving this goal and is fundamental to understanding models of complex phenomena. In this chapter, we give an overview of recent research in visualization as applied to biomedical applications, focusing on the work by researchers at the Scientific Computing and Imaging (SCI) Institute [2].

Schroeder et al. have offered the following useful definition of visualization [3]:

> Scientific visualization is the formal name given to the field in computer science that encompasses user interface, data representation and processing algorithms, visual representations, and other sensory presentation such as sound or touch. The term data visualization is another phrase to describe visualization. Data visualization is generally interpreted to be more general than scientific visualization, since it implies treatment of data sources beyond the sciences and engineering... Another recently emerging term is information visualization. This field endeavors to visualize abstract information such as hyper-text documents on the World Wide Web, directory/file structures on a computer, or abstract data structures.

The field of visualization is focused on creating images that convey salient information about underlying data and processes. In the past three decades, there has been unprecedented growth in computational and acquisition technologies, a growth that has resulted in an increased ability both to sense the physical world in precise detail and to model and simulate complex physical phenomena. As such, visualization plays a crucial role in our ability to comprehend such large and complex data—data which, in two, three, or more dimensions, convey insight into such diverse biomedical applications as understanding the bioelectric currents within the heart, characterizing white matter tracts by diffusion tensor imaging, and understanding morphology differences between different genetic mice phenotypes, among many others.

Shown in Figure 1, the "visualization pipeline" is one method of describing the process of visualization. The *filtering* step in the pipeline involves processing raw data and includes operations such as resampling, compression, and other image-processing

Figure 1 The visualization pipeline.

algorithms such as feature-preserving noise suppression. In what can be considered the core of the visualization process, the *mapping* stage transforms the pre-processed filtered data into geometric primitives along with additional visual attributes, such as color or opacity, determining the visual representation of the data. *Rendering* utilizes computer graphics techniques to generate the final image using the geometric primitives from the mapping process.

While the range of different biomedical visualization applications is vast, the scientific visualization research community has found it useful to characterize scientific visualization techniques using a taxonomy associated with the dimensionality of the (bio)physical field to visualize:

- Scalar fields (temperature, voltage, density, magnitudes of vector fields, most image data),
- Vector fields (pressure, velocity, electric field, magnetic field), and
- Tensor fields (diffusion, electrical and thermal conductivity, stress, strain, diffusion tensor image data).

We use this taxonomy to discuss the biomedical visualization techniques throughout this chapter.

2. SCALAR FIELD VISUALIZATION

Scalar data are prevalent in biomedical simulation and imaging. In biomedical computing, scalar fields represent a quantity associated with a single (scalar) number, such as voltage, temperature, the magnitude of velocity. Scalar fields are among the most common datasets in scientific visualization, and thus, they have received the most research attention (see Hansen and Johnson [4] for an overview of scalar field visualization research). Biomedical scalar data, whether derived from simulations or gathered experimentally, often have both spatial and temporal components. These components have traditionally been visualized through different software structures. However, handling both spatial and temporal components of data within a single, unified software structure is not only convenient but will tend to strengthen the underlying visualization software by providing the *same* results whether the data is measured or computed.

2.1. Direct volume rendering

Direct volume rendering is a method of displaying 3D volumetric scalar data as 2D images and is probably one of the simplest ways to visualize volume data. The individual values in the dataset are made visible by the choice of a transfer function that maps the data to optical properties, like color and opacity, which are then projected and composited to form an image. As a tool for scientific visualization, the appeal of direct volume rendering is that no intermediate geometric information need be calculated, so the process maps from the dataset "directly" to an image. This is in contrast to other rendering techniques such as isosurfacing or segmentation, in which one must first extract elements from the data before rendering them. To create an effective visualization with direct volume rendering, the researcher must find the right transfer function to highlight regions and features of interest.

A common visualization goal in volume rendering is the depiction of the interface between two different materials in a volume dataset. The material surface can usually be seen with a simple transfer function which assigns opacity only to a narrow range of values between the data values associated with each of the two materials. In datasets characterized by noise or a more complicated relationship among multiple materials, statistical analysis of the dataset values can help to guide the transfer function design process. Moreover, in cases where datasets and associated volume-rendering methods are more complex (such as volumetric fields of vector or tensor values), methods for guiding the user toward useful parameter settings, based on information about the goals of the visualization, become necessary to generate informative scientific visualizations. Figure 2 shows a maximum intensity projection (MIP) of MRA data of the cerebral vasculature. The maximum intensity projection volume rendering method is the most simple form of volume rendering and is the one most often used by radiologists for viewing volume image data.

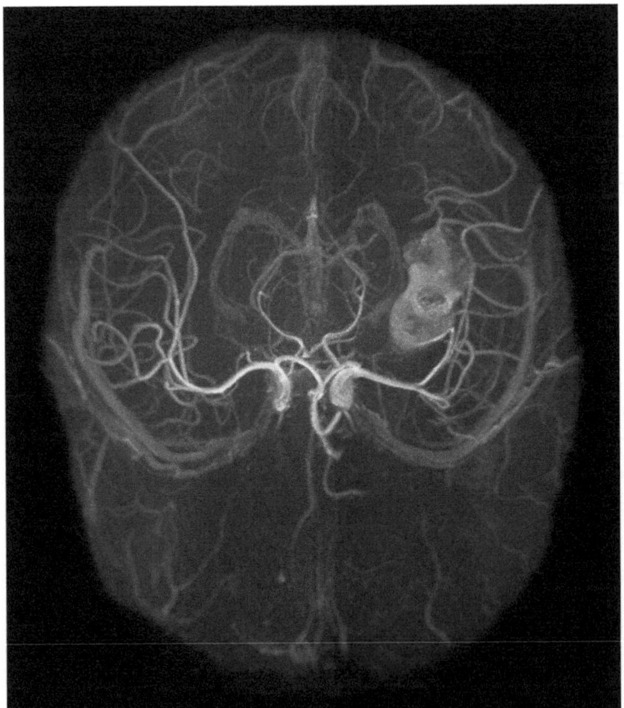

Figure 2 A maximum intensity volume rendering of magnetic resonance angiography (MRA) data showing a large fusiform aneurysm.

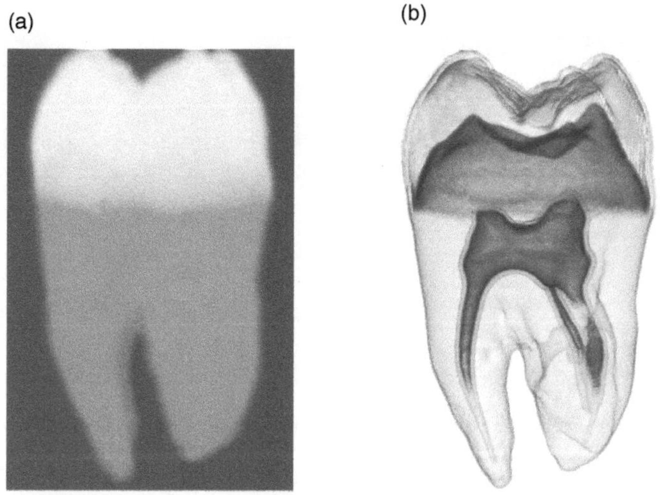

Figure 3 (a) Maximum intensity projection (MIP) volume rendering of a tooth from CT data, and (b) a full volume rendering of the same data using multi-dimensional transfer functions with SCIRun (Section 7.1). (See color plate 10).

The MIP algorithm works by projecting parallel rays (ray casting) through the volume from the viewpoint of the user. For each ray, the algorithm selects the maximum scalar value and uses that value to determine the color of the corresponding pixel on the 2D image plane. Volume rendering using MIP yields what look like "three-dimensional X-rays" in gray scales of the scalar volume data. Full volume rendering, on the other hand, traverses the rays and accumulates (integrates) color and opacity contributions along the ray. Volume rendering using full volume-rendering techniques yields an image that looks much more like what you might expect a 3D volume projection to look like in color. The differences are evident as shown below in Figure 3.

The full volume-rendered image contains more information than the MIP-rendered image and is usually more useful for understanding the 3D volume data or image. If this is true, why do most biomedical scientists use MIP and not full volume rendering? The answer is twofold. First, the MIP algorithm, since it only uses one value per voxel, is much less costly to compute than full volume rendering, which requires several values per voxel. This was the case especially before 2003, when graphic cards were not sufficiently powerful to generate fast full volume renderings. Second, full volume rendering is more difficult than the MIP algorithm to use. In particular, researchers find it difficult to relate the so-called transfer function to the scalar data values. This difficulty arose because most full volume rendering was originally done using 1D transfer functions that involved manipulating the transfer function as a sort of ramp function, as shown in Figure 4. The

Figure 4 Traditional 1D ramp transfer function.

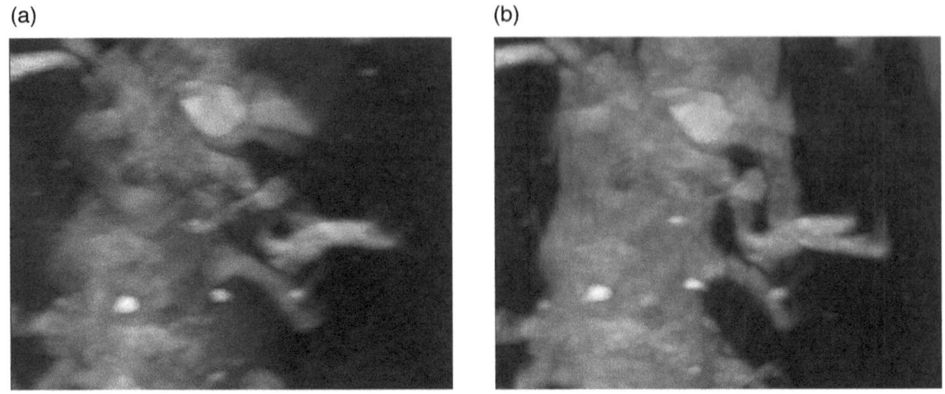

Figure 5 Maximum intensity projection (MIP) renderings of electron microscope tomography volumes: (a) an image based on the raw data that exhibits reconstruction artifacts obfuscating the boundaries of this spiney dendrite, and (b) an image based on a dataset that includes estimates of the missing views which reveals a more coherent picture of the structure.

method provided no intuitive way to choose the transfer function that might enable the user to see a particular feature in the data volume.

In some cases, volume visualization is more intimately tied to the imaging process. For instance, in electron microscope tomography (EMT), the configurations of the sample and electron beam limit the range of angles for which projections (i.e., the sinogram) are acquired. This creates reconstruction artifacts that adversely affect the quality of virtually any direct rendering or visualization strategy. By using volume rendering to fit a surface to the data [5], we are able to estimate missing parts of the sinogram and create better 3D reconstructions, as shown in Figure 5.

Finding a good transfer function is critical to producing an informative rendering, but this is a difficult task even if the only variable to set is opacity. Looking through slices of the volume dataset allows the researcher to spatially locate features of interest, and the researcher may employ a means of reading off data values from a user-specified point on the slice to help in setting an opacity function to highlight those features. However, there is no way for the researcher to know how representative of the whole 3D feature these individually sampled values are. User interfaces for opacity function specification typically allow the user to alter the opacity function by directly editing its graph, usually as a series of linear ramps joining adjustable control points. This interface does not itself guide the user toward a useful setting, as the movement of the control points is unconstrained and unrelated to the underlying data. Thus, finding a good opacity function tends to be a slow and frustrating trial-and-error process, in which apparently minor changes in an opacity function may lead to drastic changes in the rendered image. When the interaction of other rendering parameters, such as shading, lighting, and viewing angle, enters the picture, the process becomes even more confusing.

SCI Institute researchers have done substantial new research in volume rendering [6–15]. For example, Kindlmann and Durkin [15] have recently developed a method for the semi-automatic generation of transfer functions. In this method, the researcher

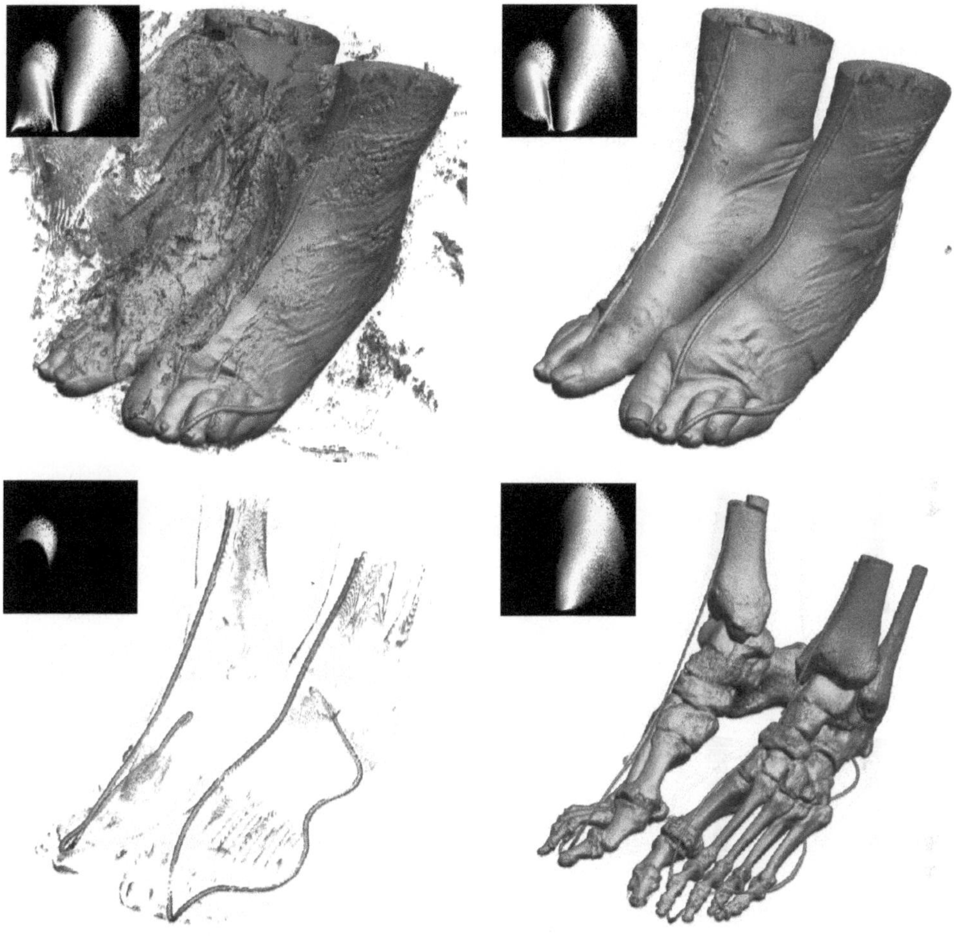

Figure 6 Exploration of a CT dataset (Visible Human project, feet of the visible woman) with semi-automatically generated multi-dimensional transfer functions. The inset black-and-white images show the opacity assignment as a function of CT data value (horizontal axis) and gradient magnitude. The image in the upper left was created through the use of default settings in the semi-automatic method. A researcher familiar with the data can identify the obscuring structures as being associated with the bag in which the specimen was placed prior to freezing and scanning. When the corresponding region is eliminated from the transfer function domain (upper right), the skin structure is clearly visible. The remaining features in the transfer function domain can be identified as the surface of the registration cord (lower left) and the bone surface (lower right). Such specificity is not possible with standard 1D transfer functions.

creates a data structure, which we call the "histogram volume structure," that captures information about the boundaries present in the volume and facilitates a high-level interface-to-opacity function creation. The user can determine which portions of the boundary will be made opaque without knowing the data values that occur in the boundary. Figure 6 illustrates an approach we have developed that is "semi-automated" in that it usefully constrains the user to a subspace of all possible transfer function while still allowing flexible depiction of surface features and characteristics [15].

Multi-dimensional transfer functions are sensitive to more than one aspect of the volume data, including, for example, both the intensity and one or more spatial gradients or other derived parameters. Such transfer functions have wide applicability in volume rendering for biomedical imaging and visualization. Even scalar datasets can benefit from multi-dimensional transfer functions. By incorporating edge-detection measures such as the first- and second-order spatial derivatives, we can easily discriminate between data values that represent material boundaries and those that represent the materials themselves. For instance, in a volume visualization of the Visible Male using CT data with the gradient magnitude and the second directional derivative in the gradient direction, we can accurately identify (classify) the sinuses, as seen in the right-hand panel of Figure 7. A method that uses only the CT data as the sole domain variable in a 1D transfer function (shown in the left-hand panel of Figure 7 cannot produce a visualization with the same specificity [6].

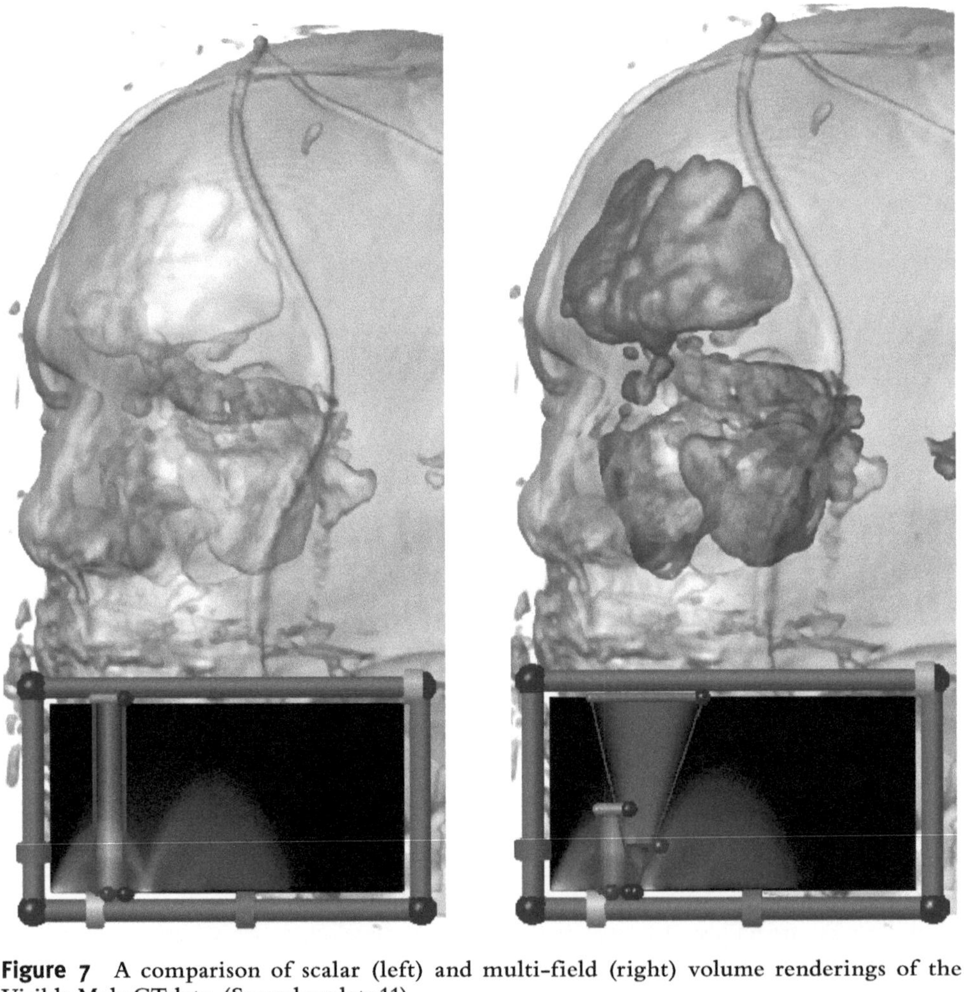

Figure 7 A comparison of scalar (left) and multi-field (right) volume renderings of the Visible Male CTdata. (See color plate 11).

Multi-dimensional transfer functions are also valuable for more general cases in which the data come from multiple variables in a physical simulation or multiple medical-imaging modalities. In such a case, the transfer function allows one to identify phenomena or features that can be described or classified based only on the basis of unique combinations of data values. Medical imaging can also benefit from multi-dimensional transfer functions when features of interest are not adequately described by any one imaging modality. For instance, CT imaging captures variations in bone structure but does not differentiate soft tissues well, whereas MR imaging differentiates soft tissue well and bone very poorly. By visualizing a co-registered CT and MRI dataset simultaneously, one can visualize features of interest in the soft tissue along with important bone landmarks.

To resolve the complexities inherent in a user interface for multi-dimensional transfer functions, we introduced a set of direct manipulation widgets that make finding and experimenting with transfer functions an intuitive, efficient, and informative process. To achieve interactive response, we exploited the fast rendering capabilities of modern graphics hardware, especially 3D texture memory and pixel texturing operations. Together, the widgets and the hardware form the basis for new interaction modes that guide the user toward transfer function settings appropriate for their visualization and data exploration interests [10,16,17].

The generalization of direct volume rendering to multi-valued fields presents at least two research challenges on which future work can focus, the first involving surface shading and the second involving the efficient evaluation of multi-dimensional transfer functions. Crucial to successful volume rendering is the calculation from the original data of the spatial gradients with which shading and surface rendering will be performed. In scalar data, the gradient has a simple mathematical definition and computation. In multi-valued fields, however, the generalization of the gradient is matrix-valued, and the surface orientation is represented as an eigenvector of this matrix. Current volume-rendering methods make little use of this information because it is computationally expensive, although advances in the abilities of GPUs increasingly support these calculations at interactive rates.

As described above, multi-dimensional transfer functions permit the visualization of structure based on combinations of values within the constituent fields. As the number of fields increases, simple implementations of multi-dimensional transfer functions become unwieldy. This observation motivated us to investigate analytic expressions for transfer function representation [7]. We have introduced a flexible framework for classifying and rendering multi-field datasets using transfer functions based on Gaussian primitives. We show in the paper that Gaussian transfer functions have a number of key features: they allow for selective classification of narrow features, can be efficiently computed on the graphics hardware, and their integral along a line segment has, under very limited assumptions, a closed-form analytic solution. Hence, they obviate the need for large precomputed lookup tables and are very memory efficient.

We expect this approach to have a large impact on the qualitative and quantitative value of volume rendering and become a significant tool for biomedical researchers. An example is shown in Figure 8.

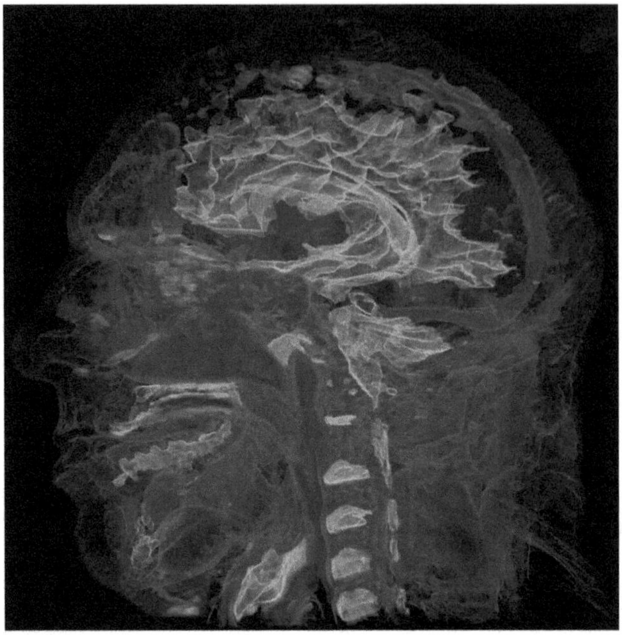

Figure 8 Example of 4D multi-field volume classified using a Gaussian transfer function. The dataset is the Visible Male Color Cryosection, courtesy of the NLM NIH. The analytical integration of the transfer function along rays creates a high-quality representation. In particular, muscles and boundaries between white and gray matter can be precisely isolated. Image courtesy of Joe Kniss [7]. (See color plate 12).

2.1.1. Example volume-rendering applications

The power of volume rendering for the interactive investigation of complex medical datasets was demonstrated in the BioPSE software system with the BioImage PowerApp [18], developed at the SCI Institute. As an example of BioImage's utility, we have used BioImage's volume-rendering tools to investigate the angiogenesis surrounding tumors in transgenic mice, as shown in Figure 9. The mouse vasculature, so clearly visible in this 3D volume rendering, could only be clearly visualized using a tool with the powerful specificity of multidimensional transfer functions. In addition, the BioImage tool that provides the interactive rendering also allows the user to crop and filter their data; it allows the user to simultaneously view slices and volume renderings of their data; and it provides the user with a set of quantitative tools for directly probing values within his data.

The same volume-rendering techniques can be used on thoracic CT datasets to investigate lung tumors, as shown in Figure 10. With the BioImage volume-rendering tools, we can interactively explore the size and shape of lung tumors without compromising the fidelity of the original high-resolution CT data.

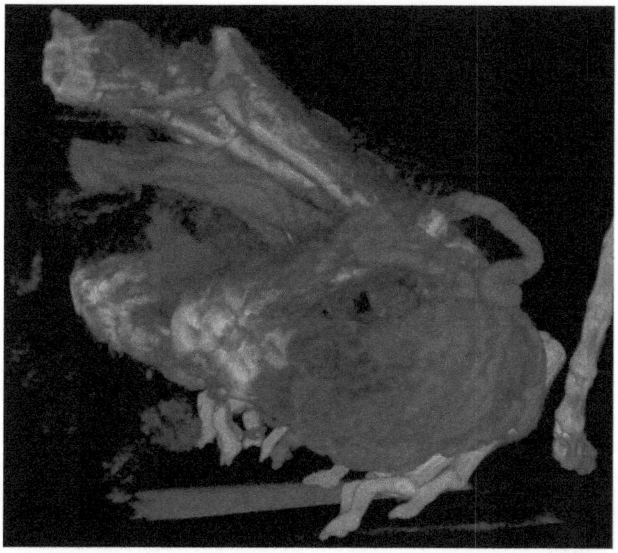

Figure 9 Mouse data visualized with BioImage's multi-dimensional transfer function volume rendering. Data courtesy of Charles Keller.

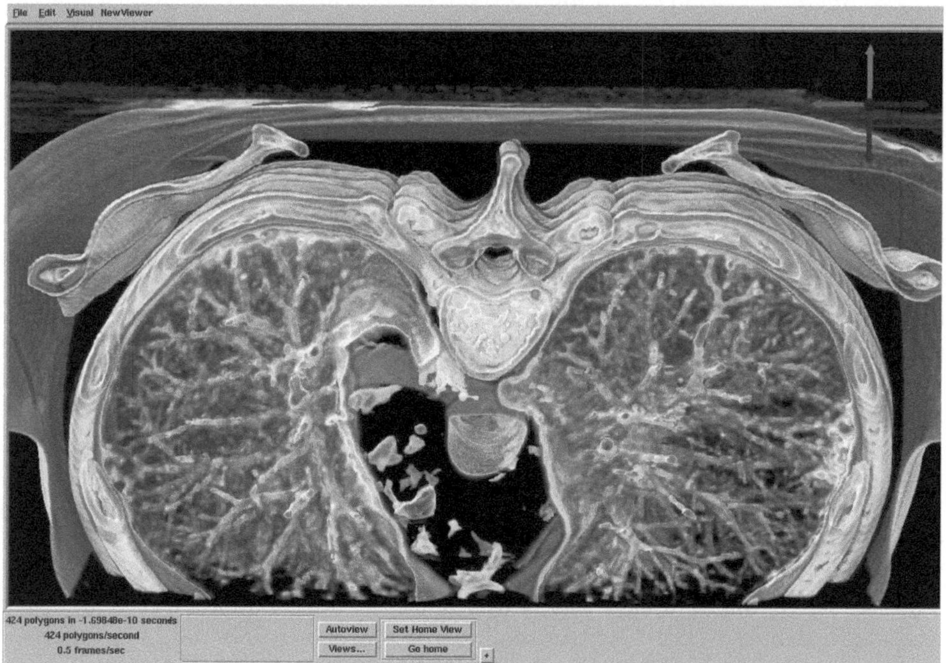

Figure 10 A volume-rendered image using multi-dimensional transfer functions. This view highlights the detailed vasculature of the lungs. Data courtesy of George Chen, MGH.

2.2. Isosurface extraction

Isosurface extraction is a powerful tool for investigating volumetric scalar fields. An isosurface in a scalar volume is a surface on which the data value is constant, separating regions of higher and lower value. Given the physical or biological significance of the data value (i.e., radio-opacity, dye concentration, fluorescence level), the position of an isosurface, as well as its relation to other neighboring isosurfaces, can provide clues to the underlying structure of the scalar field. In biomedical imaging applications, isosurfaces permit the extraction of particular anatomical structures and tissues; however, these isosurfaces are typically static in nature. A more dynamic use of isosurfaces can provide better visualization of complex space- or time-dependent behaviors in many scientific applications.

Within the last 15 years, isosurface extraction methods have advanced significantly, from an off-line, single-surface extraction process into an interactive, exploratory visualization tool. Interactivity is especially important in exploratory visualization where the user has no *a priori* knowledge of any underlying structures in the data. A typical data exploration session therefore requires the researcher to make many isovalue changes in search of interesting features. In addition, it is helpful to provide global views (to place an isosurface in the context of the entire dataset) and detailed views of small sections of interest. Maintaining interactivity while meeting these visualization goals is especially challenging for large datasets and complex isosurface geometry.

The *Marching Cubes* [19,20] method, introduced in 1986, was the first practical and most successful isosurface extraction algorithm. Its simplicity has made it the *de facto* standard extraction method even to this date. The Marching Cubes algorithm demonstrated that isosurface extraction can be reduced, using a divide and conquer approach, to solving a local triangulation problem. In addition, the marching cubes method proposed a simple and efficient local triangulation scheme that uses a lookup table. Subsequently, researchers created methods for accelerating the search phase for isosurface extraction [21–25] all of which have a complexity of $O(n)$, where n is the number of voxels in the volume. We introduced the *span space* [26] as a means for mapping the search onto a 2D space and then used it to create a *near optimal isosurface extraction* (NOISE) algorithm that has a time complexity of $O(\sqrt{n}+k)$, where k is the size of the isosurface. Cignoni et al. [27] employed another decomposition of the span space leading to a search method with optimal time complexity of $O(\log n+k)$, albeit with larger storage requirements. In addition, Bajaj et al. introduced the contour spectrum, which provides a fast isosurface algorithm and a user interface component that improves qualitative user interaction and provides real-time exact quantification in the visualization of isocontours [28,29].

We improved further on these isosurface extraction methods by using a different visibility testing approach and virtual buffer rendering to achieve a real-time, view-dependent isosurface extraction [30–33]. We also presented a progressive hardware-assisted isosurface extraction (PHASE) that is suitable for remote visualization, i.e., when the data and display device reside on separate computers. This approach works by reusing, when a view point is changed, the information and

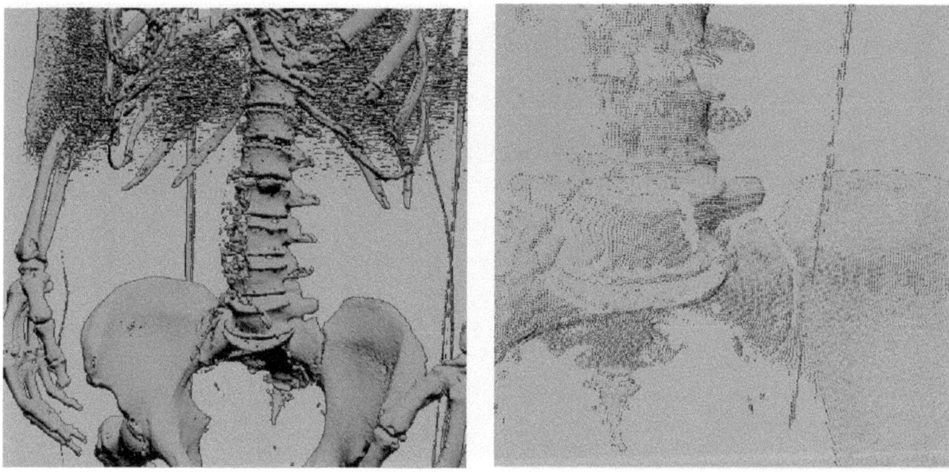

Figure 11 Isosurface extraction of the full CT data (512 × 512 × 1734, 1 mm spacing) of the NIH NLM Visible Female. **Left:** A section of the skeleton extracted by the PISA algorithm [33]. **Right:** A closeup view of the extracted points. Point shading is determined by an image-based normal computation technique that ensures high-quality results. (See color plate 13).

triangles that were extracted from the previous view point [32]. Using this approach, we can extract only newly visible sections of the isosurface and thus improve visualization performance.

Following the same view-dependent approach we have recently proposed a novel point-based approach to isosurface extraction [33]. The basic idea of our method is to address the challenge posed by the geometric complexity of very large isosurfaces by a point-based representation of sub-pixel triangles. Combined with a new fast visibility query and a robust normal estimation scheme, our method allows for the interactive interrogation of large data sets [e.g., the full NIH NLM Visible Female dataset (1 GB)] at up to 15 frames per second on a single desktop computer, see Figure 11. We also presented an extension of that algorithm that enhances the isosurface visualization by computing shadows.

2.2.1. Isosurface applications

As described above, isosurface extraction is a ubiquitous tool for exploring scalar volume datasets. The isosurface module in SCIRun software system (see the description in the Visualization Software Section 7 below) supports the standard Marching Cubes algorithm as well as the high-performance NOISE algorithm. These tools are used for a variety of SCIRun applications, including isochrone visualization, isopotential surface visualization, and cutting plane synthesis, among others.

For example, SCIRun employs the isosurface extraction algorithm to produce and visualize isochrone surfaces, as demonstrated in Figure 12. In this application, the user can interactively step through different moments in time during cardiac activation to visualize how the activation wavefront passes through the heart tissue.

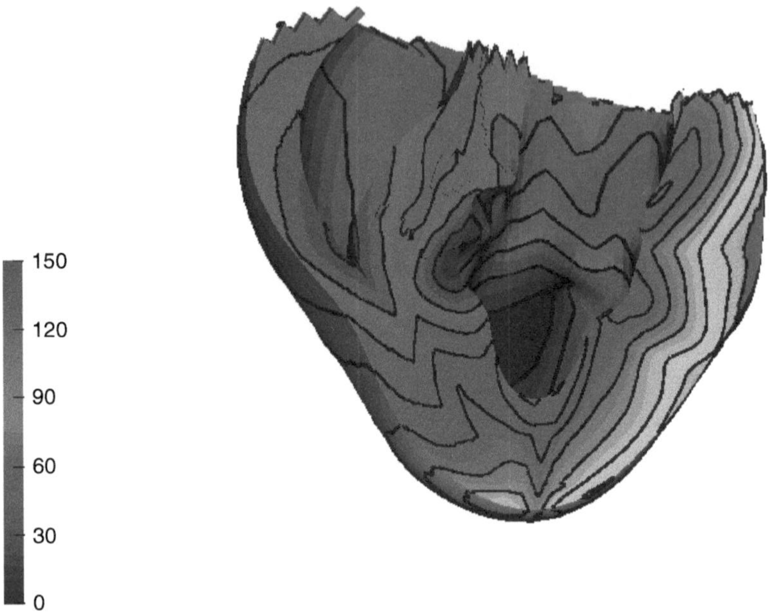

Figure 12 BioPSE visualization of the isochrones for a cardiac activation simulation. The front of the heart has been clipped away to reveal the transmural activation times. (See color plate 14).

By animating the isochrone value (time step), the electrophysiologist can investigate the specific processes and mechanisms involved in cardiac arrhythmia.

SCIRun also employs isosurface visualization to render and investigate how electrical activity on the cortical surface creates isopotential contours on the surface of the scalp. In Figure 13, we again illustrate how SCIRun can be used to visualize these data.

2.3. Time-dependent scalar field visualization

Temporal information can be incorporated and readily understood within volume rendering through the use of sequences of volume images generated across a range of time steps. The use of such sequences allows one to effectively investigate a dataset's temporal dynamics, such as wave propagation effects through a 3D scalar field. We have developed an algorithm that significantly reduces the time required to create flow animations of scalar fields while also reducing the storage requirements. This algorithm, called differential volume rendering [34], utilizes temporal and spatial coherence between consecutive time steps of simulation data to both accelerate the volume animation and to compress the volume data. The method may be used with all volume-rendering techniques and can be adapted into a variety of ray-casting paradigms that can be used to further accelerate the visualization process [35–37]. Additionally, direct volume-rendering techniques have been proposed as a methodology to visualize scalar features in a dynamic time-dependent field [38–42].

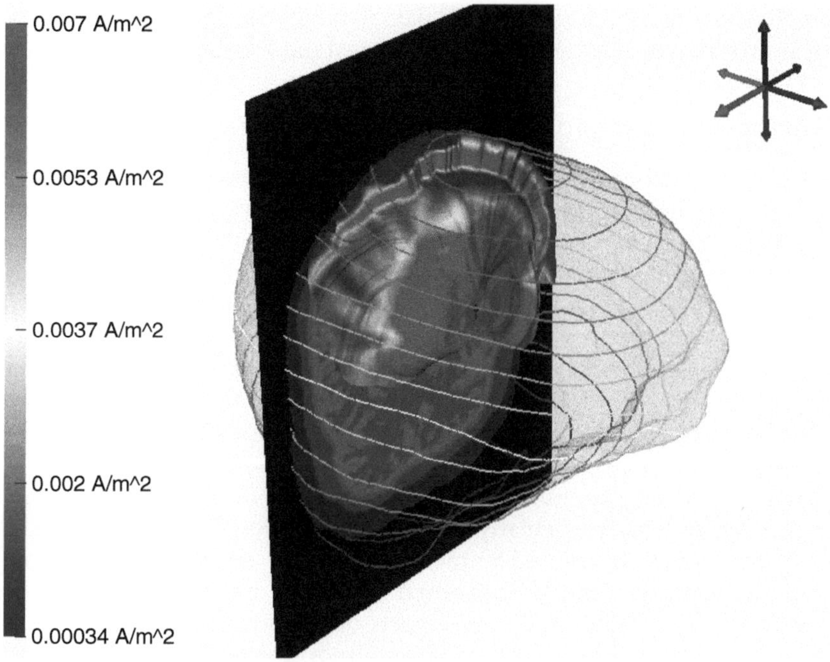

Figure 13 Visualization of the potential distribution through the cranium due to a dipolar source simulation. Potentials are color-mapped onto a stream-surface within the head volume, and evenly spaced isopotential contours are plotted on the scalp surface. Data courtesy of Scott Makeig, UCSD and Greg Worrell, Mayo Clinic. (See color plate 15).

3. VECTOR FIELD VISUALIZATION

Vector fields are a fundamental biophysical quantity that describe the underlying continuous flow structures of physiological processes. Examples of important biophysical vector fields include bioelectric fields, current densities, biomagnetic fields, as well the velocities and pressures of biofluids, and the forces associated with biomechanics. Vector-valued quantities also appear in biomedical data in the form of derivatives of scalar fields.

Common clinical examples of scalar fields associated with vector fields include electrocardiography (ECG) in cardiology, and electroencephalography (EEG) or magnetoancephalography (MEG) in neurology. Such measured clinical data are often sparse and noisy. In order to accurately approximate the associated vector fields of interest, a preprocessing stage is often needed that involves applying appropriate filtering and interpolation schemes. The derivatives of the resulting smooth quantities can then be computed to yield the vector fields. High-resolution vector fields are also directly obtained through computer simulations, including forward and inverse problems that rely on numerical approximation methods ranging from Finite Elements, Finite Differences, Finite Volumes, and Boundary

Elements. In these cases, the vector field data naturally lend themselves to scientific visualization techniques that respect the numerical models used in the computation.

3.1. Vector field methods in scientific visualization

Visualizing vector field data is challenging because no existing natural representation can visually convey large amounts of 3D directional information. Visualization methods for 3D vector fields must balance the conflicting goals of displaying large amounts of directional information while maintaining an informative and uncluttered display.

The methods used to visualize vector field datasets take their inspiration in real world experiments where a wealth of physical flow visualization techniques have been designed to gain insight into complex natural flow phenomena. To this end, external materials such as dye, hydrogen bubbles, or heat energy can be injected into the flow. As these external materials are carried through the flow, an observer can track them visually and thus infer the underlying flow structure.

Analogues to these experimental techniques have been adopted by scientific visualization researchers, particularly in the computational fluid dynamics (CFD) field. CFD practioners have used numerical methods and 3D computer graphics techniques to produce graphical icons such as arrows, motion particles, and other representations that highlight different aspects of the flow.

Among the existing flow visualization methods, the techniques relevant to the visual analysis of biomedical vector fields can be categorized as follows.

1. The simplest techniques correspond to an intuitive, straightforward mapping of the discrete vector information to so-called glyphs. Glyphs are graphical primitives that range from mere arrows to fairly complex graphical icons that display directional information, magnitude, as well as additional derived quantities such as the curl and divergence altogether.
2. The second category corresponds to the set of techniques that are based on the integration of streamlines. Streamlines are fast to compute and offer an intuitive illustration of the local flow behavior.
3. Stream surfaces constitute a significant improvement over individual streamlines for the exploration of 3D flows since they provide a better understanding of depth and spatial relationships. Conceptually, they correspond to the surface spanned by an arbitrary starting curve advected along the flow.
4. Textures and other dense representations offer a complete picture of the flow, thus avoiding the shortcomings of discrete samples. Their major application is the visualization of flows defined over a plane or a curved surface.
5. The last type of flow visualization techniques are based on the notion of flow topology. Topology offers an abstract representation of the flow and its global structure. Sinks and sources are the basic ingredients of a segmentation of the volume into regions connecting the same spots along the flow.

Next, we describe how each of these different approaches can be applied to the visualization of biomedical vector fields. Because of their lack of scalability and the strong occlusion issues that characterize them in practical applications, we do not consider glyph-based techniques in further detail.

3.2. Streamline-based techniques

Streamlines offer a natural way to interrogate a vector dataset. Given a starting position selected by the user, numerical integration over the continuous representation of the vector field yields a curve that can be readily visualized. The numerical schemes commonly used for the integration range from the first-order Euler scheme with fixed step size to Runge–Kutta methods with higher order precision and adaptive step size [43]. The choice of the appropriate method requires to take into account the complexity of the structures at play and the smoothness of the flow.

Since streamlines are unable to fill the space without visual clutter the task of selecting an appropriate set of starting points (commonly called seed points) is critical to obtaining an effective visualization. A variety of solutions have been proposed over the years to address this problem. A simple interactive solution consists in letting the user place a rack in the data volume over which seed points are evenly distributed. The orientation and spatial extent of the rack, as well as the number of seed points can be adjusted to allow for the selective exploration of a particular region of interest, as shown in Figure 14.

Another class of methods aim at automatically placing seed points throughout the domain in order to achieve an even distribution of the corresponding streamlines [44–48]. Following this approach, Sachse et al. devised a new streamline seeding technique, specifically designed for the visualization of the electrical current, that distributes streamlines in the volume based on the local value of the current density [49,50]. As a result, streamline and current densities are proportional, which produce images that engineers can easily interpret. Examples are shown in Figure 15.

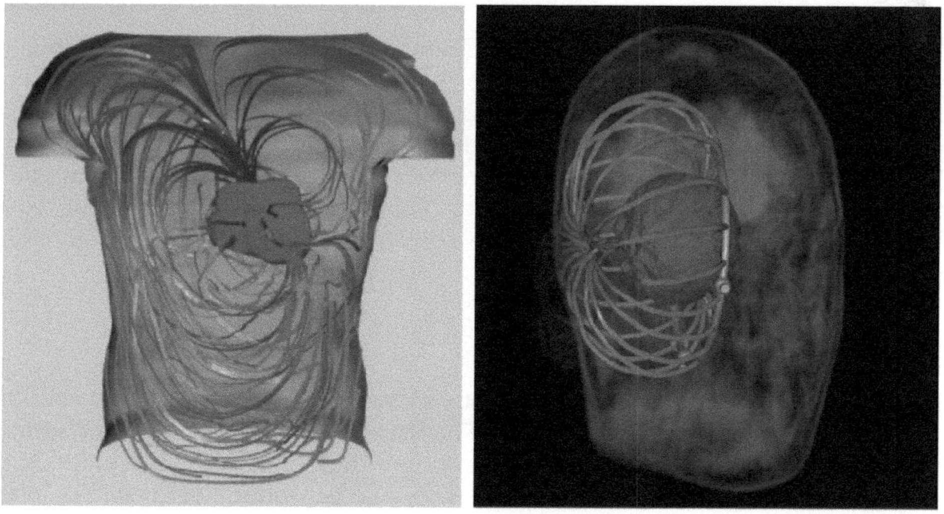

Figure 14 Applications of streamlines to Finite Element biomedical datasets. **Left:** The bioelectric field in the torso visualized through streamlines seeded randomly around the epicardium. **Right:** Use of a circular rack to seed streamlines regularly around a dipolar source in the brain. (See color plate 16).

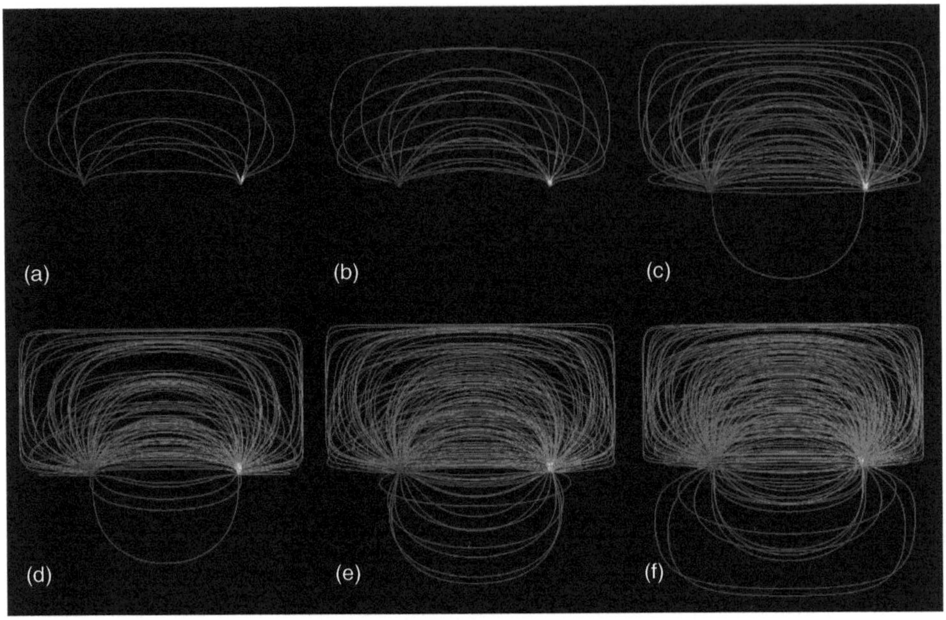

Figure 15 Streamline visualization of electrical current flow in a cube. A conductivity of 0.02 and 0.2 S/m were assigned to the lower and upper half, respectively, of the cube. (a) 10, (b) 20, (c) 50, (d) 100, (e) 150, and (f) 200 streamlines were drawn. Two Dirichlet boundary conditions were assigned representing an electrical source and sink. A raster cube consisting of 32 × 32 × 32 cubic voxels constituted the spatial domain. Images courtesy of Frank Sachse [49].

An additional limitation of flow visualizations based upon streamline techniques concerns the difficult interpretation of the depth and relative position of curves in a three-dimensional space. A solution consists in creating artificial lighting effects that emphasize curvature and assist the user in his or her perception of depth [51–53]. An alternative method that can be implemented on the graphics hardware, assigns a non-zero volume to individual streamlines. These streamlines are then depicted as tubes and filled with 3D textures to create expressive images in which various visual cues are used to enhance perception [54]. Refer to Figure 16.

3.3. Stream surfaces

The intuitive representations offered by stream surfaces make them a very valuable tool in the exploration of 3D flows. The standard method for stream surface integration is Hultquist's advancing front algorithm [55]. The basic idea is to propagate a polygonal front along the flow, while accounting for possible divergence and convergence by adapting the front resolution. Yet, this method yields triangulated surfaces of poor quality when the flow exhibits complex structures. We recently proposed a modified stream surface algorithm that improves on Hultquist's original scheme by allowing for an accurate control of the front curvature [56]. This

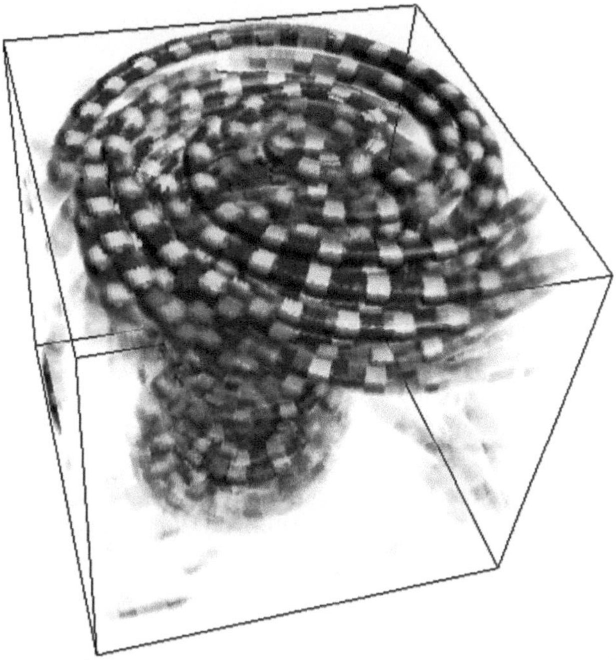

Figure 16 An extension of streamline-based Flow Visualization. The image shows a combination of streamlines and 3D textures in the visualization of a tornado data set. Textures permit to embed additional information and ease the interpretation of the spatial context. From Li et al. [54].

method creates smooth, high-quality surfaces, even for very intricate flow patterns. We applied stream surfaces to the visualization of the cardiothoracic current obtained through a Finite Elements simulation [49]. Specifically, stream surfaces were shown to permit insight into the interconnection of different regions of the epicardium through the bioelectric current inside the thorax. An illustration is shown in Figure 17.

A similar approach was used to visualize the return current in a high-resolution simulation of a realistic head model. In this case, stream surfaces proved instrumental in assessing the impact of various models of the white matter anisotropy on the current pattern and its interconnection with anatomical structures, see Figure 18 and Section 5.1.

Despite their esthetic appeal and the effective visualizations they yield, stream surfaces share with streamlines the limitation induced by their reliance on a careful seeding. Here, the seeding task consists in selecting a starting curve such that the resulting surface will exhibit the characteristic patterns of important flow structures. As with streamlines, this task can be tedious. In the case of the bioelectric current, flow structures can be defined in terms of interconnections between zones of inflow and outflow on surface enclosing dipolar sources. This is the solution that we applied in Figures 17 and 18. To visualize the cardiothoracic current isocontours of

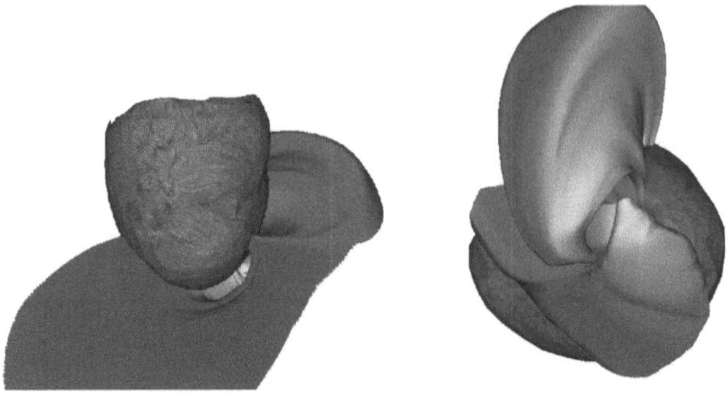

Figure 17 Visualization of the bioelectric field in the direct vicinity of epicardium with high-quality stream surfaces. The image shows the technique applied to the cardiothoracic current in a Finite Element forward computation. The surfaces capture the geometry of the current induced by the dipole equivalent cardiac source. They also provide an effective representation of the interconnections that exist between different regions on the epicardium. The seeding curves correspond to isocontours of the electric potential selected close to local extrema. A rainbow color map is used along each seeding curve to visualize flow stretch. (See color plate 17).

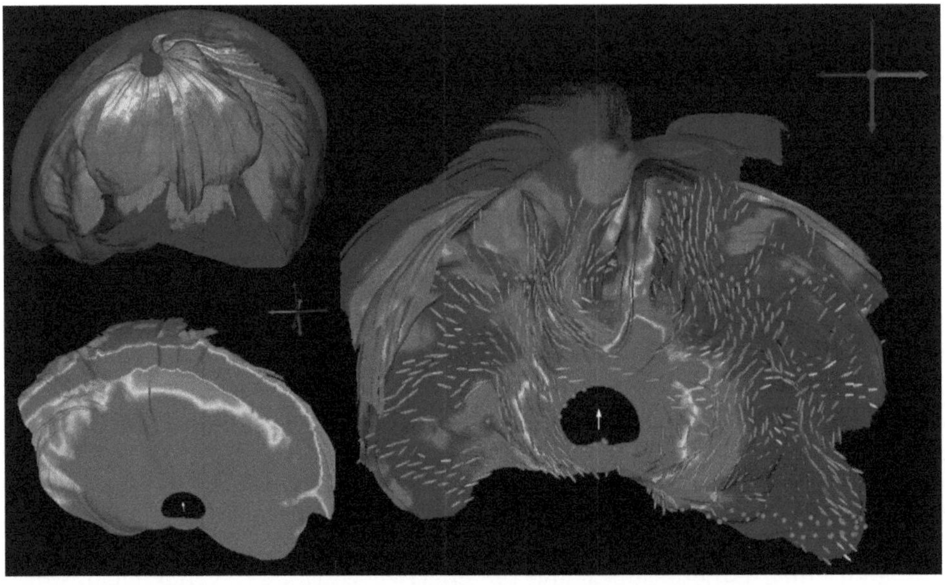

Figure 18 Stream surface visualization of bioelectric field induced by a dipolar source in left thalamus. **Left top.** Stream surfaces seeded along isocontour of electric flux on sphere enclosing the source. Culling is used to address occlusion. White matter has anisotropic conductivity. **Left bottom.** Stream surface started along circle contained in coronal slice and centered around source location. White matter is assumed isotropic. Color coding corresponds to magnitude of electric field. **Right.** Similar image obtained for anisotropic white matter. Glyphs visualize major eigenvector of conductivity tensor. Color coding shows magnitude of return current. (See color plate 18).

either the electric potential or the electric flux are used as seed curves on the epicardium, see Figure 17. In the case of the return current in the brain, however, no such obvious surface exists that encloses the source. A possible solution consists in introducing an artificial spherical surface surrounding the known source position. Refer to Figure 18, top left. Observe that culling is helpful to address the occlusion caused by the intricate shape of the stream surface.

Another solution to the seeding problem that we investigated consists in exploring the dependency of stream surfaces on a parameterization of their seeding curves. As the seeding curve changes according to some predefined mechanism, the corresponding evolution of the stream surface can then be visualized as an animation. For example, this technique can be used to allow for a better understanding of the 3D structure of the return current in the brain. Specifically, a circle of increasing radius centered around the dipolar source and lying in the transverse plane orthogonal to the dipole axis is regarded as a parameterized seeding curve. An analysis of the differences between isotropic and anisotropic conductivity of the white matter can then be made in a side by side comparison of the corresponding animations. Results are shown in Figure 19.

3.4. Texture representations

Texture-based flow visualization methods provide a unique means to address the limitations of depictions based on a limited set of streamlines. They yield an effective, dense representation which conveys essential patterns of the vector field and does not require the tedious seeding of individual streamlines to capture all the structure of interest [57]. Arguably the most prominent of those methods is Line Integral Convolution (LIC) proposed by Cabral and Leedom [58]. The basic idea is to apply a 1D low-pass filter to a white noise texture covering the 2D flow domain. The filter kernel at each pixel is aligned with streamlines of the underlying flow. Consequently, the resulting image exhibits a high correlation of the color values along the flow and little or no correlation across the flow. Hence, this method produces a dense set of streamline-type patterns that fill the domain and reveal all the flow structures that are large enough to be captured by the fixed resolution of the texture. This seminal work has inspired a number of other methods. In particular, improvements were proposed to permit the texture-based visualization of time-dependent flows [59–61], flows defined over arbitrary surfaces [62–64], and dye advection [65]. Some attempts were made to extend this visual metaphor to 3D flows [66,67].

An alternative to LIC-like visualization methods uses reaction-diffusion simulations to generate textures with shapes, sizes, and orientations driven by the underlying vector field [68]. Note that this approach to flow visualization is in fact an extension of a visualization method previously developed for diffusion tensors [14]. The human visual system naturally follows spatio-temporal patterns and can easily perceive very subtle changes affecting them. This allows the orientation and magnitude of the local vector field to be conveyed with an appropriately tuned pattern of elliptical spots. This method of visualization is based on mapping two of the vector field components, orientation and magnitude, to the diffusion kinetics in

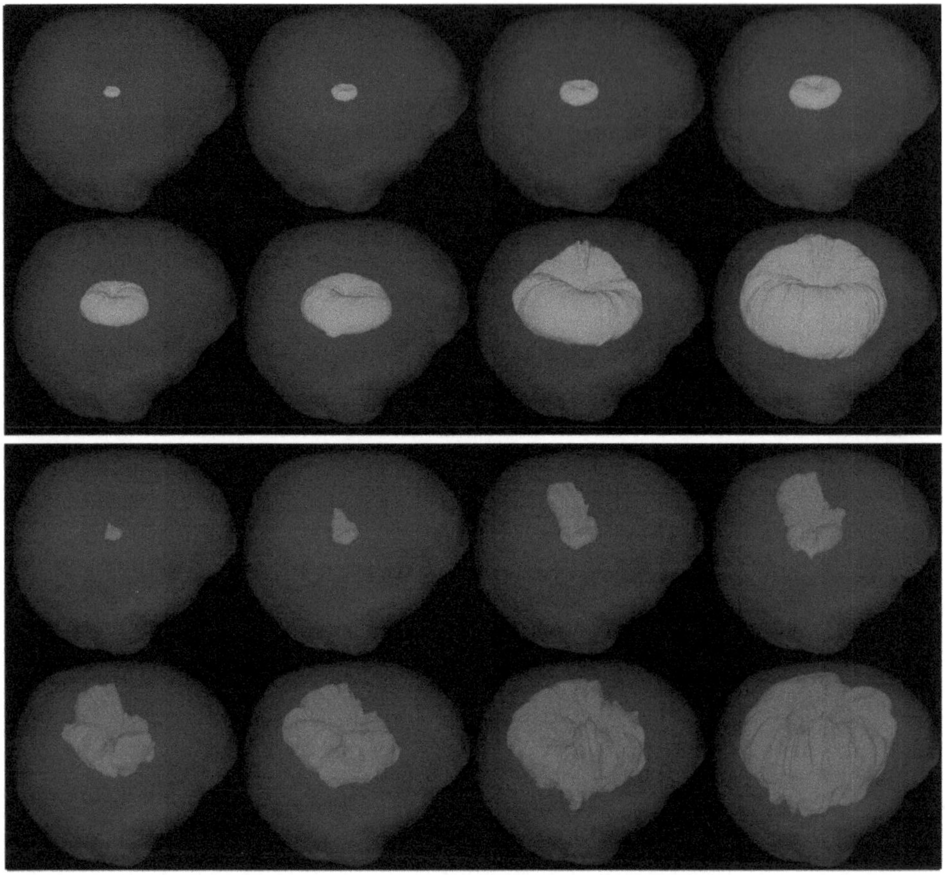

Figure 19 Evolution of a stream surface integrated along the return current with respect to the increasing radius of its seeding circle. **Top row.** Frames from an animation corresponding to isotropic white matter. **Bottom row.** Frames of the animation obtained for anisotropic white matter.

the reaction-diffusion system. Our method also supports mapping orientation *uncertainty* to the diffusion kinetics. This is further discussed in Section 5.2. The principal advantage of the reaction-diffusion model over existing flow field visualization techniques is that the pattern size and density that naturally arises from the reaction-diffusion model accurately represents the underlying vector field, based on a flexible mapping from multiple field characteristics to parameters of the diffusion kinetics. Figures 20 and 21 show the anisotropic diffusion applied to the Turing [69] and Gray and Scott [70,71] reaction-diffusion models for a vector field at 45 degrees with a random variation in the magnitude.

From the point of view of their application to biomedical vector data texture representations can be seen as the natural vector counterpart of color plots for scalar potentials. Indeed, this type of technique is best suited for the visualization of flow data defined over a plane or a curved surface. Moreover, texture-based

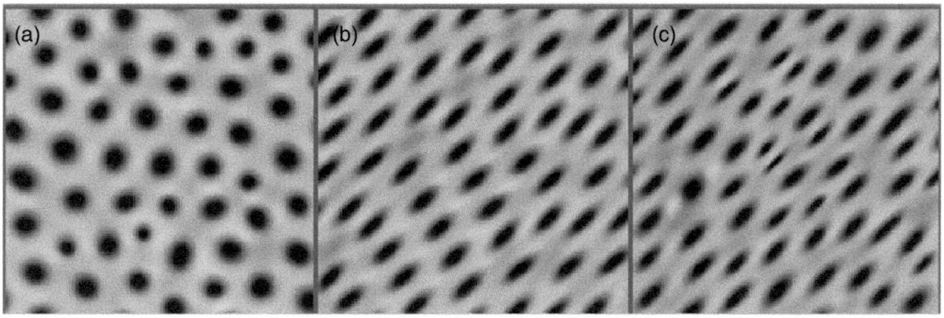

Figure 20 Turing model visualization of a vector field with (a) random magnitude, (b) constant orientation, and (c) magnitude and orientation.

Figure 21 Gray–Scott model visualization of a vector field with (a) random magnitude (b) constant orientation (c) magnitude and orientation.

visualizations can be combined with the color coding of a scalar quantity associated with the vector field to enrich the pictures.

An important property of bioelectric fields is their homogeneous Neumann boundary condition which imposes that the current cannot leave the conductive volume and therefore must be tangent to its boundary. In particular, the current is tangential to the torso and the scalp. In contrast, the epicardium surface is crossed by the current originating at sources locations contained in the cardiac tissue. As a consequence, the visualization of the restriction of the current to the heart surface by means of textures requires to first project it onto the corresponding geometry. Examples are proposed in Figure 22, left, and Figure 25.

Similar to color plots, an obvious way to address the occlusion problems encountered when applying texture visualizations to volume data is to restrict the representation to a cutting plane. To be meaningful, the choice of this plane must be made based on the symmetry of the geometry and on the known position of a bioelectric source, see Figure 22 (left) and Figure 23. However, a single plane is typically unable to convey an informative picture of the 3D current. This problem is emphasized when the conductivity of the tissue is anisotropic which breaks the

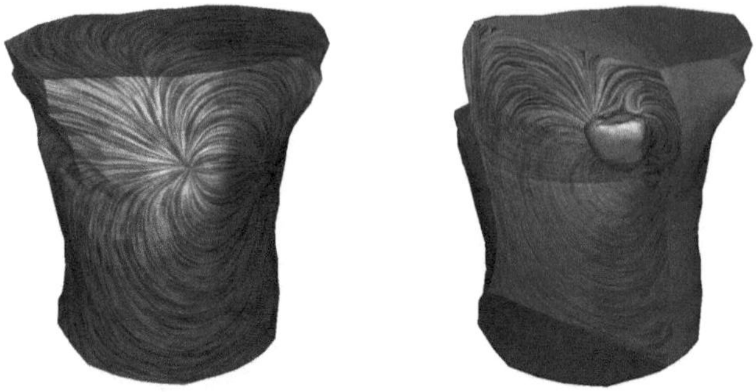

Figure 22 **Left.** LIC representation of the current on the boundary of a Finite Element grid used in forward computation. The color-coded electric potential (negative values in blue, positive values in red) are superimposed on the texture. **Right.** Two LIC textures computed over cutting planes combined by transparency. The geometry of the heart remains opaque for context. (See color plate 19).

Figure 23 LIC applied to coronal and sagittal clipping planes reveals details of the dipolar source and its interaction with the surrounding anisotropic tissue. Rather than a typical smooth, symmetric dipolar pattern, the electric current is clearly diverted by the presence of white matter tracts that lie close to the source. The field also changes direction very rapidly as it approaches the skull just beneath the surface of the head. (See color plate 20).

symmetry of the dipolar current flow pattern and increases its geometric complexity. A simple solution consists in displaying two or more such planes in combination. Additionally, transparency allows the user to see the spatial relationship between the patterns exhibited by each texture. An example is shown in Figure 22 (right).

3.5. Topology

The topological approach provides a powerful framework for flow visualization in a broad range of applications [72]. For planar vector fields, as well as vector fields defined over curved surfaces, it has established itself as a method of reference to characterize and visualize flow structures. The excessive complexity of the topology of intricate flows can be addressed by simplifying the resulting graphs while preserving essential properties in order to facilitate the analysis of large-scale flow patterns [73,74]. Refer to Figure 24.

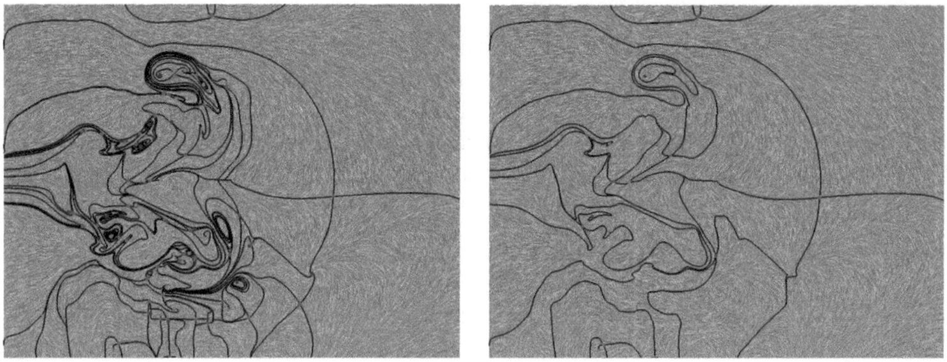

Figure 24 Topology simplification. The left image shows the original topology obtained for a CFD simulation of a streaming jet with inflow into a steady medium. Numerous small-scale structures lead to a cluttered depiction. The right image shows the same data set after topology simplification.

Topology-based methods prove interesting in the visualization of the bioelectric current on the epicardium. In this case, the automatically extracted topological features are believed to be related to what is known in cardiac electrophysiology as epicardial breakthroughs. This phenomenon occurs when an activation wave in the cardiac tissue breaks through the surface, generating a local potential minimum in the epicardial surface. Refer to Figure 25.

The usability of 3D topology for the characterization of salient patterns of the bioelectric current remains limited. The explanation comes from the structure of a bioelectric field. Indeed, bioelectric sources typically behave as dipoles, which do not exhibit an intrinsic separation surface and therefore do not allow for a natural topological segmentation of their surrounding medium. However, the application of the topological framework to the visualization of bioelectric activity is at a very early stage, and future research will be necessary to determine the full potential of this approach.

Figure 25 Topology of bioelectric field on epicardium. The images show an LIC representation of the potential gradient on the surface enhanced by the depiction of the associated topological graph. Green points correspond to potential minima, blue points mark potentials maxima. Left and middle images show an anterior view of the epicardium. Right image shows a view of the torso from the left. (See color plate 21).

Beyond its application to bioelectric data, the topological approach shows great promise to be instrumental in the structural investigation of flows in other biophysical fields as well. For instance, topology is known to permit the characterization of flow recirculation patterns in fluid dynamics [75] and can therefore provide powerful tools to analyze blood recirculation in hemodynamics. Topology could also be used to study and visualize the structures of forces in biomechanics.

4. Tensor Field Visualization

Tensors provide the language necessary to describe the intrinsic material properties of biological systems. Electrical conductivity and molecular diffusivity are examples of material properties that describe the ability of particles (such as electrons and water molecules) to pass through a given material. Strain and stress tensors, on the other hand, characterize the mechanical properties of soft tissues. Material properties are often *inhomogeneous*, that is, they vary as a function of the position within the material. Thus, proper modeling of their characteristics requires a field of tensor values sampled in three dimensions. Hence, gaining insight into the structure of 3D tensor fields is a significant and ongoing challenge in biomedical visualization.

The need for tensor visualization arises in the modeling of the material conductivity of tissue, such as in the brain or heart. Conductivity is a tensor quantity because the fiber structure of the tissue results in anisotropic conduction [76,77]; setting accurate values of fiber structure is a crucial component of a model of the cardiac propagation [78,79], or the localization of an epileptic seizure from an inverse EEG computation, tasks that require tensor field visualization capabilities.

Another application that has drawn increasing interest is diffusion tensor imaging. Diffusion tensor MRI (DT-MRI or DTI) is an imaging modality that permits, through its influence on the local diffusion of water molecules [80], the non-invasive measurement of tissue physical microstructure. In regions where the tissue has a linear organization, such as in myelinated axon bundles comprising the white matter in the brain or in muscle tissue, diffusion is preferentially directed along the fiber direction, and this phenomenon can be measured with DT-MRI. Creating meaningful images or models from diffusion tensor data is challenging because each sample point has six independent degrees of freedom.

The deformation of the heart muscle (myocardium), namely its successive contraction and expansion are described by a different tensor, the strain tensor. For clinical purposes, the analysis of the strain tensor field in the left ventricle is of great significance in the efficient screening of cardiac patients since abnormal properties of the myocardial strain can be identified before the first symptoms of a heart attack [81]. An imaging technique used to acquire the corresponding tensor data is called *Tagged MRI.* It consists in using radiofrequency pulses in an MR scanner to create dark lines in the MR image. Because these lines deform with the tissue, the motion of material points can be tracked over time to compute the strain tensor [82]. Measures obtained through MRI can be coupled with Finite Element models of the mechanics of the tissue [83] or deformable image registration [84]. The visualization of the resulting tensor data is then instrumental in producing diagnostic information for cardiac conditions.

The remainder of this section provides an overview of the different visualization techniques that have been devised to help clinicians and researchers explore and analyze the rich structural information contained in large biomedical tensor datasets.

4.1. Anisotropy and tensor invariants

The examples of tensor fields mentioned previously share the property that they correspond to real symmetric maps between vector fields. Real symmetric tensors have real eigenvalues λ_i, and the associated eigenvectors $\vec{e_i}$ are mutually orthogonal:

$$T\vec{e_i} = \lambda_i \vec{e_i}, \quad \lambda_i \in IR, \quad \forall i \neq j \;\; \vec{e_i}.\vec{e_j} = 0.$$

A standard notation consists in classifying the three eigenvectors in major, medium, and minor eigenvectors with respect to the corresponding eigenvalue: $\lambda_1 \geq \lambda_2 \geq \lambda_3$.

A tensor is called *isotropic* if it maps every vector direction uniformly. In the case of a diffusion tensor, for example, this corresponds to an equal passage of molecules through the material in every direction. An isotropic strain tensor indicates that the material deforms uniformly in all directions. This property implies that all eigenvalues are equal and individual eigenvectors in the previous definition cannot be characterized (the eigenspace is three-dimensional). When there exists a preferred direction, the tensor is called *anisotropic*. Two different anisotropic behaviors exist for three-dimensional tensors. *Linear anisotropy* corresponds to one large and two small eigenvalues: $\lambda_1 \gg \lambda_2 \approx \lambda_3$. In this case, a single preferred direction exists. *Planar anisotropy* corresponds to two large and equal eigenvalues and a small third eigenvalue: $\lambda_1 = \lambda_2 \gg \lambda_3$. In this case, the restriction of the tensor to its 2D eigenspace is isotropic. To characterize the anisotropy of a tensor, several measures have been devised that find direct applications in biomedical tensor visualization. Basser and Pierpaoli introduced the notion of *fractional anisotropy* [85] defined as:

$$\mathrm{FA} = \sqrt{\frac{3}{2}} \frac{\|T - \mu_1 I_3\|}{\|T\|},$$

where $\mu_1 = \frac{1}{3}\sum_i \lambda_i$ is the mean of the eigenvalues, I_3 is the identity map, and $\|.\|$ designates the Frobenius norm. The term $T - \mu_1 I_3$ is obtained by subtracting the isotropic part of tensor T and is called its *deviator*. FA measures the anisotropy of a given tensor but does not differentiate between linear and planar anisotropy.

To address this limitation, Westin et al. proposed three metrics that quantify linear and planar anisotropy, as well as isotropic behavior [86].

$$c_l = \frac{\lambda_1 - \lambda_2}{\lambda_1 + \lambda_2 + \lambda_2}, \qquad c_p = \frac{2(\lambda_2 - \lambda_3)}{\lambda_1 + \lambda_2 + \lambda_3}, \qquad c_s = \frac{3\lambda_3}{\lambda_1 + \lambda_2 + \lambda_3}$$

Observe that eigenvalues as well as quantities derived from them are invariant under changes of the reference frame. As such they correspond to intrinsic properties of the tensor and are of primary interest for visualization purposes.

4.2. Color coding of major eigenvector orientation

A solution to visualize the directional information associated with the major eigenvector consists in using a spherical color map. The coordinates of the major eigenvector in the reference frame of the laboratory are mapped to red, green, and blue color values. Because of the indeterminacy of an eigenvector's direction, both e_1 and $-e_1$ are mapped to the same color. Additionally, the orientation of the major eigenvector is poorly defined when the eigenvalues approach planar anisotropy. Therefore, the saturation of the color is set with respect to the value of the linear anisotropy metric c_l, which measures the significance of the major eigenvector as a means to assess the anisotropy of the tensor. Pajevic and Pierpaoli analyzed the perception issues associated with different color maps following this basic principle [87].

4.3. Tensor glyphs

At the small scale, glyphs are a staple of tensor visualization because they provide an intuitive display of individual tensor samples by mapping the eigenvalues and eigenvectors of each tensor to the shape and orientation of a geometric element such as a box or ellipsoid [3]. Observe that as opposed to diffusion or conductivity, tensors whose eigenvalues are positive by definition, strain tensors can have both positive and negative eigenvalues, corresponding to stretch and compression respectively. Therefore, Wünsche and Young [83] suggested to alternate between a red and a blue color coding mapped onto ellipsoid glyphs to disambiguate the sign of the eigenvalues. Data inspection with glyph-based methods represents the first phase of understanding structures in a tensor field. The use of traditional glyphs for diffusion tensor visualization has a number of drawbacks. While ellipsoidal glyphs are mathematically simplest because their symmetry can directly convey a tensor eigensystem, their round shape may not effectively indicate vital shape and orientation cues, a problem that may lead to ambiguous displays of disparate tensor values. Cuboid glyphs, with their sharp edges, avoid this problem, but, in the case of eigensystem symmetry (eigenvectors are not unique when two or more eigenvalues are nearly equal), they often depict an arbitrary directional information, which is easily influenced by noise.

Westin et al. proposed to address that limitation by mapping the metrics c_l, c_p, and c_s introduced previously to a line, a disk, and a sphere primitive combined in the same glyph [88]. Kindlmann recently presented an alternative strategy

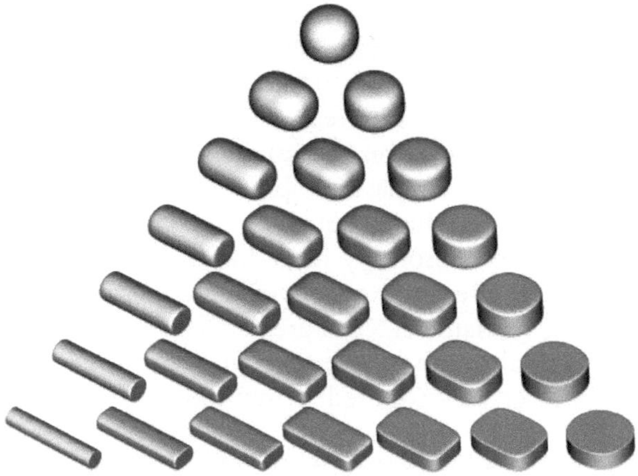

Figure 26 Superquadric tensor glyphs smoothly blend between cylinders, boxes, and spheres. From Kindlmann [90].

employing a glyph geometry that changes according to the tensor eigensystem, emphasizing directional information only where it is numerically well-defined and otherwise reverting to symmetrical configurations [89]. A technique of solid modeling called *superquadrics* allows him to represent cuboids, ellipsoids, cylinders, and all intermediate shapes. Figure 26 demonstrates the palette of super quadric glyphs, in which the extremes of spherical, linear, and planar anisotropy occur at the corners of the triangle.

Figure 27 demonstrates the differences between ellipsoidal and superquadric glyphs as used to visualize tensor field data from an axial slice of a diffusiontensor MRI dataset. In the left side of Figure 27, many of the ellipsoidal glyphs appear

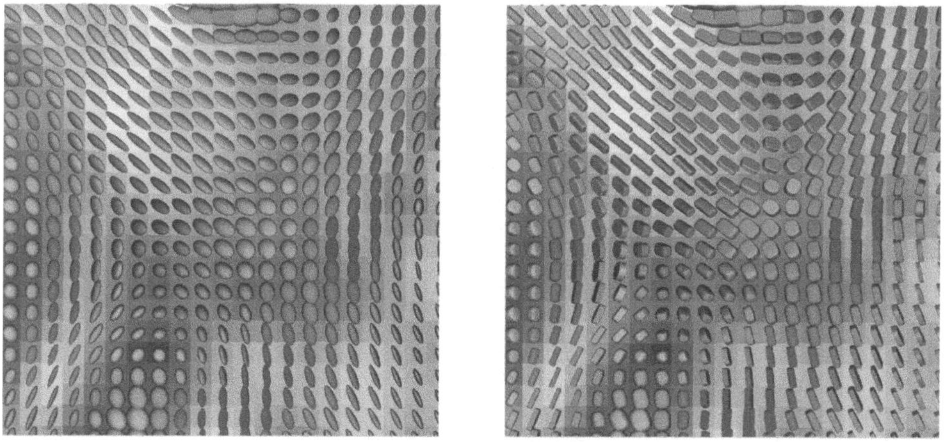

Figure 27 Ellipsoidal (left) and superquadric (right) glyph visualizations of an axial slice of a DT-MRI scan. (See color plate 22).

similar, while the shape differences are more clearly conveyed with the superquadric glyphs in the right side of Figure 27. The superquadric glyphs additionally clarify directional patterns. Ideally, the composition of multiple glyphs from across the tensor field can hint at larger scale features that may be subsequently explored and extracted with other visualization and analysis techniques.

As with vector visualization, simple attempts at indicating all the tensor variables at all sample locations rapidly produces unintelligible visual clutter. In DT-MRI of nervous tissue, the degree of anisotropy has a biological significance relating to the white matter structure, so an effective way to avoid clutter is to display only those tensors that exhibit anisotropy of a certain degree or greater. This strategy was used in Figure 28.

Alternatively, the visualization can be restricted to a single slice. For this type of representation, Laidlaw et al. proposed a method to normalize the size of the

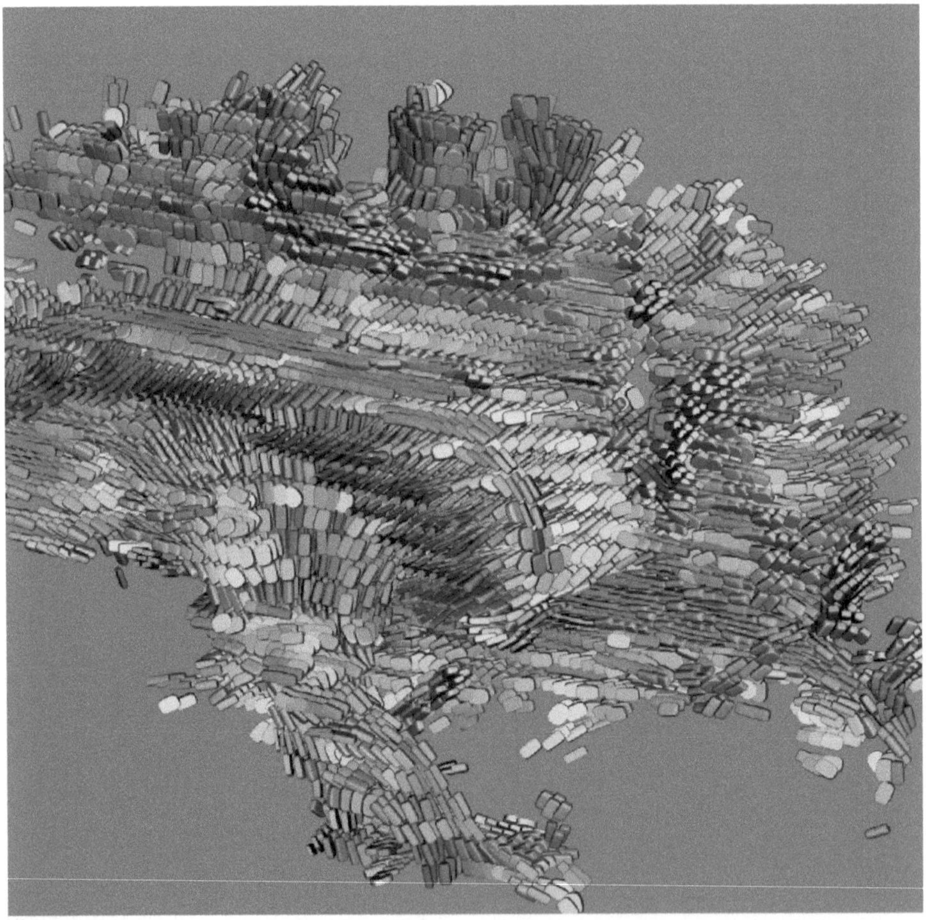

Figure 28 Visualization of half a brain DT-MRI volume using superquadrics glyphs. Glyphs are depicted only in anisotropic voxels. Red indicates left/right fiber orientation, green indicates anterior/posterior, and blue indicates superior/inferior. Image courtesy of Gordon Kindlmann. (See color plate 23).

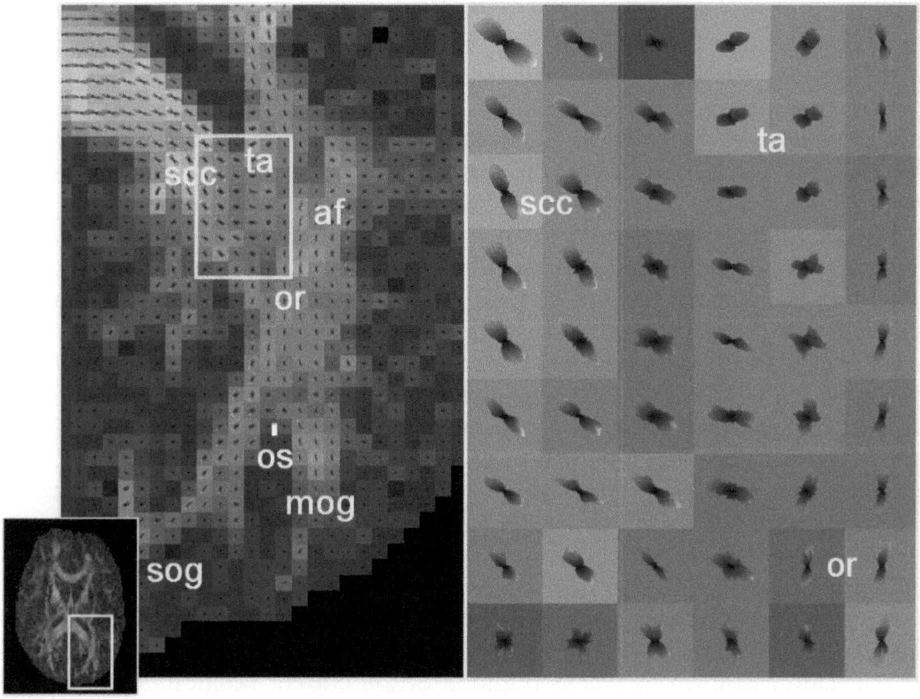

Figure 29 Q-ball image showing intravoxel fiber crossing in the major forceps. Figure courtesy of D.S. Tuch. (See color plate 24).

ellipsoids and another method that leverages techniques from oil painting to offer a complete view of the tissue anisotropy on a given slice [90].

While the symmetric tensor is a popular representation for the local diffusion, it is a simplification of the underlying biophysics. In reality, how water diffuses from a local region varies continuously over the space of directions. A tensor representation approximates that continuous distribution with an ellipsoidal shape parameterization. However, with the emerging field of high angular resolution diffusion (HARD) imaging, more complicated shapes can be measured and represented. HARD images reveal complex patterns at fiber-crossing locations within the brain. This intravoxel heterogeneity is demonstrated in Figure 29) [91,92].

In these images, the non-ellipsoidal shape of each glyph is exaggerated by subtracting the isotropic component from the local measurements.

4.4. Fiber tractography

In diffusion tensor MRI, a popular technique of feature extraction is fiber tractography [93–95]. In DT-MRI of nervous tissue, fiber tractography seeks to create models of the pathways of axon bundles in white matter tracts. In muscle tissue, such as the myocardium of the heart, tractography can illustrate the directional structure of the muscle fibers. Standard methods of tractography are essentially applications of hyperstreamlines [83,96], in which pathways are numerically integrated along the direction of the principal eigenvector of the tensor samples. The biological justification for this lies in how the direction of greatest diffusivity, the principal eigenvector of the tensor, should indicate the orientation of the fibrous microstructure of the underlying tissue. However, noise in the data, inadequate resolution, or complex structures may lead to problems in the visualization. In such cases, the direction of the principal eigenvector changes suddenly or is not numerically well defined.

To address this, several generalizations of standard fiber tractography have been proposed. *Tensorlines* [97] add an advection term to the diffusion equation in order to numerically stabilize the path integration. Zhang et al. created a visualization method that constructs tubular representations of fiber pathways in regions of strong linear anisotropy and surfaces everywhere tangential to the 2D eigenspace in regions of planar anisotropy [95]. Surfaces provide an intuitive depiction of the directional information associated with planar anisotropy while avoiding the numerical instability associated with the tracking of the ill-defined major eigenvector. The scheme computes both *streamtubes* and *streamsurfaces* everywhere in the domain. It then applies a decimation procedure that preserves the geometric structures only where they are most meaningful and prunes objects based on similarity measures. Zhukov and Barr presented a solution to the problem of tracking fiber orientation across noisy DT-MRI data [98,99]. Their approach uses a regularization method during the integration based on the technique of Moving Least Squares. It determines the local tracking direction as a combination of the orientation sampled in a neighborhood of the current position. The shape and extent of the neighborhood is determined by a gaussian weighting function that is scaled in each direction with respect to the anisotropy of the fiber direction determined in the previous iteration. The application of this method to white matter and cardiac tissue shows the robustness of this strategy and its ability to extract subtle pathways. Illustrations are proposed in Figures 30 and 31.

Alternatively, Hlawitschka and Scheuermann proposed an algorithm for fiber tracking in HARD tensor data called *HOT-Lines* [100]. Due to the higher angular

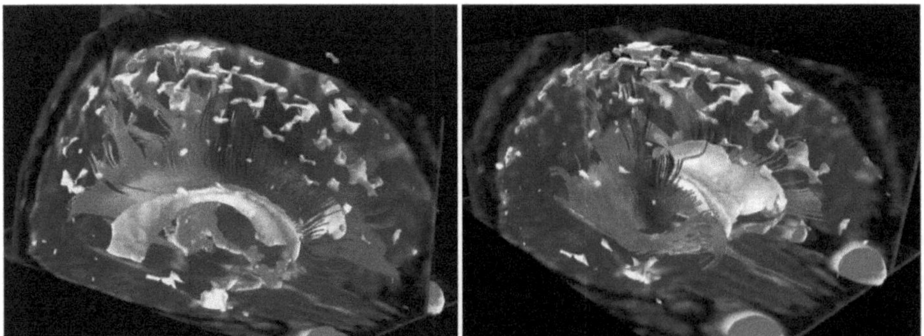

Figure 30 Brain structures: neural pathways computed using the MLS algorithm in the corpus callosum (left) and corona radiata (right) shown together with isotropic brain structures – ventricle, eye sockets, and pockets of CSF on the top of the brain. Cutting planes show isotropic c_s values. Images courtesy of Leonid Zhukov [98]. (See color plate 25).

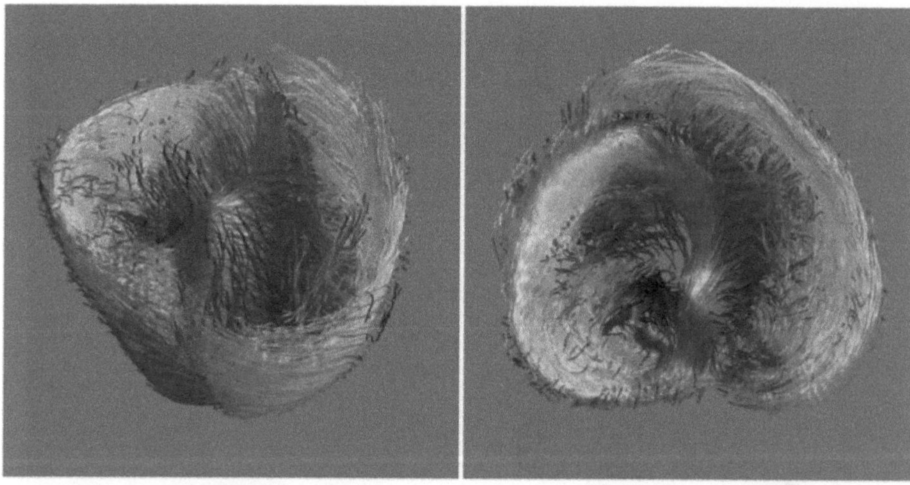

Figure 31 Reconstruction of heart muscle fibers using the MLS algorithm. The color coding changes smoothly from clockwise to counterclockwise spiral oriented fibers. Horizontal parts (very small pitch angle) of the fibers are shown in white. This coloration is consistent with observations of some heart researchers, who have described a systematic smooth variation in pitch and direction of heart muscle fibers from endocardium to epicardium. Images courtesy of Leonid Zhukov [99]. (See color plate 26).

resolution, crossings and bending of neural fibers within a single voxel can be detected which is not possible using second-order tensor models. The method computes the fiber directions by means of a gradient descent algorithm and therefore extracts fiber pathways that fully leverage the high angular accuracy available in the input data.

A natural extension of fiber tractography consists in grouping individual pathways together to yield a segmentation of the white matter into fiber bundles. From a visualization standpoint, the result of this clustering task can also be leveraged to provide a color coding of the fibers that enhance the interpretation of their spatial structure and their anatomical significance [101]. Some methods require the user to guide the clustering, typically by specifying one or several regions of interest (ROI) [94,102] in order to identify the fibers that intersect them. In the case of several ROIs, boolean logic can be applied to determine the set of fibers that connect known regions of the brain anatomy [103–105]. Clustering can also be achieved in a non-supervised way.

To that end, Brun et al. proposed to represent each fiber pathway in a high-dimensional space that embeds the mean vector of the fiber as well as its covariance matrix [106]. In that space, the authors apply so-called *normalized cuts* that iteratively split the set of fibers into halves until some homogeneity criteria are met in each cluster. The result of their method is shown in Figure 32. Alternatively, Zhang and Laidlaw used an agglomerative hierarchical clustering method [107].

4.5. Volume rendering

Beyond its traditional application to the visualization of 3D scalar fields, direct volume rendering has also been extended to tensor fields in several ways. Wenger et al. [108]

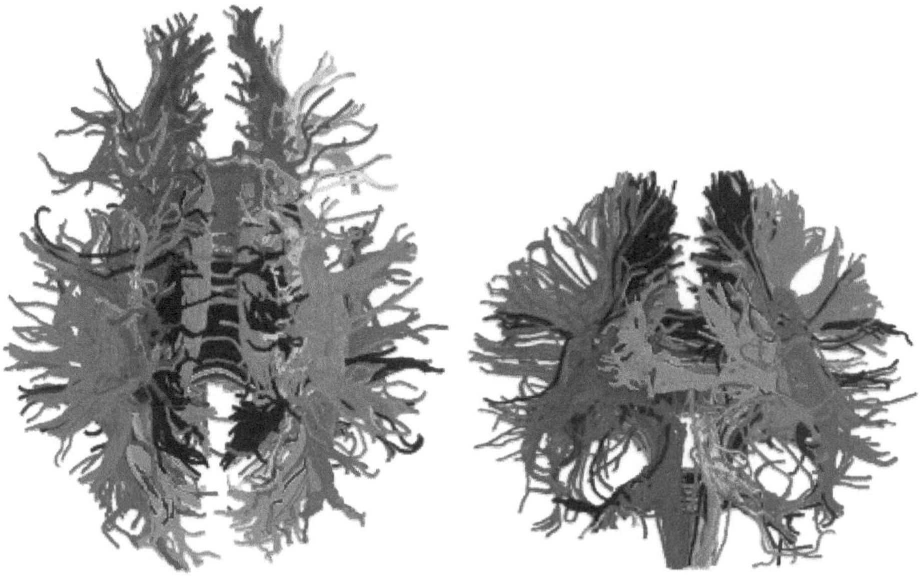

Figure 32 **Left:** Axial view of a segmentation obtained from recursive bipartitioning of the white matter fiber traces. The color of fiber traces indicate cluster membership. **Middle:** coronal view. Images courtesy of Carl-Fredrik Westin [106]. (See color plate 27).

generalized ideas previously used to visualize vector fields to achieve the volume rendering of diffusion tensor fields. The basic ingredients of their representation are threads that depict the directional information contained in the data combined with halos that enhance depth perception and whose color and opacity can be varied to encode a scalar measure of anisotropy. Following a different approach, Kindlmann and Weinstein proposed to use the barycentric coordinates of anisotropy measure introduced in Section 4.1 to control the opacity of the data volume [109,110], see Figure 33.

The additional use of a standard color map indicating the orientation of the major eigenvector helps disambiguate distinct structures of the white matter. The same authors also designed a method called *Lit-Tensors* that provides visual clues about the type of anisotropy and the corresponding orientation. Their idea consists in introducing an additional control parameter in the standard Blinn-Phong shading model that allows for a smooth transition between shading models associated with line and surface objects, respectively. In that way, regions of linear anisotropy are shaded like illuminated streamlines [111] and regions of planar anisotropy are shaded like surfaces, while intermediate anisotropy types are shaded according to a combination of these two extrema. Moreover, this method can be combined with a shading based on the gradient of an opacity function alone, which ensures a consistent shading across a feature of interest (refer to Figure 34).

In his PhD thesis, Kindlmann [112] also described a volume-rendering method based on the analytical computation of the derivatives of a tensor invariant like FA. Leveraging smooth reconstruction kernels for the computation of the tensor

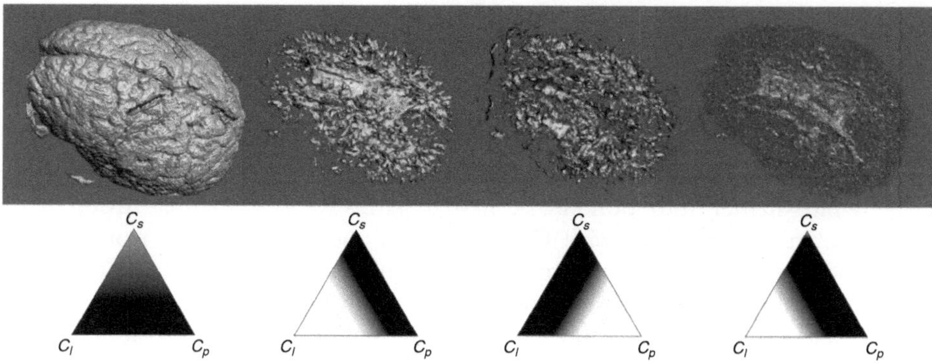

Figure 33 Examples of barycentric opacity maps and resulting volumes. Brighter regions in the triangle correspond to higher opacity assignment. From Kindlmann and Weinstein [110].

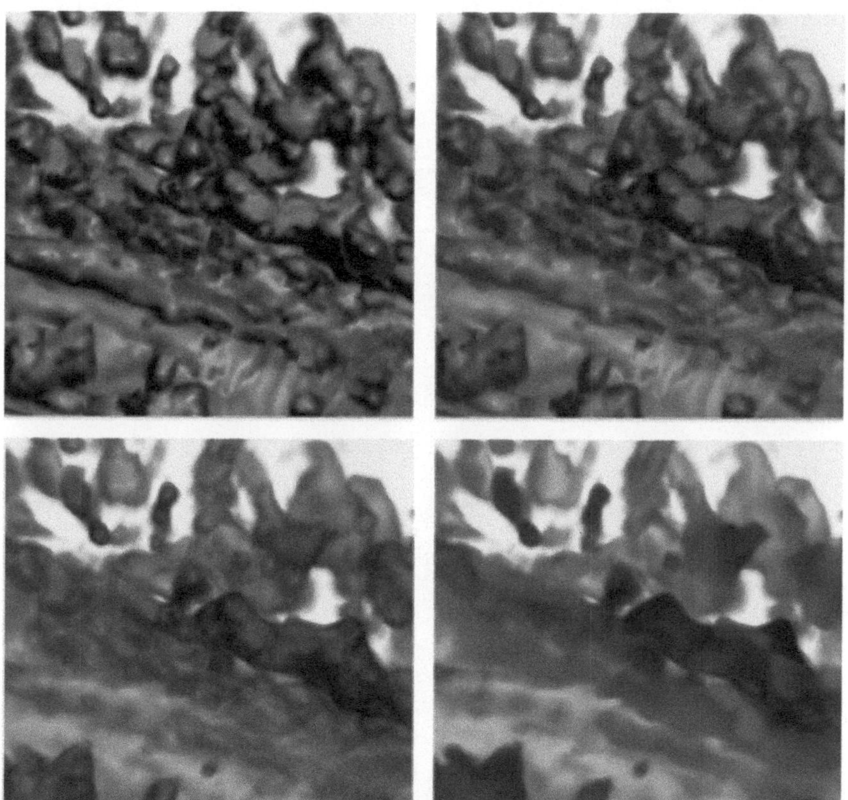

Figure 34 Mixing between lit-tensor and opacity gradient shading for a portion of brain data. Images from Kindlmann and Weinstein [110]. (See color plate 28).

invariant and its derivatives, the method is able to generate volume renderings of extremely high quality that effectively captures anatomical structures.

4.6. White matter segmentation using tensor invariants

We mentioned previously that fiber tractography can serve as a basis for segmentation of the white matter into bundles of neural pathways that exhibit strong spatial similarities. In collaboration with G. Kindlmann and C.-F. Westin from Harvard Medical School, we have recently proposed a compelling alternative to that approach [113] which does not require the potentially error-prone integration of fiber tracts in noisy tensor data. Leveraging concepts from the field of computer vision, we have devised a method that computes a segmentation of the white matter tissue through crease surfaces of the tensor invariant FA. The premise of this technique is that surfaces where the anisotropy (as measured by FA) is locally maximal (ridges) correspond to the 2D core of a fiber bundle, while surfaces where FA is locally minimal (valleys) coincide with the boundary between adjacent but distinctly oriented fiber tracts. The algorithm makes use of cubic separable reconstruction kernels so as to permit the analytical computation of the smooth second-order derivatives needed to identify crease surfaces. The results demonstrate the anatomic relevance of the structures extracted in that way. An illustration is proposed in Figure 35.

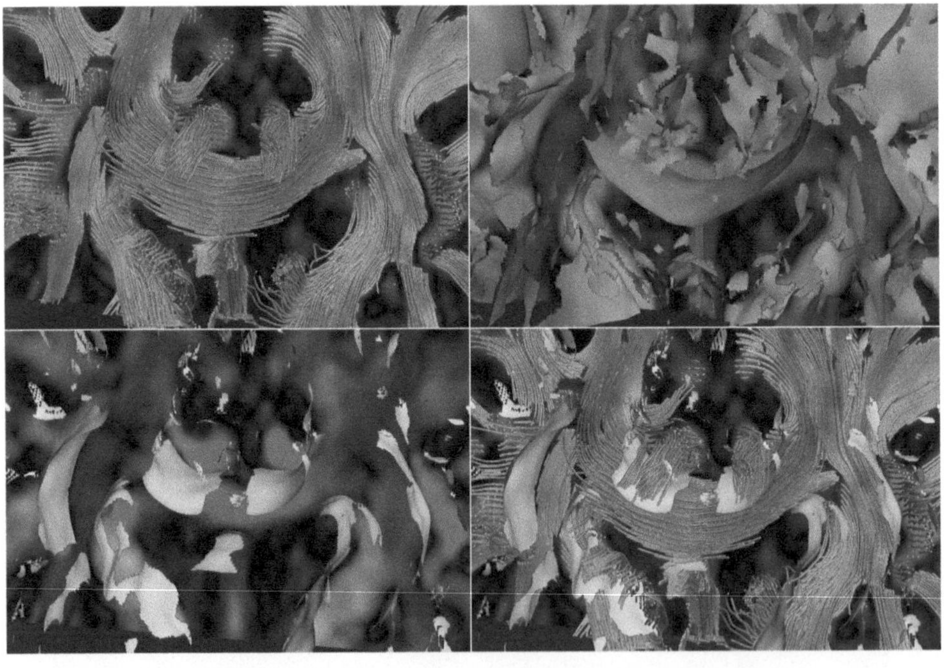

Figure 35 Anisotropy creases near the corpus callosum. **Top left**: RGB map with fibers. **Top right**: Ridge surfaces. **Bottom left**: Valley surfaces. **Bottom right**: Valleys with fibers. From Kindlmann [113]. (See color plate 29).

5. MULTI-FIELD VISUALIZATION

Computational field problems, such as occur in computational fluid dynamics (CFD), electromagnetic field simulation, and essentially any investigation of a phenomenon whose physiology and physics can be modeled effectively by ordinary and/or partial differential equations, are frequently encountered within biomedical computing. For this reason, visualization researchers have concentrated on developing effective ways to visualize large-scale computational fields. The output of such simulations might be a single field variable (such as voltage, pressure, or velocity) or a combination of variables involving a number of scalar, vector, and/or tensor fields. However, much of current and previous visualization research has focused on methods and techniques for visualizing a single computational field variable (such as the extraction of an isosurface from a single scalar field). While visualizations of a single field are sometimes sufficient for the achievement of some biomedical research goals, effective simultaneous visualizations of multiple fields would greatly enhance many projects.

For this reason, biomedical computing is ripe for research in what we will term "multi-field" visualization [114], in which combinations of the above fields are visualized together to elucidate their *interactions*. The challenges for such multi-field visualizations are many and include large-scale data, complicated geometries, and heterogeneous and anisotropic material properties. Below are two examples of multi-field visualizations. Both examples illustrate the challenges involved in providing a researcher with intuitive and useful visual feed-back.

In Figure 36, we give a simple example of a multi-field visualization from the simulation of electric current flow within an anisotropic media. The sample volume has Dirichlet (± 1 Vs) boundary conditions on the opposite sides of the cube (orthogonal to the plane of view) and Neumann zero flux boundary conditions on all remaining sides. The media is described by a single conductivity tensor with all non-zero elements. The result is very unintuitive; the isosurfaces are not parallel to the sides of the cube and current lines are not orthogonal to these isosurfaces. Electric field lines will still be orthogonal to the isosurfaces but will not be parallel to the sides of the cube.

In another example of multi-field visualization, Figure 37 shows the results of a large-scale finite element simulation of the distribution of electric current flow and voltage within an inhomogeneous model of the human head and brain. The image shows a combination of an isovoltage surface and streamlines of current flow within the context of the magnetic resonance image scans and a geometric head model.

In Figure 38, we see a multi-field visualization from a computational fluid flow simulation.

Another multi-field visualization idea is to use a combination of scalar and vector field visualization using glyphs for time-dependent data as shown in Figures 39 and 40, which illustrate the electrical potential along with the current on the cardiac surface during excitation and repolarization, highlighting regions of in- and outflow.

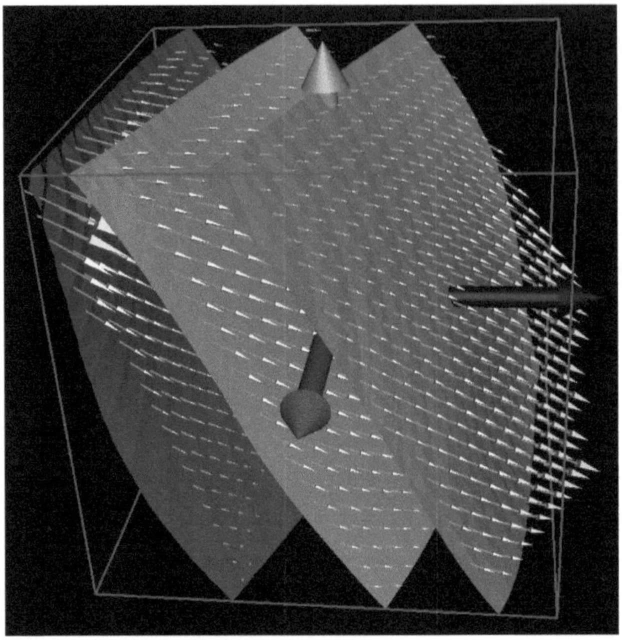

Figure 36 Electric current flow within an anisotropic media. Note how current is not orthogonal to the isovoltage surfaces because of the anisotropic conductivity.

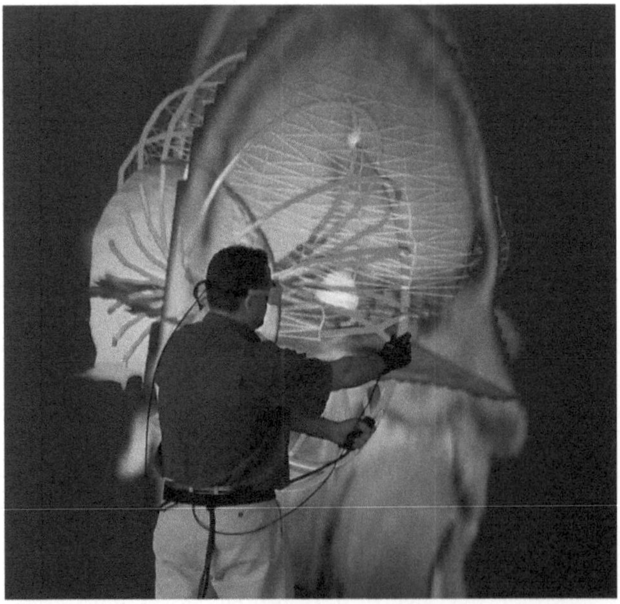

Figure 37 Electric current flow within the brain arising from a localized source, visualized in an immersive environment. (See color plate 30).

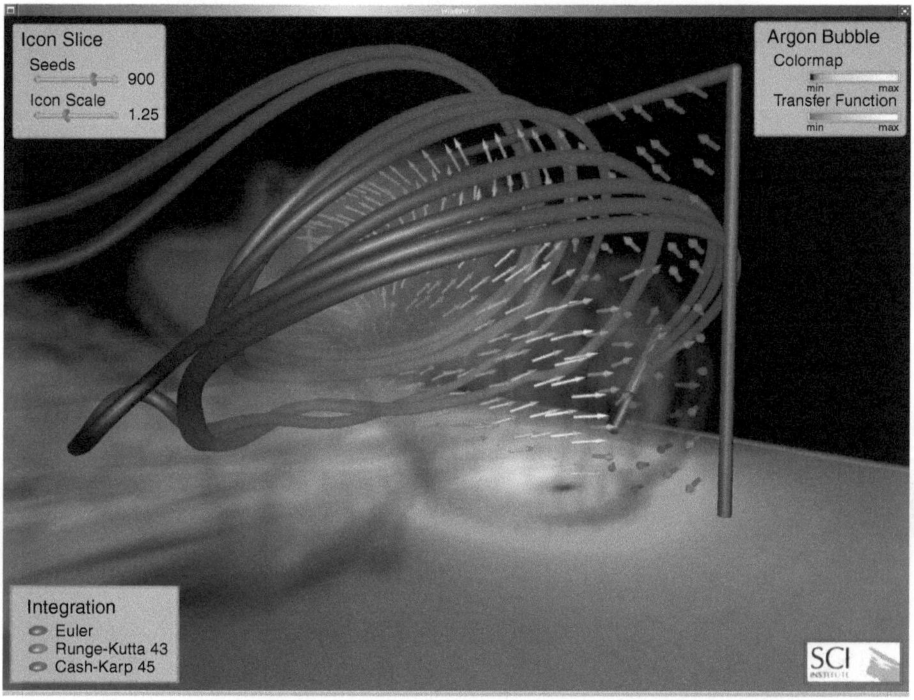

Figure 38 This figure shows a multi-field visualization from a computational fluid flow simulation. Here we are visualizing two scalar fields by using volume rendering and the cutting plane in the *x-y* plane. We are also visualizing two vector fields using streamlines and arrows in the *x-z* cutting plane. Data courtesy of Lawrence Berkeley National Laboratory.

Another compelling example of the need and use of multi-field visualization arises from EEG inverse simulations. One of the open questions in inverse EEG source localization is what effect the anisotropy of white matter has on source localization accuracy. We applied a combination of stream surface visualization with simple tensor field visualization to look at the effects of including anisotropy within a realistic head model for the EEG source localization simulation. Figure 41 illustrates the effects of white matter anisotropy using these visualization techniques. We can observe a correlation between the primary direction of the conductivity structure of the white matter fiber bundles and the direction of the return currents. The visualization of return currents in bioelectric field problems can reveal important details about the distribution of sources, interactions at conductivity boundaries, and the effect of geometric distortion on bioelectric fields. By integrating the stream surfaces with a visualization of the diffusion tensors representing the white matter, we can better understand the relationship between structural, spatial relationships [115,116].

For generations, artists have attempted to convey complex and multiple ideas and images via limited channels. Painters, for example, use different brush strokes and texture in layers to add depth and structure to oil paintings. In a similar way, we

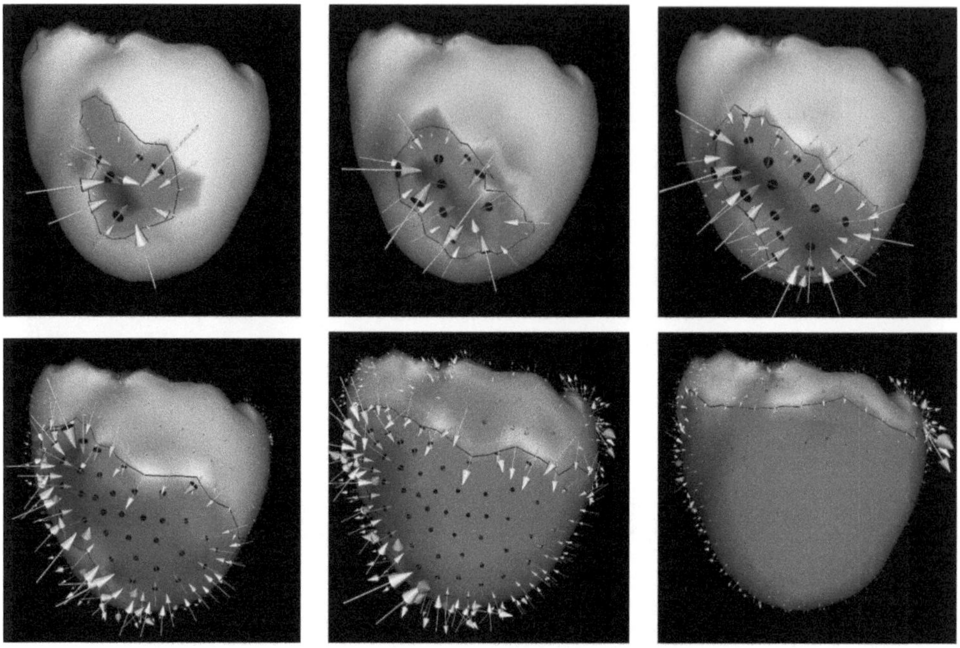

Figure 39 Combined scalar and vector visualization I. The image shows the electrical potential and the electrical current over the cardiac surface during excitation, at 10, 20, 30, 40, 50, and 60 ms, respectively. The black lines indicate the boundary between regions of in- and outflow of the electrical current. Data and image courtesy of Bruno Taccardi and Frank Sachse of the Cardiovascular Research and Training Institute. (See color plate 31).

will employ creative methods to increase the number of channels that can be presented clearly in visualization. A successful example of the application of a painterly method applied to a scientific dataset is shown in Figure 42. The authors, Kirby et al. [117], use a combination of discrete and continuous visual elements arranged in multiple layers to represent the data and additional mathematically derived quantities such as derivatives or eigenvalues and eigenvectors of a diffusion tensor. Figure 42 shows a visualization of an airflow simulation based on this layered approach, which presents six different quantities in a compact way. In the bottom layer, the area of the ellipses represents divergence, while eccentricity encodes shear. The arrow layer conveys the velocity and speed via the direction and area, respectively, of the arrows. Finally, the ellipses encode the vorticity through their color, texture, and contrast.

These are just a few examples of our initial ideas and research in the emerging field of multi-field visualization. From a scientific perspective, multi-field visualization is valuable because it can generate novel hypotheses about the relationships between the underlying physical or biological quantities. Successful multi-field visualization requires visual transparency in all the constituent field visualizations. In the 3D domains of biomedical researchers, it will be a challenge to generate such transparency, especially through techniques that are also intuitive and interactive.

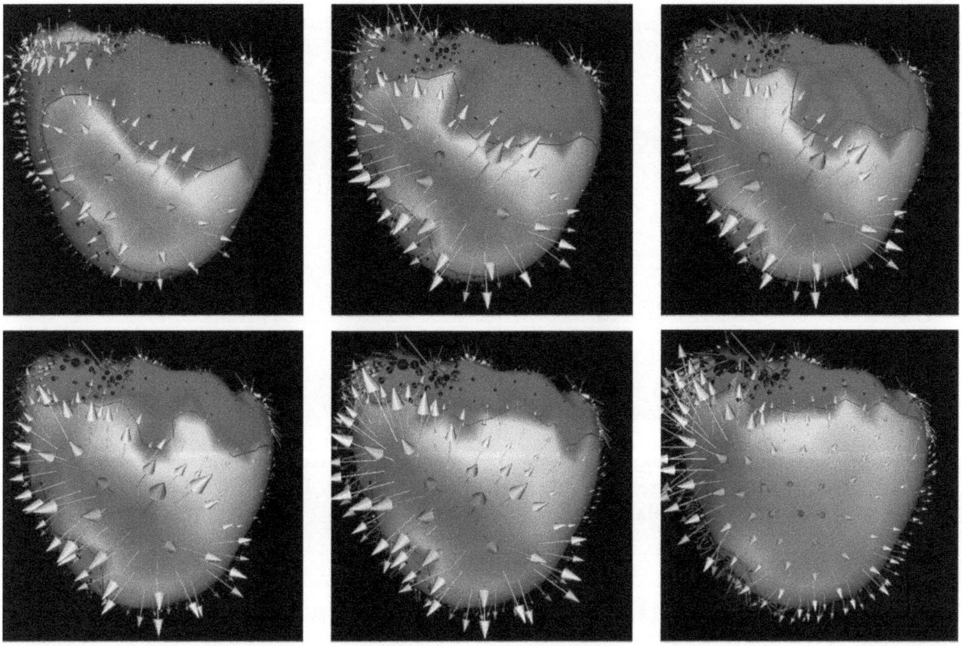

Figure 40 Combined scalar and vector visualization II. Electrical potential and electrical current over the cardiac surface are shown during repolarization, at 110, 130, 150, 170, 190, and 210 ms, respectively. Data and image courtesy of Bruno Taccardi and Frank Sachse of the Cardiovascular Research and Training Institute. (See color plate 32).

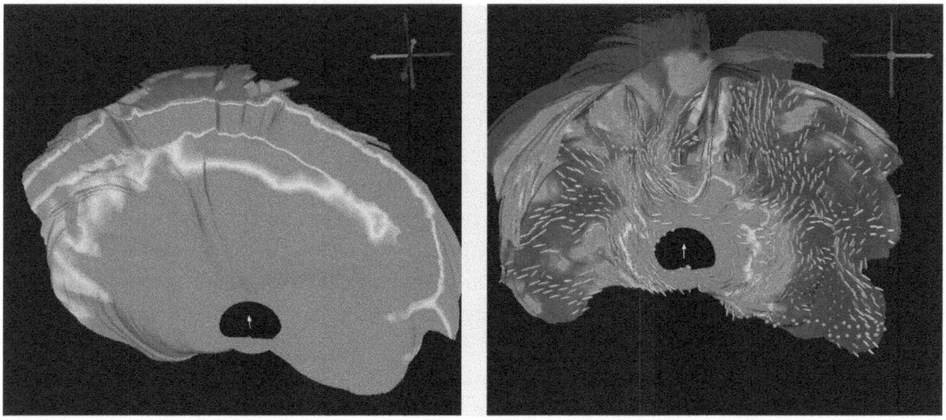

(a) (b)

Figure 41 Visualization of return current surfaces from an EEG simulation using a deep thalamic source. The two visualizations show the results of the simulation using an isotropic model (a) and a model including anisotropic white matter (b). (See color plate 33).

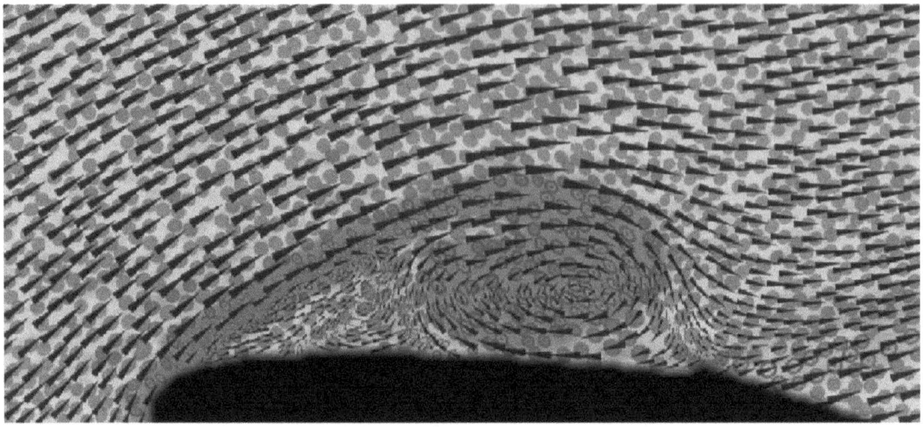

Figure 42 Visualization of experimental 2D flow past an airfoil, indicating velocity, speed, vorticity, rate of strain, divergence, and shear. Image courtesy R.M. Kirby and H. Marmanis and D.H. Laidlaw. (See color plate 34).

6. Error and Uncertainty Visualization

With few exceptions, visualization research has ignored the visual representation of errors and uncertainty for 3D visualizations. Certainly, this lack can be attributed partly to the inherent difficulty in defining, characterizing, and controlling comparisons between different datasets and partly to the corresponding error and uncertainty in the experimental, simulation, and/or visualization processes. In addition, visualization researchers have developed few methods that allow for easy comparison and representation of error and uncertainty in data for visualizations. To make current visualization techniques and software more useful to biomedical computing (and other) researchers, we need to incorporate visual representations of error and uncertainty [118].

A few visualization researchers have started thinking about how to create effective 3D visual representations of errors and uncertainties, the sources of which can include uncertainty in acquisition (instrument measurement error, numerical analysis error, statistical variation), uncertainty in the model (both in mathematical and in geometric models), uncertainty in transformation (where errors may be introduced from resampling, filtering, quantization, rescaling), and uncertainty in visualization. (A useful overview of uncertainty definitions can be found in Taylor and Kuyatt [119].) Pang et al. [120] have summarized a variety of techniques that might be used for uncertainty visualization. These techniques range from adding or modifying the geometry with, for example, a bump map or altered lighting attributes, to using textures. Perhaps the most interesting technique proposed is the use of blurring, as shown in Figure 43. Grigoryan and Rheingans [121], on the other hand, used point-based primitives instead of blurring to create a fuzzy surface that achieves similar results (Figure 44) Blurring is a natural cue to the eye that something is amiss. This technique can easily be applied to a variety of visualization techniques from particle tracing to isosurfacing, as we describe later on.

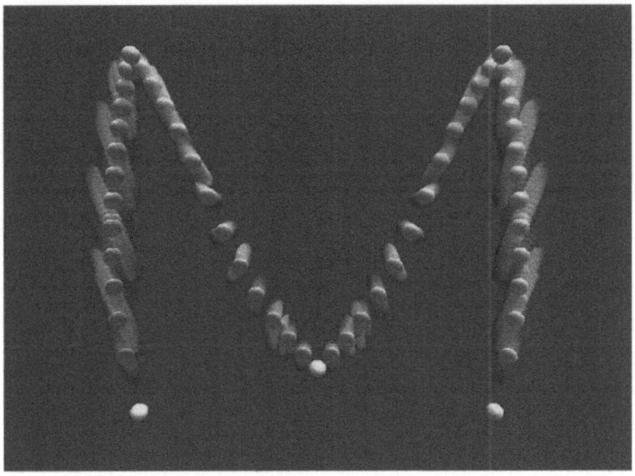

Figure 43 Particle tracing using blurring to show the uncertainty in the path. Used with permission of the author. (See color plate 35).

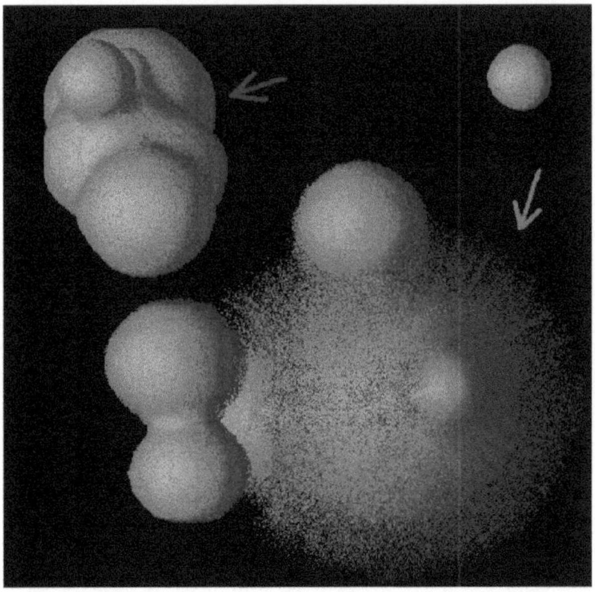

Figure 44 Point-based primitives used to create a fuzzy surface to show uncertainty. Used with permission of the author.

Our reaction-diffusion visualization technique described in Section 3.4 can be modified to create a way to represent uncertainty in the vector field. By allowing the amount of anisotropy in the reaction-diffusion model to vary, we embed another variable that can be visually mapped and controlled. For example, let us assume that we have a uncertainty in the vector field associated with orientation.

We can use the anisotropy tensor within the reaction diffusion equation to help create a visual mapping of uncertainty to the resulting spot pattern. When the amount of anisotropy within the reaction diffusion equation is small (corresponding to a high value of uncertainty), we can cause the spot to take an almost circular shape, with the ratio of the semi-axes being approximately one. However, when the anisotropy within the reaction diffusion equation is high (corresponding to a low value of uncertainty), we can cause the spot to take the shape of an ellipse, deforming at times so extremely that it appears almost as a thick line i.e., the ratio of the semi-axes becomes very large. This creates a visual difference very well suited to mapping an orientation uncertainty. When the orientation uncertainty is very small, the spot is elliptical, reflecting a precise orientation. When the uncertainty is very high, the spot is more circular, reflecting the uncertainty in the orientation. This is demonstrated in Figure 45 on a synthetic circular field where the uncertainty in the direction is a function of the angle [68,122].

One simple example of the error and uncertainty visualization techniques combines isosurface methods with volume-rendering methods. For example, one can represent the average value of a scalar field with an isosurface and then represent the error or uncertainty of the scalar field using volume rendering, as shown in Figures 46 and 47.

Most recently we have proposed an alternative approach to encoding uncertainty in volume rendering visualizations [123]. Our method offers an intuitive means to interactively select the transfer function for a precomputed fuzzy classified

Figure 45 Reaction diffusion vector field model visualization of orientation uncertainty. A more elliptical pattern represents low orientation uncertainty, while a more circular pattern represents a higher uncertainty in the orientation of the vector field.

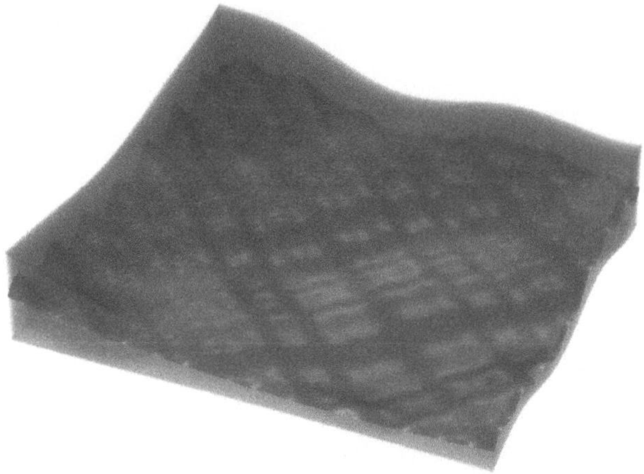

Figure 46 An isosurface of a synthetic dataset (in gray) is bracketed with volume-rendered regions (red and green) indicating uncertainty around the isovalue. (See color plate 36).

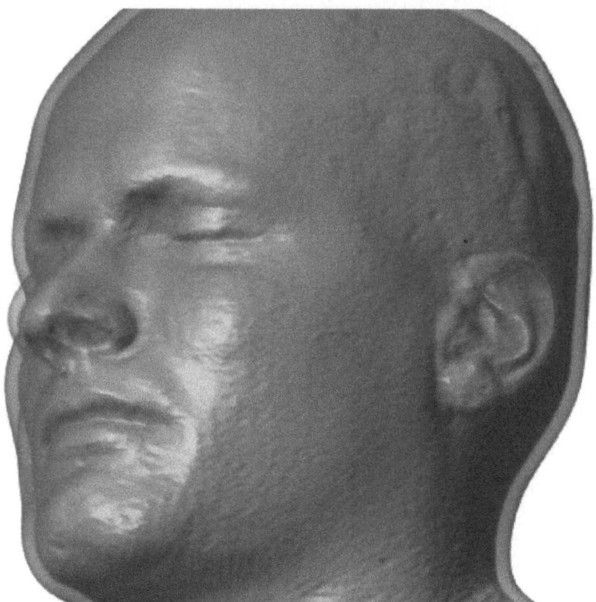

Figure 47 An isosurface of an MRI dataset (in orange) is surrounded by a volume-rendered region of low opacity (in green) to indicate uncertainty in surface position. (See color plate 37).

volume data. One key feature of this new method is to permit the manipulation of the transfer function directly in the feature space. The user can therefore interact with the visualization at a semantic level. Moreover, the uncertainty present in the fuzzy classification as a result of the mixture of neighboring material

Figure 48 A Comparison of transfer function-based classification versus data-specific probabilistic classification. Both images are based on T1 MRI scans of a human head and show fuzzy classified white matter, gray matter, and cerebro-spinal fluid from Kniss et al. [123]. **Left**: Results of classification using a carefully designed 2D transfer function based on data value and gradient magnitude. **Right**: Visualization of the data classified using a fully automatic, atlas-based method that infers class statistics using minimum entropy, non-parametric density estimation [124]. (See color plate 38).

intensities at each sample point can be properly integrated in the volume-rendering representation to draw user attention to the lack of a clear-cut boundary between homogeneous materials. The benefits of this approach are illustrated in Figure 48.

7. Visualization Software

There are a variety of commercially available and research-based general visualization systems that may be useful for biomedical visualization (see Hansen and Johnson [4] for an overview of visualization systems). While certainly not an exhaustive list, examples of these systems are

Amira: Amira is a professional image segmentation, reconstruction, and 3D model generation application produced by Mercury Computer Systems GmbH (www.amiravis.com). It is used by research and development groups in chemistry, biology, medicine, material science, etc. Amira is designed to handle confocal microscopy, MRI, or CT data. It uses the tcl language as a command interface for scripting and is built on top of the OpenGL and Open Inventor toolkits. Modules can be developed to extend the Amira system and can use parallelization techniques if the developer so desires.

IDL: Research Systems Inc.'s Interactive Data Language (IDL) (www.rsinc.com/idl) is used for data analysis and visualization and to rapidly develop algorithms and interfaces. IDL is a "Matlab"-like tool that interprets the commands entered by the user.

Micro View: Micro View is an open-source, freely-distributed 3D volume viewer. It is written in Python and uses VTK for its graphical capabilities.

OsiriX: OsiriX is an open-source image-processing software package dedicated to DICOM images produced by medical equipment (MRI, CT, PET, PET-CT, etc.) and confocal microscopy designed to run under OSX on the Mac. It offers many visualization modes including surface and volume rendering for 2D and 3D interrogation of the data. It is multi-threaded and takes advantage of graphic board acceleration.

VTK: VTK, The Visualization Toolkit (www.kitware.com), is a public domain visualization package that is widely used in both classroom settings and research laboratories. It provides general visualization capabilities for scalars, vectors, tensors, textures, and volumetric data. Written in C++, VTK includes Tcl, Python, and Java bindings for application development and prototyping. VTK contains some built-in parallelization pieces for both threading and MPI.

AVS: Advanced Visual Systems (www.avs.com) (AVS) was the first commercially available dataflow visualization system. AVS/Express provides general visualization capabilities for scalar and vector visualization. The Manchester England AVS Center provides user contributed models that add functionality to the commercial system. AVS/Express allows for C++/C/Fortran legacy codes to be integrated using a modular approach.

IBM Open Visualization Data Explorer: IBM Open Visualization Data Explorer, OpenDX (www.research.ibm.com/dx), is a public domain dataflow visualization system similar to AVS/Express. OpenDX as it is a community-based open software project.

Iris Explorer: Iris Explorer (www.nag.com) is another commercially available dataflow visualization system. The newest version allows modules to be optionally (off-line) compiled into a single executable. This allows the user to run the group as a single process thus using less CPU time and memory than required by the original separate modules. Individual modules may be written to use parallel processing, but this is not directly facilitated by Iris Explorer.

ANALYZE: The ANALYZE software system (developed at the Mayo Clinic) is a comprehensive commercially available software packages for 3D biomedical image visualization and analysis.

In this next section, we describe open source visualization software tools that have been developed at the SCI Institute.

7.1. SCIRun/BioPSE visualization tools

The SCIRun system has been a focus of research and development at the SCI Institute since 1995 [125,126]. Its framework for visualization, modeling, and simulation has been the test bed for much of our fundamental research in visualization techniques and their applications to real-world scientific problems. As such, it supports all the standard techniques of scientific visualization and our improvements upon them. This includes efficient extraction of isosurfaces (unstructured and structured grids) and volume rendering of scalar fields, streamlines, and hedge-hogs for vector fields, as well as of tensorlines and glyphs for tensor fields. The visualization can be interactively created and modified with data probe widgets located within the field geometry. In 2000, the SCIRun Problem-Solving Environment (PSE) was made available to the research community as open source software [127].

We show an example of the SCIRun/BioPSE software system for an electroencephalography (EEG) source localization application in Figures 49 and 50.

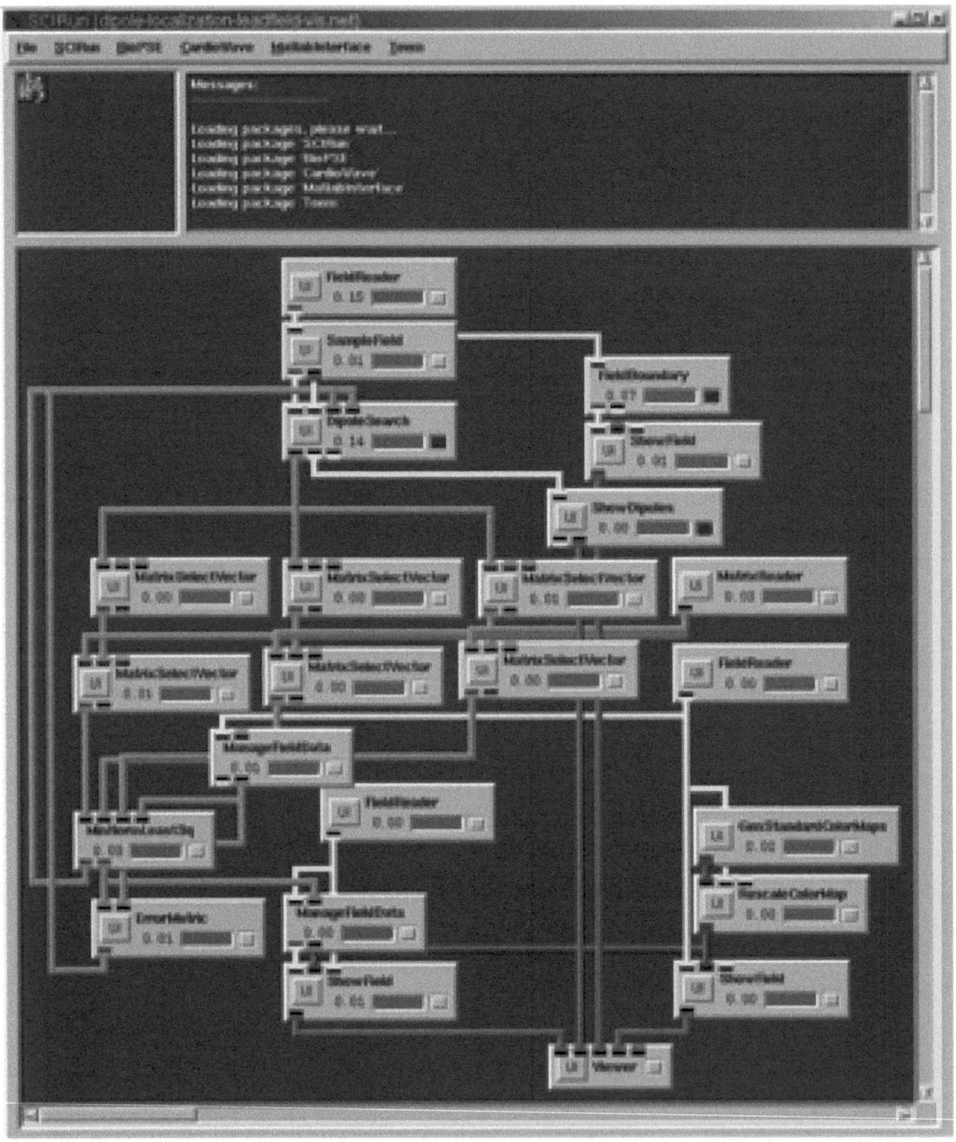

Figure 49 SCIRun/BioPSE modules combined for EEG modeling (unstructured mesh generation), simulation (finite element simulation, parallel linear system solves, and inverse source localization), and visualization (mesh visualization, isosurface extraction, and vector field visualization.

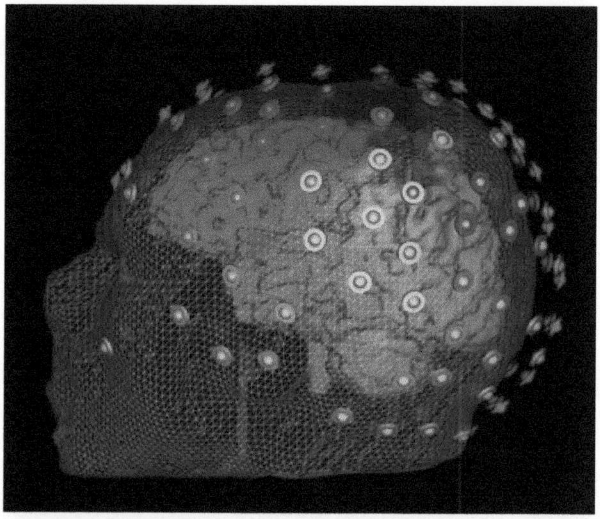

Figure 50 Visualization of simulation results of an EEG simulation localizing a neural source. (See color plate 39).

Figure 49 contains the dataflow network that implements an inverse EEG application. At the top of the network, the input data files are loaded; these include the finite element mesh that defines the geometry and conductivity properties of the model and a precomputed lead-field matrix that encodes the relationship between electric sources in the domain and the resulting potentials that would be measured at the electrodes. Further down in the network, we have a set of modules that optimize the dipole location in order to minimize the misfit between the measured potentials from the electrodes and the simulated potentials due to the dipole. Finally, we have visualization and rendering modules, which provide interactive feedback to the user.

7.1.1. SCIRun PowerApps

One of the major hurdles to SCIRun becoming a practical tool for the scientists and engineers has been SCIRun's dataflow interface. While visual programming is natural for computer scientists and some engineers, who are accustomed to writing software and building algorithmic pipelines, it is sometimes overly cumbersome for application scientists. Even when a dataflow network implements a specific application (such as the bioelectric field simulation network provided with BioPSE and detailed in the BioPSE Tutorial), the user interface (UI) components of the network are presented to the user in separate UI windows, without any semantic context for their settings. For example, SCIRun provides file browser user interfaces for reading in data. However, on the dataflow network, all of the file browsers have the same generic presentation. Historically, there has not been a way to present the filename entries in their semantic context, for example, to indicate

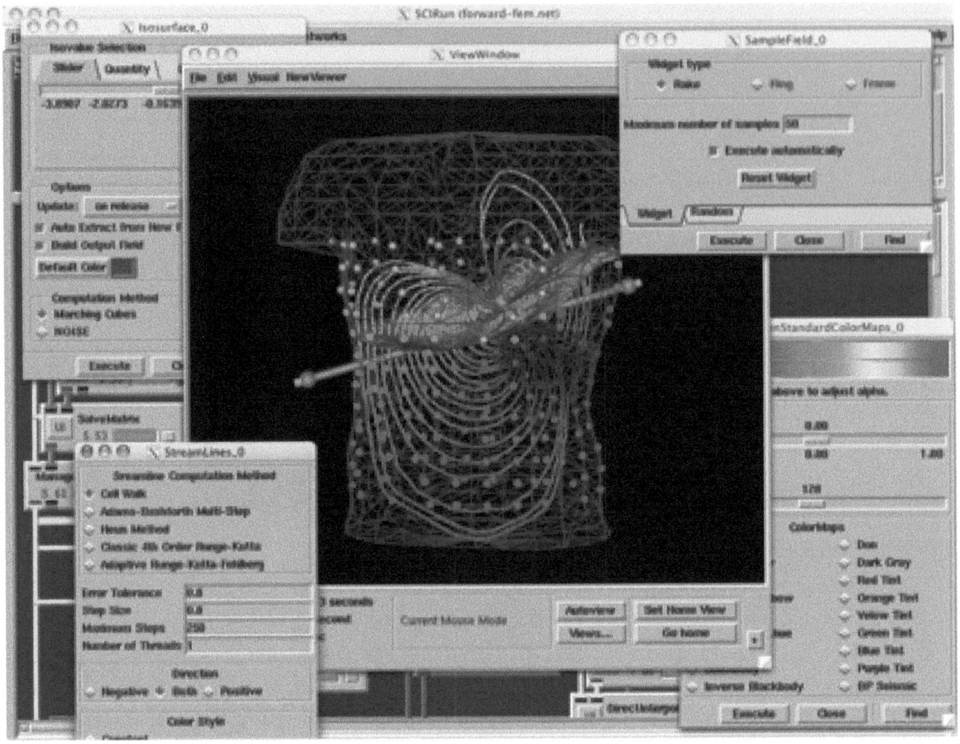

Figure 51 BioPSE dataflow interface for a bioelectric field application. (See color plate 40).

that one entry should identify the electrodes input file and another should identify the finite element mesh file. While this interface shortcoming has long been identified, it has only recently been addressed. With a recent release of BioPSE/SCIRun, we introduced PowerApps. A PowerApp is a customized interface built atop a dataflow application network. The dataflow network controls the execution and synchronization of the modules that comprise the application, but the generic user interface windows are replaced with entries that are placed in the context of a single application-specific interface window. A comparison of the dataflow version and the PowerApp version is shown in Figures 51 and 52.

7.2. map3d

A special area of visualization that arises in many biomedical domains is displaying sets of time signals that represent scalar values at numerous locations in space. Examples come from electrophysiology of the heart [128–136] and nervous system [137–139] and many others in which scientists and engineers deal with high spatial and temporal resolution of their measured or simulated data. The challenges of this form of visualization include representing both

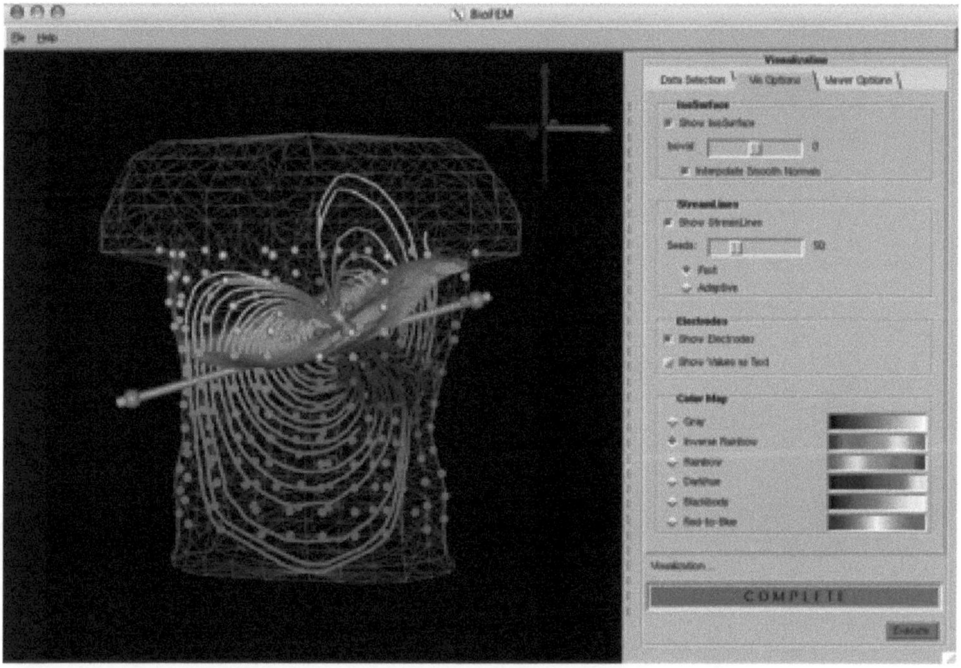

Figure 52 The BioFEM PowerApp custom interface. Though the application is functionality equivalent to the dataflow version shown in Figure 51, this PowerApp version provides an easier-to-use custom interface. Everything is contained within a single window; the user is lead through the steps of loading and visualizing the data with the tabs on the right; and generic control settings have been replaced with contextually appropriate labels; and application-specific tooltips (not shown) appear when the user places the cursor over any user interface element. (See color plate 41).

geometric shape and the associated data in a manner that is clear to the user, establishing quantitative, flexible coding and scaling of the data for detailed interactive analysis, and preserving efficiency so that a user may sort through very large (tens to hundreds of MBytes) datasets.

One approach to these visualization goals is *map3d*, an open-source software project that has been developed over the past 10 years for use in a wide range of applications in biomedical research [128,129,131]. *map3d* is written using the OpenGL standard graphics library, and there are versions for all the major platforms (Windows, MacOSX, Linux, SGI Irix). The motivating application for *map3d* was visualizing results from simulations and experiments in cardiac electrophysiology of the whole heart, and although this remains the dominant application, it has also found use for displaying skin temperature profiles, scalp electric potentials from electroencephalography, and even concentration profiles in the nephrons of the kidney. More generally, *map3d* contains elements and capabilities that are typical of other solutions to the need to view this type of data.

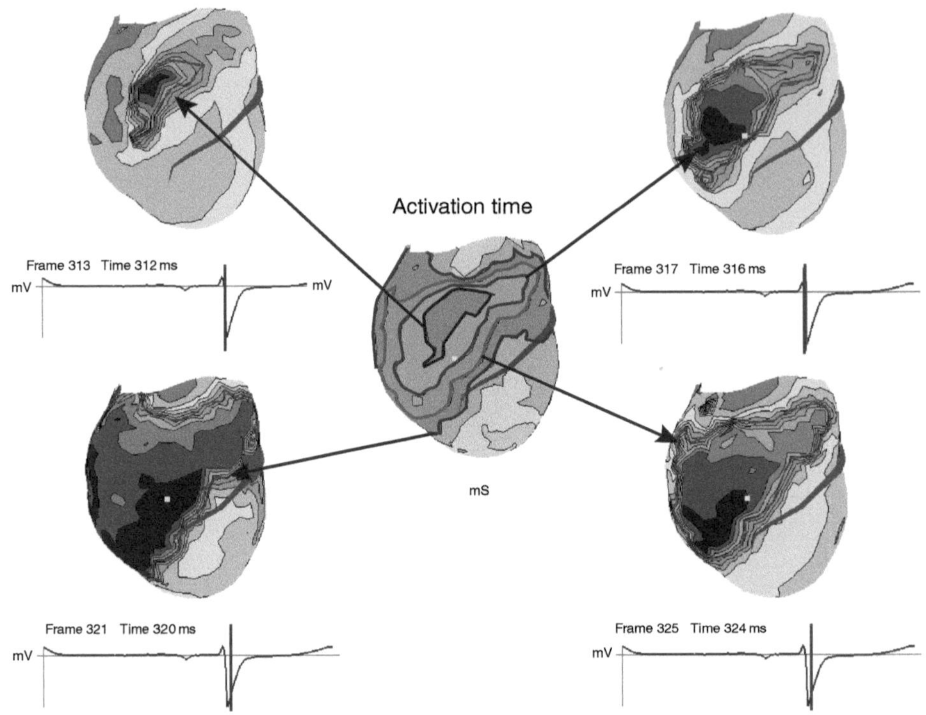

Figure 53 Example of *map3d* visualization of cardiac potentials and activation times.

The essential structure of *map3d* is a set of locations in space (nodes), usually connected to form surfaces of triangles and time signals associated with each of these nodes. The user must identify the nodes and may use *map3d* to connect them and must provide some form of mapping between channels of time signal and the nodes. From this information, *map3d* provides a highly interactive environment with which to view multiple sets of data associated with the same or multiple sets of nodes/triangles.

To illustrate *map3d* functionality, Figure 53 shows an example of potentials recorded from the surface of a dog heart using a 600-electrode sock array. The figure also shows the capability to display not just scalar values from time series but also quantities derived from those time series. In the figure, the center image contains what is known as the "activation" time, i.e., the time instant at which the cardiac wave of excitation passes each location. The specific case in Figure 53 shows a heart beat that initiated in the center of the ventricles, indicated by the thick dark innermost ellipse in the image. Contours here indicate discrete values of activation time, spaced at 5-ms intervals, and show the pathway of a single activation sequence i.e., the electrical sequence that leads to the contraction phase of the heart beat. The surrounding maps contain renderings of electric potential at a sequence of time instants, as indicated by the red vertical bar on the time signal below each frame. *map3d* offers the capability of rendering not only the activation time in its own window but also the activation contour that corresponds to each instant in time within the potential maps – the arrows in the figure indicate the relationship between the activation time and the thick red band (refer to color map) in each of the four potential maps corresponding to 312, 316, 320, and 324 ms, respectively from the start of the measurement. This figure also shows the unique support *map3d* provides for displaying visual queues or "landmarks", in this case the red tubes (refer to color map) lying on the surface of the heart that show the location of the coronary arteries.

The importance of this type of visualization lies in its ability to incorporate multiple information sources of quite different nature into the same visualization. There are many quantities one can derive from bioelectric time signals that have meaning on their own, just as activation time provides a compact representation of the trajectory of the excitation wave in the heart. The challenge becomes then to display this mixed information in a way that is intuitive and simple for the user. *map3d* now supports complex file formats that contain all the information derived from a set of time signals as well as the time signals themselves so that once constructed (the task of other programs), the user can quickly evaluate a large volume of information with *map3d*.

Another example of *map3d* visualization of multiple data sources is shown in Figure 54, which contains a sequence of electric potential maps recorded from the outer surface (epicardium) and inner surface (left-ventricular endocardium) of a canine heart. *map3d* offers many options for quantitative examination of datasets such as this, including the ability to select any point in the surface rendering and see the associated time signal; an enormous array of scaling options, including dynamically coupling the scaling across multiple surfaces; selecting new reference signals (important in evaluating EEG data); and even marking of data values directly on the surface.

A final example of *map3d* functionality that provides additional visual references is shown in Figure 55. Here, the entire display of geometry, potentials, time signals, and scaling bars is superimposed over a related image of the source of the data. The user can manually scale and align the geometry or prescribe the parameters for consistent alignment across multiple instances of the program.

7.2.1. Application domains of map3d

Typical applications of *map3d* – and the origins of the program – are in mapping the spatial and temporal features of cardiac electrophysiology, and for this purpose there remains, to our knowledge, no suitable alternative. Other application areas require the same basic functionality, i.e., whenever there exist time signals associated with locations on a surface in 3D space, there are is a similar set of visualization needs. Some examples from biomedicine include bioelectric signals from the head, skeletal muscle, or even the eye. However, one might also visualize temperature variations over the surface of the body or even the distribution of RNA expression over a 2D tissue preparation.

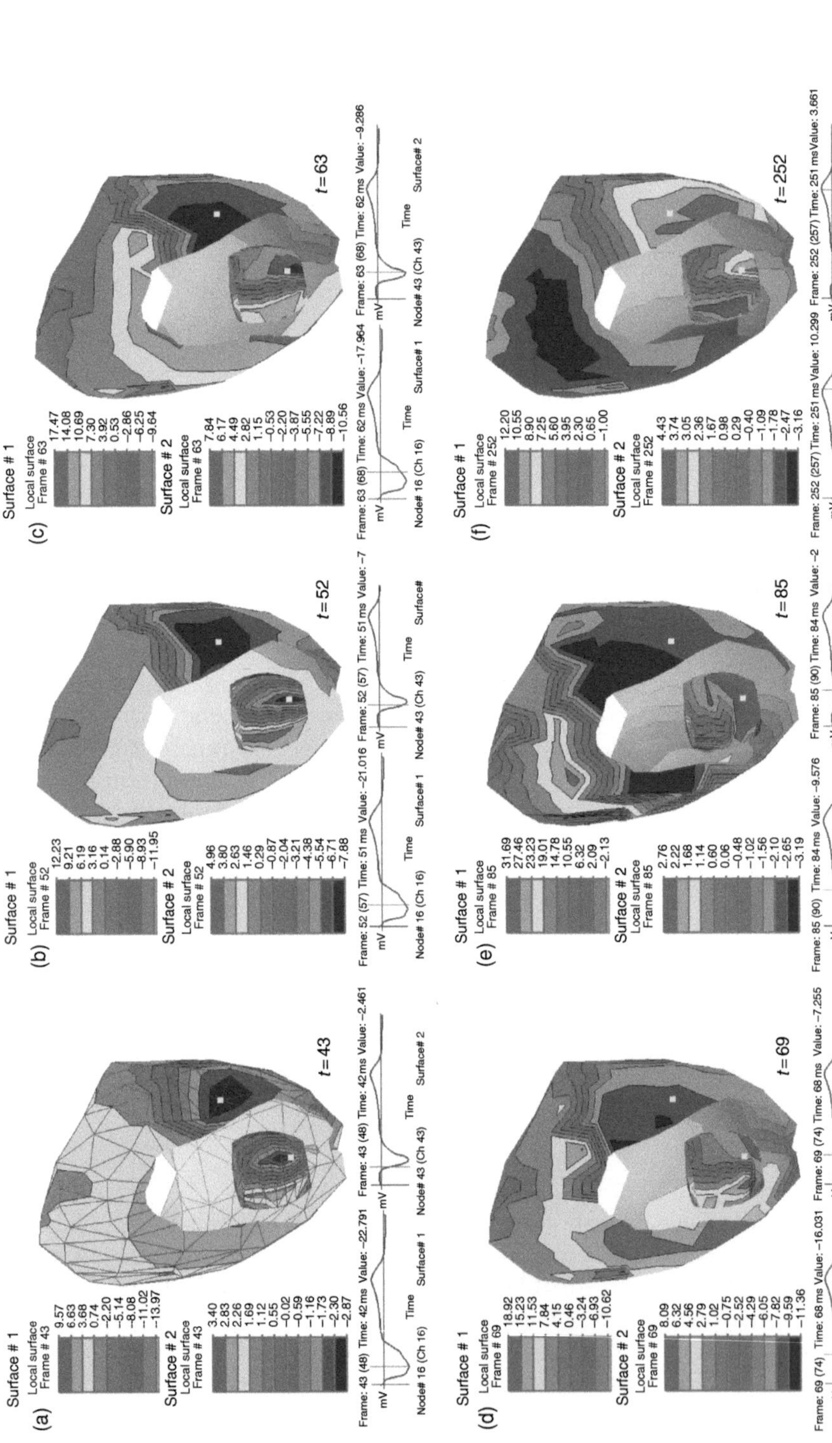

Figure 54 Sequence of potentials from the outer and inner surfaces of the heart. Images show the use of clipping planes to reveal data from two concentric surfaces simultaneously. The first image in the sequence shows the underlying measurement meshes for both surfaces (in red) and the time values marked in each panel indicate the time after stimulation of the heart from the epicardium. (See color plate 42).

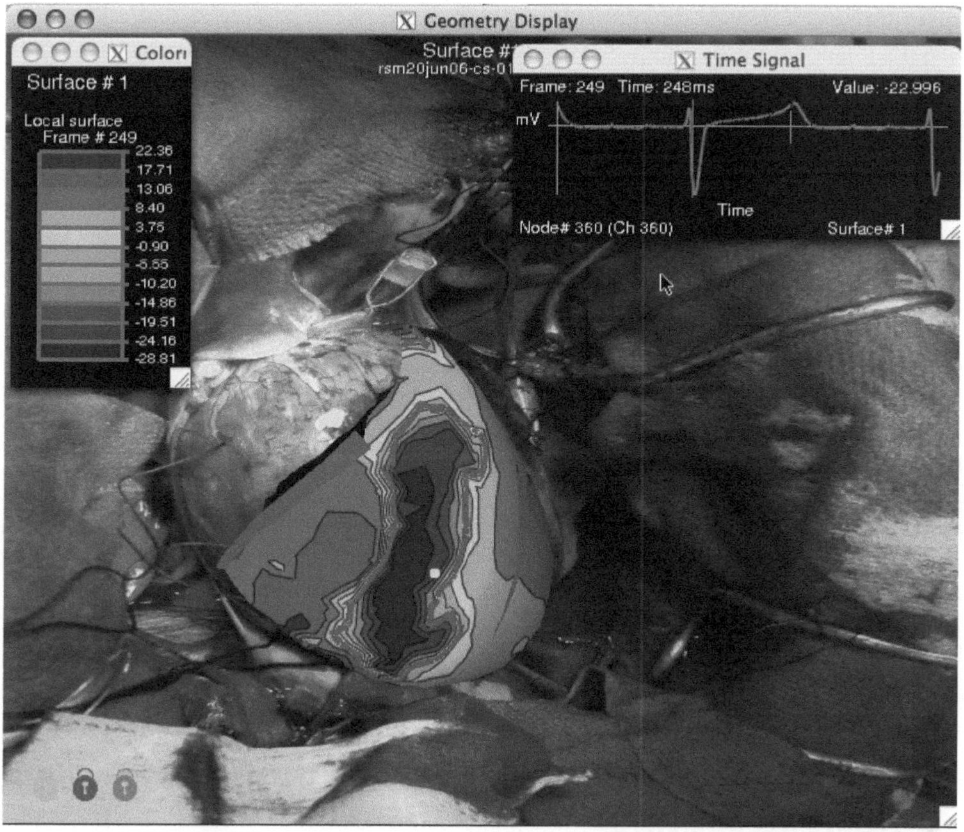

Figure 55 Visualization of data from the surface of the heart superimposed on an image of the same heart during the experiment. Color coding is for surface potentials with the think red contour indicating the location of activation. The time signal in the upper right corner of the figure shows a vertical red mark indicating the time instant captured in the surface map. (See color plate 43).

8. Summary and Conclusion

New imaging modalities, more accurate simulation models, and continued growth in computational power all contribute to confronting biomedical researchers and engineers with an unprecedented volume of information to further their understanding of biological systems and improve clinical practice. As the size and complexity of the resulting data explode, the tools created by Scientific Visualization research become crucial to gaining insight into the underlying biophysical phenomena. In this chapter, we have offered a selected overview of recent advances in biomedical visualization, while emphasizing promising avenues for future research.

Following a natural decomposition, techniques were presented that address the specific challenges posed by the exploration and analysis of large-scale scalar, vector, and tensor biophysical datasets. We showed that, in each case, both effective general purpose and application specific methods have been devised by the Scientific Visualization community over the last decade. The proposed algorithms address the need for effective visual representations while ensuring an efficient processing of large data that supports user interaction. We also discussed the question of the depiction of the error and uncertainty present in the visualization pipeline. While the importance of this aspect has long been underestimated, it is now regarded as an essential component of a useful visual analysis within a scientific study. We concluded our presentation with an overview of some of the existing visualization software packages that feature the methods that we discussed. In particular, we gave a more detailed description of *BioPSE* and *map3d* that the Scientific Computing and Imaging Institute has developed and made publicly available for the benefit of the biomedical community.

Finally, we mention two important additional visualization research areas that we did not cover in this chapter that could have entire chapters devoted to them. Information visualization is an active research field concerned with the visual representation of abstract data with no natural spatial embedding. For example, it provides compelling tools to support analysis in bioinformatics [140,141]. Another biomedical visualization area that has received a significant interest over the last few years is molecular visualization [142–145].

ACKNOWLEDGMENTS

The authors thank the many people who contributed to this article including Charles Hansen, Gordon Kindlmann, Joe Kniss, Rob MacLeod, Steve Parker, C.F. Westin, Ross Whitaker, and Leonid Zhukov. The authors acknowledge support from NIH NCRR Grant P41-RR12553-07, and additional support from NIH, NSF, and DOE.

REFERENCES

[1] C.R. Johnson, R. Moorhead, T. Munzner, H. Pfister, P. Rheingans, and T.S. Yoo. NIH-NSF Visualization Research Challenges Report, 2006. http://tab.computer.org/vgtc/vrc/index.html.

[2] Scientific Computing and Imaging Institute (SCI), 2007, University of Utah, www.sci.utah.edu.

[3] W. Schroeder, K. Martin, and B. Lorensen. *The Visualization Toolkit: An Object Oriented Approach to Graphics*, chapter 6. Kitware, Inc., Clifton Park, New York, 2003.

[4] C.D. Hansen and C.R. Johnson. *The Visualization Handbook*. Elsevier, 2005.

[5] V. Elangovan and R.T. Whitaker. From sinograms to surfaces: A direct approach to the segmentation of tomographic data. In *Proceedings of the 4th Int. Conf. Medical Image Computing and Computer-Assisted Intervention*, pages 213–223. Springer-Verlag, 2001.

[6] J. Kniss, G. Kindlmann, and C. Hansen. Multidimensional transfer functions for interactive volume rendering. In *IEEE Transactions on Visualization and Computer Graphics*, 8(3):270–285, 2002.
[7] J.M. Kniss, S. Premoze, M. Ikits, A.E. Lefohn, C.D. Hansen, and E. Praun. Gaussian transfer functions for multi-field volume visualization. In *Proceedings IEEE Visualization 2003*, IEEE Computer Society Press, pages 497–504, October 2003.
[8] G. Kindlmann, R. Whitaker, T. Tasdizen, and T. Möller. Curvature-based transfer functions for direct volume rendering: Methods and applications. In *Proceedings IEEE Visualization 2003*, IEEE Computer Society Press, page 67, October 2003.
[9] M. Ikits, J.M. Kniss, A.E. Lefohn, and C. Hansen. Volume rendering techniques. In R. Fernando, editor, *GPU Gems: Programming Techniques, Tips and Tricks for Real-Time Graphics*, pp. 667–692. Addison Wesley, 2004.
[10] J.M. Kniss, J.P. Schulze, U. Wossner, P. Winkler, U. Lang, and C.D. Hansen. Medical applications of multi-field volume rendering. In *Data Visualization 2004*, A.K. Peters Ltd., pages 249–254, 2004.
[11] J.M. Kniss, S. Premoze, C. Hansen, and D. Ebert. Interactive translucent volume rendering and procedural modeling. In *Proceedings IEEE Visualization 2002*, IEEE Computer Society Press, pages 109–116, 2002.
[12] J. Kniss, P. McCormick, A. McPherson, J. Ahrens, J. Painter, A. Keahey, and C. Hansen. Interactive texture-based volume rendering for large datasets. In *IEEE Computer Graphics and Applications*, 21(4):52–61, July/August 2001.
[13] J.M. Kniss, P. McCormick, A. McPherson, J. Ahrens, J. Painter, A. Keahey, and C.D. Hansen. T-rex, texture-based volume rendering for extremely large datasets. In *IEEE Computer Graphics and Applications*, 21(4):52–61, 2001.
[14] G.L. Kindlmann, D.M. Weinstein, and D. Hart. Strategies for direct volume rendering of diffusion tensor fields. In *IEEE Transactions on Visualization and Computer Graphics*, 6(2):124–138, April–June 2000.
[15] G. Kindlmann and J. Durkin. Semi-automatic generation of transfer functions for direct volume rendering. In *IEEE Symposium on Volume Visualization*, pages 79–86. IEEE Press, October 1998.
[16] J.M. Kniss, G. Kindlmann, and C.D. Hansen. Multidimentional transfer functions for volume rendering. In C.D. Hansen and C.R. Johnson, editors, *The Visualization Handbook*, pages 189–210. Elsevier, 2005.
[17] J. Kniss, G. Kindlmann, and C. Hansen. Interactive volume rendering using multi-dimensional transfer functions and direct manipulation widgets. In *Proceedings IEEE Visualization 2001*, IEEE Computer Society Press, pages 255–262, October 2001.
[18] BioPSE: Problem Solving Environment for modeling, simulation, image processing, and visualization for biomedical computing applications. Scientific Computing and Imaging Institute (SCI), 2007, Download from: http://www.sci.utah.edu/cibc/software/.
[19] W.E. Lorensen and H.E. Cline. Marching cubes: A high resolution 3D surface construction algorithm. *Computer Graphics*, 21(4):163–169, 1987.
[20] G. Wyvill, C. McPheeters, and B. Wyvill. Data structure for soft objects. *The Visual Computer*, 2:227–234, 1986.
[21] R.S. Gallagher. Span filter: An optimization scheme for volume visualization of large finite element models. In *Proceedings of Visualization '91*, pages 68–75. IEEE Computer Society Press, Los Alamitos, CA, 1991.
[22] T. Itoh, Y. Yamaguchi, and K. Koyyamada. Volume thining for automatic isosurface propagation. In *Visualization '96*, pages 303–310. IEEE Computer Society Press, Los Alamitos, CA, 1996.
[23] J.S. Painter, P. Bunge, and Y. Livnat. Case study: Mantle convection visualization on the cray t3D. In *Visualization '96*, pages 409–412. IEEE Computer Society Press, Los Alamitos, CA, 1996.
[24] H. Shen and C.R. Johnson. Sweeping simplicies: A fast iso-surface extraction algorithm for unstructured grids. In *Proceedings of IEEE Visualization '95*, IEEE Computer Society Press, pages 143–150, 1995.
[25] H.W. Shen, C.D. Hansen, Y. Livnat, and C.R. Johnson. Isosurfacing in span space with utmost efficiency (ISSUE). In *Visualization '96*, pages 287–294. IEEE Computer Society Press, 1996.
[26] Y. Livnat, H. Shen, and C.R. Johnson. A near optimal isosurface extraction algorithm using the span space. *IEEE Transaction on Visualization and Computer Graphics*, 2(1):73–84, 1996.

[27] P. Cignoni, C. Montani, E. Puppo, and R. Scopigno. Optimal isosurface extraction from irregular volume data. In *Proceedings of IEEE 1996 Symposium on Volume Visualization*. ACM Press, 1996.
[28] C.L. Bajaj, V. Pascucci, and D.R. Schikore. Fast isocontouring for improved interactivity. In *Proceedings of 1996 Symposium on Volume Visualization*, IEEE Computer Society Press, pages 39–46, 1996.
[29] C.L. Bajaj, V. Pascucci, and D.R. Schikore. The contour spectrum. In *Proceedings IEEE Visualization 1997*, pages 167–173, 1997.
[30] Y. Livnat, C.D. Hansen, and C.R. Johnson. Isosurface extraction for large-scale data sets. In F. Post, editor, *Data Visualization: The State of the Art*, 77–94, 2003.
[31] Y. Livnat, X. Cavin, and C. Hansen. PHASE: Progressive hardware assisted isosurface extraction framework. SCI Institute Technical Report UUSCI-2002-001, University of Utah, 2002.
[32] Y. Livnat and C. Hansen. Dynamic view dependent isosurface extraction. SCI Institute Technical Report UUSCI-2003-004, University of Utah, 2003.
[33] Y. Livnat and X. Tricoche. Interactive point based isosurface extraction. In *Proceedings IEEE Visualization 2004*, IEEE Computer Society Press, pages 457–464, 2004.
[34] H.W. Shen and C.R. Johnson. Differential volume rendering: A fast alogrithm for scalar field animation. In *Visualization 94*, pages 180–187. IEEE Press, 1994.
[35] R. Yagel and A. Kaufman. Template-based volume viewing. In *Proceedings of EUROGRAPHICS '92*, volume 11, pages 153–157. Blackwell, Cambridge, England, September 1992.
[36] R. Yagel and Z. Shi. Accelerating volume animation by space-leaping. In *Proceedings of Visualization '91*, pages 62–69. IEEE Computer Society Press, Los Alamitos, CA, October 1993.
[37] K.L. Ma, J. Painter, C. Hansen, and M. Krogh. Parallel volume rendering using binary-swap compositing. *IEEE Computer Graphics and Applications*, 14(4):59–68, 1994.
[38] C. Wang, J. Gao, L. Li, and H.W. Shen. A multiresolution volume rendering framework for large-scale time-varying data visualization. In *International Workshop on Volume Graphics*, IEEE Computer Society Press, pages 11–19, 2005.
[39] H.W. Shen, L.J. Chiang, and K.L. Ma. Time-varying volume rendering using a time-space partitioning tree. In *IEEE Visualization 99*, IEEE Computer Society Press, pages 371–377, 1999.
[40] A. Kaufman. *Volume Visualization*. IEEE CS Press, Los Alamitos, CA, 1990.
[41] R. Crawfis and N. Max. Direct volume visualization of three-dimensional vector fields. In *Proceedings of 1992 Workshop on Volume Visualization*, pages 55–60. IEEE Computer Society Press, Los Alamitos, CA, 1992.
[42] P.G. Swann and S.K. Semwal. Volume rendering of flow-visualization point data. In *Proceedings of Visualization '91*, pages 25–32. IEEE Computer Society Press, Los Alamitos, CA, 1991.
[43] W.H. Press, S.A. Teukolsky, W.T. Vetterling, and B.P. Flannery. *Numerical Recipes*, 2nd Edition. Cambridge University Press, Cambridge, 1992.
[44] G. Turk and D. Banks. Image-guided streamline placement. *Computer Graphics (SIGGRAPH)*, 30 (Annual Conference Series), pages 453–460, 1996 http://citeseer.ist.psu.edu/article/turk96imageguided.html.
[45] B. Jobard and W. Lefer. Creating evenly-spaced streamlines of arbitrary density. In *Proceedings of the Eurographics Workshop on Visualization in Scientific Computing*, pages 45–55, 1997.
[46] V. Verma, D. Kao, and A. Pang. A flow-guided streamline seeding strategy. In *VISUALIZATION '00: Proceedings of the 11th IEEE Visualization 2000 Conference (VIS 2000)*. IEEE Computer Society, 2000.
[47] A. Mebarki, P. Alliez, and O. Devillers. Farthest point seeding for efficient placement of streamlines. In *Proceedings of IEEE Visualization 2005*, IEEE Computer Society Press, pages 479–486, October 2005.
[48] X. Ye, D. Kao, and A. Pang. Strategy for seeding 3d streamlines. In IEEE Computer Society Press, editor, *IEEE Visualization Proceedings 2005*, pages 471–478, October 2005.
[49] F.B. Sachse and B. Taccardi. Visualization of electrical current flow with a new streamline technique: Application in mono- and bidomain simulations of cardiac tissue. In *Engineering in Medicine and Biology Society 2004*, IEEE EMBS 26th Annual International Conference of the IEEE, Volume 1, pages 1846–1849, 1–5 September 2004, Volume 3.

[50] E. Hughes, B. Taccardi, and F.B. Sachse. A heuristic streamline placement technique for visualization of electrical current flow. *Journal of Flow Visualization and Image Processing 13*, pages 53–66, 2006.

[51] D.C. Banks. Illumination in diverse codimensions. In *SIGGRAPH 94 Proceedings*, volume 28, pages 327–334, 1994.

[52] M. Zöckeler, D. Stalling, and H.-C. Hege. Interactive visualization of 3D-vector fields using illuminated streamlines. In *IEEE Visualization 96 Proceedings*, pages 107–113, 1996.

[53] O. Mallo, R. Peikert, C. Sigg, and F. Saldo. Illuminated streamlines revisited. In *Proceedings of IEEE Visualization 2005*, pages 19–26, October 2005.

[54] G. Li, U.D. Bordoloi, and H. Shen. Chameleon: An interactive texture-based rendering framework for visualizing three-dimensional vector fields. In *Proceedings of IEEE Visualization 2003*, page 32, October 2003.

[55] J.P.M. Hultquist. Constructing stream surfaces in steady 3D vector fields. In *Proceedings of IEEE Visualization 1992*, pages 171–178, October 1992.

[56] C. Garth, X. Tricoche, T. Salzbrunn, T. Bobach, and G. Scheuermann. Surface techniques for vortex visualization. In *Proceedings of Joint Eurographics–IEEE TCVG Symposium on Visualization*, pages 155–164, May 2004.

[57] G. Erlebacher and D. Weiskopf. Flow textures: High-resolution flow visualization. In C.D. Hansen and C.R. Johnson, editors, *Visualization Handbook*, pages 279–294. Elsevier Academic Press, 2005.

[58] B. Cabral and C. Leedom. Imaging vector fields using line integral convolution. In SIGGRAPH *'93: Proceedings of the 20th Annual Conference on Computer Graphics and Interactive Techniques*, pages 263–270, New York, NY, USA, ACM Press, 1993, http://doi.acm.org/10.1145/166117.166151.

[59] H.W. Shen and D.L. Kao. A new line integral convolution algorithm for visualizing unsteady flows. *IEEE Transactions on Visualization and Computer Graphics*, 4(2), pages 98–108, 1998.

[60] B. Jobard, G. Erlebacher, and Y. Hussaini. Lagrangian Eulerian advection of noise and dye textures for unsteady flow visualization. *IEEE Transactions on Visualization and Computer Graphics*, 8(3):211–222, 2002.

[61] G.-S. Li, X. Tricoche, and C.D. Hansen. Gpuflic: Interactive and dense visualization of unsteady flows. In *Data Analysis 2006: Proceedings of Joint IEEE VGTC and EG Symposium on Visualization (Euro Vis) 2006*, Eurographics Press, pages 29–34, May 2006.

[62] H. Battke, D. Stalling, and H.-C. Hege. Fast line integral convolution for arbitrary surfaces in 3d. In Springer Berlin, editor, *Visualization and Mathematics Experiments, Simulations and Environments,* Springer-Verlag, New York Inc., pages 181–195, 1997.

[63] B. Laramee, J.J. van Wijk, B. Jobard, and H. Hauser. ISA and IBFVS: Image space based visualization of flow on surfaces. *IEEE Transactions on Visualization and Computer Graphics*, 10(6):637–648, November 2004.

[64] D. Weiskopf and T. Ertl. A hybrid physical/device space approach for spatiotemporally coherent interactive texture advection on curved surfaces. In GI '04: *Proceedings of Graphics Interface 2004*, pages 263–270, 2004.

[65] M.S. Shephard, J.E. Flaherty, H.L. de Cougny, C.L. Bottasso, and C. Ozturan. Parallel automatic mesh generation and adaptive mesh control. In M. Papadrakakis, editor, *Solving Large Scale Problems in Mechanics: Parallel and Distributed Computer Applications*, pages 459–493. John Wiley & Sons, Ltd, Chichester, 1997.

[66] V. Interrante and C. Grosch. Visualizing 3d flow. *IEEE Computer Graphics and Applications*, 8(4):49–53, 1998.

[67] C. Resz-Salama, P. Hastreiter, C. Teitzel, and T. Ertl. Interactive exploration of volume line integral convolution based on 3d-texture mapping. In *Proceedings of IEEE Visualization 1999*, IEEE Computer Society Press, pages 233–240, October 1999.

[68] A. Sanderson, C.R. Johnson, and M. Kirby. Display of vector fields using a reaction diffusion model, In *Proceedings IEEE Visualization 2004*, IEEE Computer Society Press, pages 115–122, 2004.

[69] A.M. Turing. The chemical basis of morphogenesis. *Philosophical Transactions of the Royal Society London*, B237:37–72, 1952.

[70] P. Gray and S.K. Scott. Sustained oscillations and other exotic patterns of behaviour in isothermal reactions. *The Journal of Physical Chemistry*, 59:22–32, 1985.

[71] P. Gray and S.K. Scott. *Chemical Oscillations and Inestability: Non-linear Chemical Kinetics*. Oxford University Press, 1990.

[72] G. Scheuermann and X. Tricoche. Topological methods for flow visualization. In C.D. Hansen and C.R. Johnson, editors, *The Visualization Handbook*, pages 341–356. Elsevier, 2005.

[73] X. Tricoche, G. Scheuermann, and H. Hagen. A topology simplification method for 2D vector fields. In *Proceedings of IEEE Visualization 2000*, IEEE Computer Society Press, pages 359–366, 2000.

[74] X. Tricoche, G. Scheuermann, and H. Hagen. Continuous topology simplification of planar vector fields. In *Proceedings of IEEE Visualization 2001*, IEEE Computer Society Press, pages 159–166, October 2001.

[75] C. Garth, X. Tricoche, and G. Scheuermann. Tracking of vector field singularities in unstructured 3D time-dependent data sets. In *Proceedings of IEEE Visualization 2004*, IEEE Computer Society Press, pages 329–336, 2004.

[76] L.V. Corbin and A.M. Scher. The canine heart as an electrocardiographic generator. *Circulation Research*, 41:58–67, 1977.

[77] D.E. Roberts and A.M. Scher. Effect of tissue anisotropy on excitation potential fields in canine myocardium in situ. *Circulation Research*, 50(3):342–351, March 1982.

[78] P. Colli Franzone, L. Guerri, C. Viganotti, E. Macchi, S. Baruffi, S. Spaggiari, and B. Taccardi. Potential fields generated by oblique layers modeling excitation wavefronts in the anisotropic myocardium. *Circulation Research*, 51:330–346, 1982.

[79] C.S. Henriquez, A.L. Muzikant, and C.K. Smoak. Anisotropy, fiber curvature, and bath loading effects on activation in thin and thick cardiac tissue preparations. *Journal of Cardiovascular Electrophysiology*, 7:424–444, 1996.

[80] P.J. Basser, J. Mattiello, and D. Le Bihan. Estimation of the effective self-diffusion tensor from the NMR spin-echo. *Magnetic Resonance*, 247–254, 1994.

[81] M.A. Guttman, E.A. Zerhouni, and E.R. McVeigh. Analysis of cardiac function from mr images. *IEEE Computer Graphics and Applications*, 17(1):30–38, 1997.

[82] E.A. Zerhouni, D.M. Parish, W.J. Rogers, A. Yang, and E.P. Shapiro. Human heart: Tagging with MR imaging – a method for noninvasive assessment of myocardial motion. *Radiology*, 169(1):59–63, 1988.

[83] B. Wünsche and A.A. Young. The visualization and measurement of left ventricular deformation using finite element models. *International Journal of Visual Languages and Computing*, 14(4):299–326, 2003.

[84] A.I. Veress, G.T. Gullberg, and J.A. Weiss. Measurement of strain in left ventricle during diastole with cine-MRI and deformable image registration. *ASME Journal of Biomechanical Engineering*, page to appear, February 2006. 127(7): pp. 1195–1207, 2005.

[85] P.J. Basser and C. Pierpaoli. Microstructural and physiological features of tissues elucidated by quantitative-diffusion-tensor MRI. *Journal Magnetic Resonance B*, 111(3):209–219, 1996.

[86] C.F. Westin, S. Peled, H. Gubjartsson, R. Kikinis, and F. A. Jolesz. Geometrical diffusion measures for MRI from tensor basis analysis. In *Proceedings 5th Annual ISMRM*, 1997. International Society for Magnetic Resonance in Medicine.

[87] S. Pajevic and C. Pierpaoli. Color schemes to represent the orientation of anisotropic tissues from diffusion tensor data: Application to white matter fiber tract mapping in the human brain. *Magnetic Resonance in Medicine*, 42(3):526–540, 1999.

[88] C.-F. Westin, S.E. Meier, H. Mamata, A. Nabavi, F.A. Jolesz, and R. Kikinis. Processing and visualization for diffusion tensor MRI. *Medical Image Analysis*, 6:93–108, 2002.

[89] G. Kindlmann. Superquadric tensor glyphs. Data Visualization 2004: Proceedings of The Joint Eurographics – IEEE TCVG Symposium on Visualization, Eurographics Press, pages 147–154, May 2004.

[90] D.H. Laidlaw, E.T. Ahrens, D. Kremers, M.J. Avalos, R.E. Jacobs, and C. Readhead. Visualizing diffusion tensor images of the mouse spinal cord. In *Proceedings IEEE Visualization 98*, IEEE Computer Society Press, pages 127–134, 1998.

[91] D.S. Tuch, T.G. Reese, M.R. Wiegell, and V.J. Wedeen. Diffusion MRI of complex neural architecture. *Neuron*, 40(5):885–895, 2003.

[92] D.S. Tuch. Q-ball imaging. *Magnetic Resonance in Medicine*, 52:1358–1372, 2004.

[93] P.J. Basser. Fiber-tractography via diffusion tensor MRI (DT-MRI). In *Proceedings of the 6th Annual Meeting ISMRM*, International Society for Magnetic Resonance in Medicine, page 1226, 1998.
[94] P.J. Basser, S. Pajevic, C. Pierpaoli, J. Duda, and A. Aldroubi. In vivo fiber tractograpy using DT-MRI data. *Magnetic Resonance in Medicine*, pages 625–632, 2000.
[95] S. Zhang, C. Demiralp, and D.H. Laidlaw. Visualizing diffusion tensor MR images using streamtubes and streamsurfaces. *IEEE Transactions on Visualization and Computer Graphics*, 9(4):454–462, October 2003.
[96] T. Delmarcelle and L. Hesselink. A unified framework for flow visualization. In R.S. Gallagher, editor, *Computer Visualization: Graphics Techniques for Scientific and Engineering Analysis*, pages 129–170, CRC Press, 1995.
[97] D. Weinstein, G. Kindlmann, and E. Lundberg. Tensorlines: Advection-diffusion based propagation through diffusion tensor fields. In *Proceedings IEEE Visualization 1999*, IEEE Computer Society Press, pages 249–253, 1999.
[98] L. Zhukov and A. Barr. Oriented tensor reconstruction: Tracing neural pathways from diffusion tensor MRI. In *Proceedings of IEEE Visualization 2002*, IEEE Computer Society Press, pages 387–394, 2002.
[99] L. Zhukov and A. Barr. Heart-muscle fiber reconstruction from diffusion tensor MRI. In *Proceedings of IEEE Visualization 2003*, IEEE Computer Society Press, pages 597–602, 2003.
[100] M. Hlawitschka and G. Scheuermann. Hot-lines: Tracking lines in higher-order tensor fields. In *Proceedings of IEEE Visualization 2005*, IEEE Computer Society Press, pages 27–34, October 2005.
[101] A. Brun, H.-J. Park, H. Knutsson, and C.-F. Westin. Coloring of DT-MRI fiber traces using laplacian eigenmaps. *Computer Aided Systems Theory – EUROCAST 2003*, Lecture Notes in Computer Science 2809, Springer, page 2809, 2003.
[102] M. Cantani, R.J. Howard, S. Pajevic, and D.K. Jones. Virtual in vivo interactive dissection of white matter fasciculi in the human brain. *NeuroImage*, 17:77–94, 2002.
[103] S. Wakana, H. Jiang, L.M. Nagae-Poetscher, P.C. van Zijl, and S. Mori. Fiber tract-based atlas of human white matter anatomy. *Radiology*, 230(1):77–87, 2004.
[104] D. Akers, A. Sherbondy, R. Mackenzie, R. Dougherty, and B. Wandell. Exploration of the brain's white matter pathways with dynamic queries. In *Proceedings of IEEE Visualization 2004*, IEEE Computer Society Press, pages 377–384, 2004.
[105] J. Blaas, C.P. Botha, B. Peters, F.M. Vos, and F.H. Post. Fast and reproducible fiber bundle selection in dti visualization. In *Proceedings of IEEE Visualization 2005*, IEEE Computer Society Press, pages 59–64, 2005.
[106] A. Brun, H. Knutsson, H.J. Park, M.E. Shenton, and C.-F. Westin. Clustering fiber tracts using normalized cuts. In *Seventh International Conference on Medical Image Computing and Computer-Assisted Intervention (MICCAI'04)*, Lecture Notes in Computer Science, pages 368–375, Rennes – Saint Malo, France, September 2004.
[107] S. Zhang and D.H. Laidlaw. DTI fiber clustering and cross-subject cluster analysis. In *Proceedings of ISMRM International Society for Magnetic Resonance in Medicine*, 2005.
[108] A. Wenger, D.F. Keefe, S. Zhang, and D.H. Laidlaw. Interactive volume rendering of thin thread structures within multivalued scientific datasets. *IEEE Transactions on Visualization and Computer Graphics*, 10(6):664–672, 2004.
[109] G.L. Kindlmann and D.M. Weinstein. Hue-balls and lit-tensors for direct volume rendering of diffusion tensor fields. In *Proceedings of the IEEE Visualization 99*, pages 183–189, 1999.
[110] G.L. Kindlmann and D.M. Weinstein. Strategies for direct volume rendering of diffusion tensor fields. *IEEE Transactions on Visualization and Computer Graphics*, IEEE Computer Society Press, 6(2):124–138, 2000.
[111] D. Stalling, M. Zöckler, and H.-C. Hege. Fast display of illuminated field lines. *IEEE Transactions on Visualization and Computer Graphics*, 3(2):118–128, 1996.
[112] G. Kindlmann. *Visualization and Analysis of Diffusion Tensor Fields*. Ph.D. Thesis, University of Utah, 2004.
[113] G. Kindlmann, X. Tricoche, and C.-F. Westin. Anisotropy creases delineate white matter structure in diffusion tensor MRI. In *Proceedings of Medical Imaging Computing and Computer-Assisted Intervention, MICCAI '06, Lecture Notes in Computer Science 3749*, Springer, 2006.

[114] C.R. Johnson. Top scientific visualization research problems. *IEEE Computer Graphics and Applications: Visualization Viewpoints*, 24(4):13–17, July/August 2004.

[115] C.H. Wolters, A. Anwander, X. Tricoche, D. Weinstein, M.A. Koch, and R.S. MacLeod. Influence of tissue conductivity anisotropy on EEG/MEG field and return current computation in a realistic head model: A simulation and visualization study using high-resolution finite element modeling. *NeuroImage*, 2006 Apr 15; 30(3):813–26. December 2005.

[116] D. Gullmar, J.R. Reichenbach, A. Anwander, T. Knosche, C.H. Wolters, M. Eiselt, and J. Haueisen. Influence of anisotropic conductivity of the white matter tissue on eeg source reconstruction – an fem simulation study. Biomedical Engineering, IEEE Transactions on 53(9):1841–1850, September 2006, 7(1):108–110, 2005.

[117] R.M. Kirby, H. Marmanis, and D.H. Laidlaw. Visualizing multivalued data from 2D incompressible flows using concepts from painting. In *Proceedings of the IEEE Visualization 99*, IEEE Computer Society Press, 1999.

[118] C.R. Johnson and A.R. Sanderson. A next step: Visualizing errors and uncertainty. *IEEE Computer Graphics and Applications*, 23(5):6–10, September/October 2003.

[119] B.N. Taylor and C.E. Kuyatt. Guidelines for expressing and evaluating the uncertainty of NIST experimental results. Technical report, NIST Technical Note 1297, 1994.

[120] A.T. Pang, C.M. Wittenbrink, and S.K. Lodha. Approaches to uncertainty visualization. *The Visual Computer*, 13(8):370–390, November 1997.

[121] G. Grigoryan and P. Rheingans. Probabilistic surfaces: Point based primitives to show surface uncertainty. In *IEEE Visualization '02*, pages 147–153. IEEE Press, 2002.

[122] A.R. Sanderson, M. Kirby, C.R. Johnson, and L. Yang. Advanced reaction-diffusion models for texture synthesis. *Journal of Graphics Tools*, 11(3):47–71, 2006.

[123] J.M. Kniss, R. Van Uitert, A. Stephens, G. Li, and T. Tasdizen. Statistically quantitative volume visualization. In *IEEE Visualization 2005*, 2005.

[124] T. Tasdizen, S.P. Awate, R.T. Whitaker, and N. Foster. MRI tissue classification with neighborhood statistics: A nonparametric, entropy-minimizing approach. In *Proceedings of the 8th International Conference on Medical Image Computing and Computer Assisted Intervention (MICCAI)*, Lecture Notes in Computer Science 3749, Springer, pages 517–525, 2005.

[125] S.G. Parker and C.R. Johnson. SCIRun: A scientific programming environment for computational steering. In *Supercomputing '95*. IEEE Press, 1995.

[126] C.R. Johnson and S.G. Parker. Applications in computational medicine using SCIRun: A computational steering programming environment. In H.W. Meuer, editor, *Supercomputer '95*, pages 2–19. Springer-Verlag, 1995.

[127] SCIRun: A Scientific Computing Problem Solving Environment, Scientific Computing and Imaging Institute (SCI), 2007, Download from: http://software.sci.utah.edu/scirun.html.

[128] R.S. MacLeod, C.R. Johnson, and M.A. Matheson. Visualization tools for computational electrocardiography. In *Visualization in Biomedical Computing*, pages 433–444, Bellingham, Washington, 1992. Proceedings of the SPIE #1808.

[129] R.S. MacLeod, C.R. Johnson, and M.A. Matheson. Visualization of cardiac bioelectricity – a case study. In *Proceedings of the IEEE Visualization 92*, pages 411–418. IEEE CS Press, 1992.

[130] R.S. MacLeod, C.R. Johnson, and M.A. Matheson. Visualizing bioelectric fields. IEEE Computer Graphics and Applications, 13(4):10–12, 1993.

[131] R.S. MacLeod and C.R. Johnson. Map3d: Interactive scientific visualization for bioengineering data. pages 30–31. IEEE Press, 1993. http://software.sci.utah.edu/map3d.html.

[132] B. Taccardi, B.B. Punske, R.L. Lux, R.S. MacLeod, P.R. Ershler, T.J. Dustman, and Y. Vyhmeister. Useful lessons from body surface potential mapping. 9:773–786, 1998.

[133] B.P. Punske, Q. Ni, R.L. Lux, R.S. MacLeod, P.R. Ershler, T.J. Dustman, Y. Vyhmeister, and B. Taccardi. Alternative methods of excitation time determination on the epicardial surface. Engineering in Medicine and Biology Society, 2000. Proceedings of the 22nd Annual International Conference of the IEEE Volume 2, Issue, 888–890, 2000.

[134] B. Taccardi, B.B. Punske, R.S. MacLeod, and Q. Ni. Extracardiac effects of myocardial electrical anisotropy. *Biomedizinische Technik*, 46(Supp 2):216–218, 2001.

[135] B.B. Punske, Q. Ni, R.L. Lux, R.S. MacLeod, P.R. Ershler, T.J. Dustman, M.J. Allison, and B. Taccardi. Spatial methods of epicardial activation time determination. Annals of Biomedical Engineering, 31(7):781–792(12), July 2003, 31:781–792, 2003.

[136] B. Taccardi, B.B. Punske, R.S. MacLeod, and Q. Ni. Visualization, analysis and physiological interpretation of three-dimensional cardiac electric fields. pages 1366–1367. IEEE EMBS, IEEE, 2002.

[137] F.H. Lopes da Silva. A critical review of clinical applications of topographic mapping of brain potentials. *Journal of Clinical Neurophysiology*, 7(4):535–551, 1990.

[138] J.S. George, C.J. Aine, J.C. Mosher, D.M. Schmidt, D.M. Ranken, H.A. Schlitt, C.C. Wood, J.D. Lewine, J.A. Sanders, and J.W. Belliveau. Mapping function in the human brain with magnetoencephalography, anatomical magnetic resonance imaging, and functional magnetic resonance imaging. *Journal of Clinical Neurophysiology*, 12(5):406, 1995.

[139] F. Lopes da Silva. Functional localization of brain sources using EEG and/or MEG data: Volume conductor and source models. *Magnetic Resonance Imaging*, 22(10):1533–1538, December 2004.

[140] J. Slack, K. Hildebrand, T. Munzner, and K. St. John. SequenceJuxtaposer: Fluid navigation for large-scale sequence comparison in context. In *Proceedings of German Conference in Bioinformatics*, pages 37–42, 2004.

[141] J. Slack, K. Hildebrand, and T. Munzner. Partitioned rendering infrastructure for scalable accordion drawing (extended version). *Information Visualization* 5(2):137–151, 2006.

[142] C. Bajaj, P. Djeu, V. Siddavanahalli, and A. Thane. Interactive visual exploration of large flexible multi-component molecular complexes. In *Proceedings of IEEE Visualization '04*, IEEE Computer Society Press, pages 243–250, 2004.

[143] C. Bajaj, J. Castillon-Candas, V. Siddavanahalli, and Z. Xu. Compressed representations of macromolecular structures and properties. *Structure*, 13(3):463–471, March 2005.

[144] C.H. Lee and A. Varshney. Representing thermal vibrations and uncertainty in molecular surfaces. In *Proceedings SPIE Conference on Visualization and Data Analysis*, January 20–25, 2002.

[145] C.H. Lee and A. Varshney. Computing and displaying intermolecular negative volume for docking. In G.M. Nielson G.-P. Bonneau, T. Ertl, editors, *Scientific Visualization: The Visual Extraction of Knowledge from Data*. Springer-Verlag, ISBN 3-540-26066-8, 2005.

Index

Absorbance or absorption, 148
  bands, 148–149
  spectrum, 148–149
ABS *see* Average beat subtraction (ABS)
Adaption coefficient *see* Step size
AF *see* Atrial fibrillation (AF)
Algorithms:
  batch, 113, 133
  block *see* batch
  on-line, 112–113
  off-line, 112–113
  recursive *see* on-line
  stochastic, 113
  windowed *see* batch
Anisotropy:
  fractional, 235
  linear, 235
  planar, 235
Artficial organs, 2, 11
Artifacts, ocular, 124–125, 133–138
Atrial activity (AA), 115
Atrial fibrillation (AF), 38, 115, 117
Atria *see* Heart, parts
Atrio–ventricular (AV) node, 102, 115–116
Auto-fluorescent proteins, 145–174
Average beat, 120–121, 123
Average beat subtraction (ABS), 123

Barycentric coordinates, 241–242
Beer-Lambert law, 148–149, 151
Bioelectric currents, 210, 226–229, 233
Biomedical devices, 2
Blind source separation (BSS), 97, 114, 123
  algorithms, 126–127, 131
  assumptions, 125
  indeterminacies, 127
  whitening, 131
  see also Independent component analysis (ICA)
Brain, 105
  cortical lobes, 105–106
  electrical model, 106
  functions, 105
BSS *see* Blind source separation (BSS)

Circulatory system, 11, 15, 19, 35, 101–102
Contrast function, 129–132
  maximization *see* Blind source separation (BSS), algorithms
Correlation, cross-correlation function, 121–123
Covariance matrix, 112, 121–123, 129–132, 241
  sample estimate, 112
Crease surfaces, 244
Cumulants, 128
  cross-cumulants, 128, 130
  fourth-order, 130
  higher-order, 127, 131
  marginal, 128, 130–131
  multilinearity, 130–131
  sample, 133
  second-order, 128,

Detector, 176–180, 184
  collimator, 177–179, 184
  design, 186
    in time-of-flight PET, 184
  ring detector, 178–179, 195–196
  scintillating crystal, 177–178
Diffusion tensor imaging, 210, 234

Dipole, 98
in a heterogeneous medium, 100
in a homogeneous medium, 99
moment, 99, 101, 104
realistic models, 100
Direct volume rendering, 211, 222, 241–242
3D texture memory, 217
boundary, 214–216
cerebral vasculature, 212
classification, 217
classify, 216
CT, 216
derivatives, 216
direct manipulation widgets, 217
edge-detection, 216
eigenvector, 217
electron microscope tomography (EMT), 214
features of interest, 211–212
Gaussian transfer functions, 217
gradient direction, 216
gradient magnitude, 216
gradients, spatial, 216
graphic cards, 213–214
graphics hardware, 217
interactive response, 217
lookup tables, 217
matrix, 217
maximum intensity projection (MIP), 212
memory efficient, 217
MR imaging, 217
multi-dimensional transfer functions, 216
opacity function, 214
optical properties, 211–212
ramp function, 213–214
reconstruction artifacts, 214
rendering techniques, 211–212
isosurfacing, 211–212
segmentation, 211–212
semi-automatic generation of transfer functions, 214–215
transfer function, 211–212
transfer function design, 212
user interface, 217
Visible Male, 216
X-rays, 213
Dosimetry, 186, 200
DsRed, 165–166

ECG *see* Electrocardiogram (ECG)
EEG *see* Electroencephalogram (EEG)
Eigenvalue decomposition (EVD), 130–132
Eigenvalues, 235
invariant, 236
Eigenvectors, 235
major, 235
medium, 235
minor, 235
spherical color map, 236
Electrocardiogram (ECG), 103
fetal, 109
P wave, 103*t*
QRS complex, 103*t*
T wave, 103*t*
Electroencephalogram (EEG), 106
artifacts, 108
Emission tomography *see* Single photon emission computed tomography (SPECT); Positron emission tomography (PET)
small animal imaging, 179, 194
Epilepsy, 106, 123–124
ERP *see* Event-related potential (ERP)
Error and uncertainty visualization, 250
fuzzy classification, 252–254
fuzzy surface, 250
uncertainty definitions, 250
EVD *see* Eigenvalue decomposition (EVD)

Event-related potential (ERP), 106, 108*t*
Extinction coefficient, 148–149, 151, 157–159, 161, 166

Fiber tractography, 239, 244
 clustering, 241
 hyperstreamlines, 239–240
 local tracking, 240
 moving least squares, 240
 muscle fibers, 239–240
 normalized cuts, 241
 pathways of axon bundles, 239–240
 principal eigenvector, 239–240
 regions of interest, 241
 streamtubes, 240
 tensorlines, 240
Filtering:
 optimal *see* Wiener
 spatial, 96–97, 111, 113, 119
 spatio-temporal, 113
Finite element, 63
 basics, 63
 effective use, 68
 stent design, 66
Fluorescence, 145–174
 anisotropy, 155
 brightness, 151, 160, 163–165
 decay, 152–153
 delayed, 150
 emission, 149
 excitation, 149
 non-radiative decay, 149
 quenching, 157
Fluorescence correlation spectroscopy (FCS), 166–167
Fluorescence lifetime, 151
 measurement, 153
  frequency domain, 154
  time domain, 153
 modulation, 154–155
 phase, 154–155
Fluorescence lifetime imaging microscopy (FLIM), 167–169
Fluorescence loss in photobleaching (FLIP), 166–167
Fluorescence recovery after photobleaching (FRAP), 166–167
Fluorescence resonance energy transfer (FRET), 158
 efficiency, 166–169
 rate, 158–159
Fluorescent proteins, 160–161, 165
 see also Green fluorescent protein (GFP)
Fluorophore, 151–153, 155–165, 167
FMRI *see* Magnetic resonance imaging (MRI), functional
Förster radius, 158–159
Franck–Condon principle, 148, 150
Frank system *see* Leads
Frobenius norm, 235–236

Gamma camera, 177–178, 189
GATE, 188
 digitization, 193
 features, 188
 time management, 192
GEANT4, 182, 187–189, 191–192, 194
Givens rotation, 132
Gradient, 23–24, 27, 32, 36, 68, 112–113, 216–217, 240–242
Gram–Schmidt orthogonalization, 132
Green fluorescent protein (GFP), 161
 chromophore, 161
 structure, 161
 variants, 163

Heart, 8, 34, 101
 dipole moment, 104
 electrical model, 104

Heart (*Continued*)
electric heart vector, 104
parts, 101
High angular resolution diffusion (HARD), 239–241
HOS *see* Statistics, higher-order

ICA *see* Independent component analysis (ICA)
IED *see* Interictal epileptic discharge (IED)
Illuminated streamlines, 242
Image degradation factors, 185
Image reconstruction, 88, 179, 195
see also Nuclear medicine imaging, data acquisition
Independence, 97, 126–127
see also Uncorrelation
Independent component analysis (ICA), 129–130, 135–136,
see also Blind source separation (BSS)
Independent subspace analysis (ISA), 138
Information visualization, 210, 224, 264
Interictal epileptic discharge (IED), 123
Internal conversion, 149–150
Intersystem crossing, 149–150
ISA *see* Independent subspace analysis (ISA)
Isosurface, 220
accelerating, 220
applications, 221
cardiac arrhythmia, 221–222
cortical surface, 222
cutting plane, 221
high-performance, 221
isochrone, 221
isopotential, 221
scalp, 222
SCIRun, 221
complexity, 220
contour spectrum, 220
divide and conquer, 220
lookup table, 220
Marching Cubes, 220
point-based approach, 221
real-time, 220
remote visualization, 220–221
shadows, 221
span space, 220
triangulation, 220
view-dependent, 220–221
visibility, 221
Visible Female, 221

Jabłoński diagram, 147
Jacobi diagonalization, 132

Kurtosis *see* Cumulants, fourth-order

Leads:
10–20 International System, 106
Frank system, 104
orthogonal, 104, 117
X, Y, Z *see* orthogonal
Least mean squares (LMS), 113
Least squares (LS), 121–122
LMS *see* Least mean squares (LMS)
LS *see* Least squares (LS)
Luminescence, 146

Magnetic resonance imaging (MRI), 75, 106, 124, 140, 176–177, 234, 237–240, 255
functional, 124, 140
Magnetoencephalogram (MEG), 106, 140, 223–224
Map3d, 258
"activation" time, 260–261
cardiac wave, 260–261
coronary arteries, 260–261
electroencephalography, 259

electrophysiology, 258–259
heart beat, 260–261
ventricles, 260–261
Material properties, 65
stainless steel, 66
Maximum likelihood (ML), 122–123, 129, 211
Medical robots, 43
MEG *see* Magnetoencephalogram (MEG)
Methodology, 68–69, 72–73, 75–78, 81, 84, 89, 98, 117, 222
Methods *see* Algorithms
Micro-CT, 64
microtomography, 89
MI *see* Mutual information (MI)
Mixing matrix, 119, 121, 126–127, 129–130, 133–134, 136, 138–139
Mixtures, 35–36, 97, 110, 117, 119, 129, 135, 138–139
overdetermined, 126–127, 135
underdetermined, 138–139
ML *see* Maximum likelihood (ML)
Model:
blind, 121
brain, 106
heart, 104
physical-based, 121
Moments, sample, 133
Monte Carlo simulation, 180, 184, 187
applications, 186
calculation time, 194
variance reduction techniques, 194–195
definition, 180
particle transport, 182
patient data, 184–185, 196–197
photon interactions, 180, 182, 194–195
probability distribution function, 181–183
random number generator, 180
sampling techniques, 181
distribution function, 181
mixed Monte Carlo method, 181
non-analog sampling, 181
rejection function, 181
scatter compensations, 182–185
scoring, 183
significance, 184, 200
validation of, 194
MRI *see* Magnetic resonance imaging (MRI)
Multi-field visualization, 245
computational field problems, 245
ordinary and/or partial differential equations, 245
return currents, 247
source localization, 247
Multimodal processing, 140
Mutual information (MI), 129

Noise, 13, 97, 100, 108–110, 112, 114, 195, 210–212, 229, 236, 239–240
Non-Gaussianity, 126–127
Normal sinus rhythm (NSR), 102, 115–117
NSR *see* Normal sinus rhythm (NSR)
Nuclear medicine imaging, 176
data acquisition, 179–180
correction methods, 185, 195
simulation, 180

PCA *see* Principal component analysis (PCA)
PET *see* Positron emission tomography (PET)
Phenotypes, 210
Phosphorescence, 146, 149–150
Photoluminescence, 146
Photon transport modeling, 182
PHV *see* Piecewise homogeneous volume (PHV)

Piecewise homogeneous volume (PHV), 100
Polarization, 155
Positron emission tomography (PET), 106, 176, 178
  annihilation photons, 178–179, 185, 189–190, 192, 195–196
  depth-of-interaction, 184–185
  modeling, 184–185
  time-of-flight PET, 184, 192
PowerApps, 257
  pipelines, 257–258
  user interface (UI), 257–258
Principal component analysis (PCA), 129
Prior information, 135–136, 138, 140
Projection:
  oblique, 126–127
  orthogonal, 119

Quantum yield, 151–153, 157, 158–161

Radiopharmaceutical, 176–180, 196–197
  3D distribution of, 177–178
Receiver operating characteristic (ROC), 186
Red fluorescent protein (RFP), 165
  monomeric, 166
Reference:
  spatial, 136
  temporal, 136
Regression, 119, 121–122
Rhodamine 6G, 153, 156–157, 161
Ridges, 244
Rotational correlation time, 155–157

Scalar fields, 211, 222, 223–224, 241– 242, 252, 255–256
Scientific visualization, 210–212, 223–224, 255–256
SCIRun/BioPSE, 255
  interactive feedback, 256–257
  problem-solving environment (PSE), 255–256
Separating matrix, 126–127, 129, 131–132
Simulator *see* GATE
Single photon emission computed tomography (SPECT), 106, 176
  modeling, 180, 184–185, 194, 200
  spatial resolution, 177–178, 185, 193
  system matrix, 197–198
Single photon imaging, 177
Singular value decomposition (SVD), 118
  economical, 132
Sino–atrial (SA) node, 102
Source signals, 136
Sparseness *see* Sparsity
Sparsity, 126, 138–139
Spatio-temporal cancellation, 115, 117
SPECT *see* Single photon emission computed tomography (SPECT)
Spherical color map, 236
Standardized variable *see* Whitening, whitened observations
Stationarity, 98, 110–111
Statistics, 127
  higher-order, 127
  second-order, 97, 130, 139–140
Stent modeling, 68
  alternative methods, 84
  cylindrical balloon, 74
  folded balloon, 78
  neglect of the balloon, 69
  validation, 81
Step size, 113, 225
Stokes equation, 155–156
Stokes shift, 149–150
SVD *see* Singular value decomposition (SVD)

TCSPC (time-correlated single photon counting), 153, 167–168
Tensors:
anisotropic, 235
deviator, 235–236
FA, 235–236
isotropic, 235
Tensor fields, 211, 234, 245, 247, 255–256
Tensor field visualization, 234
conductivity, 234
diffusion tensor MRI (DT-MRI or DTI), 234
epileptic seizure, 234
heart attack, 234–235
inhomogeneous, 234
inverse EEG computation, 234
left ventricle, 234–235
material properties, 234
molecular diffusivity, 234
myelinated axon bundles, 234
myocardial strain, 234–235
myocardium, 234–235
strain, 234
stress, 234
tagged MRI, 234–235
Tensor glyphs, 236
cuboid, 236
cylinders, 236–237
ellipsoid, 236
fiber-crossing, locations, 239
high angular resolution diffusion (HARD), 239
nervous tissue, 238
spherical anisotropy, 236–237
superquadrics, 236–237
Time-dependent scalar field visualization, 222
differential volume rendering, 222
ray-casting, 222
Time-of-flight PET (TOF-PET), 184, 192, 195–196
TOF-PET *see* Time-of-flight PET (TOF-PET)
Tomographic reconstruction, 179
Monte Carlo-based, 185, 197
Topology, 232
biomechanics, 234
cardiac tissue, 233
dipoles, 233
electrophysiology, 233
epicardial breakthroughs, 233
epicardial surface, 233
flow recirculation, 234
forces, 234
graphs, 232
hemodynamics, 234
separation surface, 233

Uncorrelation, 97, 112, 117–118, 120–121, 128, 130, 134–135, 139–140
see also Independence
User interface, 2, 62, 214, 217, 220, 257–258

Valleys, 244
VCG *see* Vectorcardiography (VCG)
Vectorcardiography (VCG), 104
Vector fields, 211, 223–225, 229–232, 235, 245, 255–256
Vector field visualization:
bioelectric fields, 223
biofluids, 223
biomagnetic fields, 223
biomechanics, 223
boundary elements, 223–224
cardiology, 223–224
current densities, 223
electrocardiography, 223–224
electroencephalography (EEG), 223–224
finite differences, 223–224

Vector field visualization: (*Continued*)
finite elements, 223–224
finite volumes, 223–224
flow, 223
forward and inverse problems, 223–224
magnetoancephalography (MEG), 223–224
neurology, 223–224
physiological, 223
scientific visiualization, methods, 224
3D computer graphics, 224
computational fluid dynamics, 224
dense representations, 224
glyphs, 224
graphical icons, 224
streamlines, 224
stream surfaces, 224
textures, 224
topology, 224
streamline-based techniques, 225
3D textures, 226
current density, 225
electrical current, 225
Euler, 225
lighting effects, 226
numerical integration, 225
perception, 226
perception of depth, 226
rack, 225
region of interest, 225
Runge–Kutta, 225
seed points, 225
streamline seeding, 225
tubes, 226
stream surfaces, 226
advancing front, 226–227
anatomical structures, 227
animations, 229
anisotropic conductivity, 229
anisotropy, 227
brain, 227–229
curvature, 226–227
dipolar sources, 227–229
epicardium, 226–227
flow patterns, 226–227
inflow, 227–229
isotropic conductivity, 229
occlusion, 227–229
outflow, 227–229
realistic head model, 227
source position, 227–229
thorax, 226–227
texture-based flow visualization, 229
filter kernel, 229
homogeneous Neumann boundary condition, 231
line integral convolution (LIC), 229
low-pass filter, 229
reaction-diffusion simulations, 229–230
torso, 231
uncertainty, 229–230
white noise, 229
Ventricles *see* Heart, parts
Ventricular activity (VA), 117, 119
Vibrational relaxation, 147–148, 149–150
Visualization pipeline, 210–211, 264
Visualization software, 254
Amira, 254
OpenGL, 254
ANALYZE, 255
AVS, 255
dataflow, 255
IBM Open Visualization Data Explorer, 255
IDL, 254
Iris Explorer, 255

Micro View, 255
open-source, 255
Python, 255
VTK, 255
OsiriX, 255
VTK, 255
C++, 255
Java, 255
MPI, 255
Python, 255
Tcl, 255
Volume-rendering applicatications, 218
angiogenesis, 218
BioImage PowerApp, 218
BioPSE, 218
lung tumors, 218
thoracic CT, 218
transgenic mice, 218
tumors, 218

White matter tracts, 210, 239–240
Whitening, 131
matrix, 132
whitened observations, 130
Wiener, 109–110, 136, 139–140
canceller, 111, 114
Wiener–Hopf solution, 112

X, Y, Z leads *see* Leads, orthogonal

FSC
www.fsc.org
MIX
Papier | Fördert gute Waldnutzung
FSC® C083411

Kolibri 360 GmbH
Ferdinand-Jühlke-Straße 7
99095 Erfurt, Deutschland
produktsicherheit@kolibri360.de